Communications in Computer and Information Science 505

Commenced Publication in 2007
Founding and Former Series Editors:
Alfredo Cuzzocrea, Dominik Ślęzak, and Xiaokang Yang

More information about this series at http://www.springer.com/series/7899

Pavel Braslavski · Nikolay Karpov
Marcel Worring · Yana Volkovich
Dmitry I. Ignatov (Eds.)

Information Retrieval

8th Russian Summer School, RuSSIR 2014
Nizhniy Novgorod, Russia, August 18–22, 2014
Revised Selected Papers

 Springer

Editors
Pavel Braslavski
Ural Federal University
Yekaterinburg
Russia

Nikolay Karpov
National Research University Higher School
of Economics
Nizhniy Novgorod
Russia

Marcel Worring
Intelligent Systems Laboratory
University of Amsterdam
Amsterdam
The Netherlands

Yana Volkovich
Barcelona Media Research Foundation
Barcelona
Spain

Dmitry I. Ignatov
National Research University Higher School
of Economics
Moscow
Russia

ISSN 1865-0929   ISSN 1865-0937 (electronic)
Communications in Computer and Information Science
ISBN 978-3-319-25484-5   ISBN 978-3-319-25485-2 (eBook)
DOI 10.1007/978-3-319-25485-2

Library of Congress Control Number: 2015951408

Springer Cham Heidelberg New York Dordrecht London

Printed on acid-free paper

Springer International Publishing AG Switzerland is part of Springer Science+Business Media
(www.springer.com)

Preface

The 8th Russian Summer School in Information Retrieval (RuSSIR 2014) was held on August 18–22, 2014 in Nizhniy Novgorod, Russia.[1] The school was co-organized by the National Research University Higher School of Economics[2] and the Russian Information Retrieval Evaluation Seminar (ROMIP).[3]

The RuSSIR school series started in 2007 and has developed into a renowned academic event with solid international participation. Previously, RuSSIRs took place in Yekaterinburg, Taganrog, Petrozavodsk, Voronezh, Saint Petersburg, Yaroslavl, and Kazan. RuSSIR courses were taught by many prominent international researchers in information retrieval (IR) and related areas.

The 2014 RuSSIR program featured a track of courses focusing on visualization for information retrieval along with other topics related to IR. The program led to fruitful discussions among participants coming from different domains and allowed students to learn cross-disciplinary competencies. The school program consisted of two invited lectures, six courses running in two parallel sessions, two sponsor talks, and the RuSSIR 2014 Young Scientist Conference.

The school welcomed 91 participants selected based on their applications. The majority of students came from Russia, but there were also 12 students from the European Union and 8 students from the rest of the world. The RuSSIR audience comprised of undergraduate, graduate, and doctoral students, as well as young academic faculty and industrial developers. The total number of attendees was 119 including students, sponsor representatives, lecturers, and organizers.

School participation was free of charge thanks to the sponsorship support. In addition, 20 accommodation grants were awarded to Russian participants and nine of the European-based students received travel support from the European Science Foundation (ESF)[4] through the ELIAS network[5]. Travel expenses of three school teachers from Europe were also funded through the ELIAS/ESF grant.

The RuSSIR program was compiled based on reviewing of submitted course proposals by the Program Committee. Each course proposal was reviewed by at least six PC members. In total, 17 course proposals were submitted, six of which were selected for the school program. Additionally, there were two invited lectures. Each of the six courses consisted of five 90-minute lectures taught during five subsequent days. The invited lectures ran as plenary sessions, the other six courses ran in two parallel sessions.

---

[1] http://romip.ru/russir2014/

[2] http://www.hse.ru/en/

[3] http://romip.ru/en/

[4] http://www.esf.org/

[5] http://www.elias-network.eu/

- Seeking Simplicity in Search User Interfaces – Marti Hearst, UC Berkeley, USA
- Multimedia Analysis and Multimedia Visualization – Marcel Worring, University of Amsterdam, The Netherlands
- Large Scale Information Retrieval – Katja Hofmann, Microsoft Research, Cambridge, UK
- Web as a Corpus: Going Beyond the n-gram – Preslav Nakov, Qatar Computing Research Institute, Doha, Qatar
- Visualization and Data Mining for High-Dimensional Data – Alfred Inselberg, Tel Aviv University, Israel and Pei Ling Lai, Southern Taiwan University of Science and Technology, Tainan, Taiwan
- Introduction to Formal Concept Analysis and Its Applications in Information Retrieval and Related Fields – Dmitry I. Ignatov, National Research University Higher School of Economics, Moscow, Russia
- Document Analysis and Retrieval in Scientific Digital Libraries: Case Studies in Applying Machine Learning for Information Retrieval – Sujatha Das G., Institute for Infocomm Research, Agency for Science and Technology Research, Singapore
- Author Profiling and Plagiarism Detection – Paolo Rosso, Technical University of Valencia, Spain

Sponsoring organizations made two scientific presentations additionally to the school program. Ludmila Ostroumova (Yandex) presented an overview of efficient approaches for crawling and indexing newly created web pages such as news, and posts in blogs and forums. Dmitry Solovyov (Mail.Ru) gave a talk on application of self-organizing maps to search engines analytics.

For the eight time the RuSSIR Young Scientist Conference has been held within the school program. The conference provided a platform for a dialog between young researchers from different areas such as mathematics, computer science, and linguistics as well as social and media sciences. The conference ran over two consecutive evenings and consisted of two parts, oral presentations and poster sessions.

There were two types of submissions: full papers that went the thorough reviewing process and short poster notes. Out of 22 submitted full papers, 8 papers were accepted for oral presentation at the conference and published in the school proceedings.

At the poster sessions all participants had an opportunity to discuss and exchange their research results and ideas. In total about 70 posters were displayed. As in the previous years, the Young Scientist Conference was one of the main highlights of the school program.

The volume features two sections: six tutorial notes and eighth revised papers from the associated Young Scientist Conference.

The 8th Russian Summer School in Information Retrieval was a successful event: It brought together researchers and students with different background, and facilitated interdisciplinary ideas exchange. Students had an unique opportunity to learn new material that is not usually presented in university curricula and got feedback from peers and teachers during the poster sessions and informal communications. The event contributed to supporting a lively IR community in Russia and establishing ties with international colleagues. We received very positive feedback from attendees on all the different aspects of the school.

We thank all the local Organizing Committee members (namely, Alexey Malafeev, Dmitry Zelonkin, Cyril Sherstnev, and Julia Baranova) for their commitment, which made the school possible, all the Programme Committee members for their time and efforts ensuring a high level of quality for the RuSSIR 2014 program and, in particular, all the lecturers and students who came to Nizhny Novgorod and made the school such a success. We also thank the student volunteers who contributed to the school organization on-site. Our special gratitude goes to Maxim Gubin, who was responsible for legal and financial matters.

We appreciate the generous financial support from our sponsors: National Research University Higher School of Economics[6] (main organizer), Yandex[7] and Mail.Ru[8] (gold level), Google[9] and ABBYY[10] (bronze level). We are also grateful to the ELIAS network[11] of the European Science Foundation, and Springer representatives, namely, Alfred Hofmann and Aliaksandr Birukou, for their support.

February 2015

Pavel Braslavski
Nikolay Karpov
Marcel Worring
Yana Volkovich
Dmitry I. Ignatov

---

[6] http://www.hse.ru/

[7] http://yandex.com

[8] http://go.mail.ru/

[9] http://google.com/

[10] http://abbyy.com/

[11] http://www.elias-network.eu/

Organization

The conference was organized by the National Research University Higher School of Economics (Nizhniy Novgorod, Russia).

Program Committee Chairs

Pavel Braslavski	Ural Federal University/Kontur Labs, Russia
Marcel Worring	University of Amsterdam, The Netherlands

Organizing Chair

Nikolay Karpov	National Research University Higher School of Economics, Russia

Proceedings Chair

Dmitry I. Ignatov	National Research University Higher School of Economics, Russia

Young Scientific Conference Program Committee Chairs

Yana Volkovich	Barcelona Media, Spain/Cornell Tech, USA
Svitlana Volkova	Johns Hopkins University, USA

Steering Committee

Pavel Braslavski	Ural Federal University/Kontur Labs, Russia
Nikolay Karpov	National Research University Higher School of Economics, Russia
Nikita Zhiltsov	Kazan Federal University, Russia
Dmitry Chalyy	Yaroslavl Demidov State University, Russia
Alexander Goncharov	CVisionLab, Russia
Maxim Gubin	Google, USA
Ksenia Rogova	Katholieke Universiteit Leuven, Belgium
Alexander Sychev	Voronezh State University, Russia
Natalia Vassilieva	HP Labs, Russia

Organizing Committee

Valery Kalygin	National Research University Higher School of Economics, Russia
Fedor Vitugin	National Research University Higher School of Economics, Russia

Julia Baranova	National Research University Higher School of Economics, Russia
Alexey Malafeev	National Research University Higher School of Economics, Russia
Dmitry Zelonkin	National Research University Higher School of Economics, Russia
Cyril Sherstnyv	National Research University Higher School of Economics, Russia

Program Committee

Paolo Boldi	Università degli Studi di Milano, Italy
Peter Brusilovsky	University of Pittsburgh, USA
Claudio Carpineto	Fondazione Ugo Bordoni, Italy
Sergey Chernov	Yandex, Russia
Nicola Ferro	University of Padua, Italy
Maxim Gubin	Google, USA
Matthias Hagen	Bauhaus University Weimar, Germany
Dmitry I. Ignatov	National Research University Higher School of Economics, Russia
Hideo Joho	University of Tsukuba, Japan
Robert Jäschke	Leibniz Universität Hannover, Germany
Jaap Kamps	University of Amsterdam, The Netherlands
Daan Odijk	University of Amsterdam, The Netherlands
Stefan Rüger	The Open University, UK
Konstantin Savenkov	Moscow State University, Russia
Tobias Schreck	University of Konstanz, Germany
Pavel Serdyukov	Yandex, Russia
Denis Turdakov	Institute for System Programming of RAS, Russia
Natalia Vassilieva	HP Labs, Russia
Alexey Voropaev	Mail.Ru, Russia
Ingmar Weber	Qatar Computing Research Institute, Qatar
Michael Yudelson	Carnegie Learning, Inc., USA

Young Scientist Conference Program Committee

Mikhail Ageev	Moscow State University/National Research University Higher School of Economics, Russia
Ismail Altingovde	Middle East Technical University, Turkey
Paolo Boldi	Università degli Studi di Milano, Italy
Pavel Braslavski	Ural Federal University/Kontur Labs, Russia
Claudio Carpineto	Fondazione Ugo Bordoni, Italy
Sergey Chernov	Yandex, Russia
Nicola Ferro	University of Padua, Italy
David Gleih	Purdue University, USA
Maxim Gubin	Google, USA

Matthias Hagen	Bauhaus University Weimar, Germany
Dmitry I. Ignatov	National Research University Higher School of Economics, Russia
Robert Jäschke	Leibniz Universität, Germany
Jaap Kamps	University of Amsterdam, The Netherlands
Nattiya Kanhabua	Leibniz Universität, Germany
Nikolay Karpov	National Research University Higher School of Economics, Russia
Maxim Khalilov	bmmt GmbH, Germany
Natalia Konstantinova	University of Wolverhampton, UK
Daan Odijk	University of Amsterdam, The Netherlands
Alexander Panchenko	Université catholique de Louvain, Belgium
Stefan Rüger	The Open University, UK
Pavel Serdyukov	Yandex, Russia
Denis Turdakov	Institute for System Programming of RAS, Russia
Antti Ukkonen	Aalto University, Finland
Natalia Vassilieva	HP Labs, Russia
Ingmar Weber	Qatar Computing Research Institute, Qatar
Marcel Worring	University of Amsterdam, The Netherlands
Michael Yudelson	Carnegie Learning, Inc., USA

Partners and Sponsoring Institutions

Gold sponsors	ABBYY	Yandex
Bronze sponsors	Google	Mail.Ru
Partners	ELIAS Network	European Science Foundation

Contents

Tutorial Papers

Document Analysis and Retrieval Tasks in Scientific Digital Libraries 3
    Sujatha Das Gollapalli, Cornelia Caragea, Xiaoli Li, and C. Lee Giles

Online Experimentation for Information Retrieval 21
    Katja Hofmann

Introduction to Formal Concept Analysis and Its Applications
in Information Retrieval and Related Fields 42
    Dmitry I. Ignatov

Visualization and Data Mining for High Dimensional Data: –With
Connections to Information Retrieval 142
    Alfred Inselberg and Pei Ling Lai

Web as a Corpus: Going Beyond the *n*-gram 185
    Preslav Nakov

Author Profiling and Plagiarism Detection 229
    Paolo Rosso

Young Scientists Conference Papers

Transformation of Categorical Features into Real Using Low-Rank
Approximations 253
    Alexander Fonarev

A Comparative Evaluation of Statistical Part-of-Speech Taggers for Russian . . . 263
    Rinat Gareev and Vladimir Ivanov

Recommendation of Ideas and Antagonists for Crowdsourcing
Platform Witology 276
    Dmitry I. Ignatov, Maria Mikhailova, Alexandra Yu. Zakirova,
    and Alexander Malioukov

Modelling Movement of Stock Market Indexes with Data from Emoticons
of Twitter Users 297
    Alexander Porshnev, Ilya Redkin, and Nikolay Karpov

ImSe: Exploratory Time-Efficient Image Retrieval System 307
    Ksenia Konyushkova and Dorota Glowacka

Semantic Clustering of Russian Web Search Results: Possibilities
and Problems   320
    Andrey Kutuzov

A Large-Scale Community Questions Classification Accounting
for Category Similarity: An Exploratory Study   332
    Galina Lezina and Pavel Braslavski

Towards Crowdsourcing and Cooperation in Linguistic Resources.   348
    Dmitry Ustalov

Author Index   359

Tutorial Papers

Document Analysis and Retrieval Tasks
in Scientific Digital Libraries

Sujatha Das Gollapalli[1]([✉]), Cornelia Caragea[2], Xiaoli Li[1], and C. Lee Giles[3]

[1] Institute for Infocomm Research,
Agency for Science and Technology Research, Singapore, Singapore
{gollapallis,xlli}@i2r.a-star.edu.sg
[2] Computer Science and Engineering, University of North Texas, Denton, USA
ccaragea@unt.edu
[3] Information Sciences and Technology, Computer Science and Engineering,
The Pennsylvania State University, State College, USA
giles@ist.psu.edu

Abstract. Machine Learning (ML) algorithms have opened up new possibilities for the acquisition and processing of documents in Information Retrieval (IR) systems. Indeed, it is now possible to automate several labor-intensive tasks related to documents such as categorization and entity extraction. Consequently, the application of machine learning techniques for various large-scale IR tasks has gathered significant research interest in both the ML and IR communities. This tutorial provides a reference summary of our research in applying machine learning techniques to diverse tasks in Digital Libraries (DL). Digital library portals are specialized IR systems that work on collections of documents related to particular domains. We focus on open-access, scientific digital libraries such as CiteSeer^x, which involve several crawling, ranking, content analysis, and metadata extraction tasks. We elaborate on the challenges involved in these tasks and highlight how machine learning methods can successfully address these challenges.

Keywords: Classification · Focused crawling · PageRank · Citations · Topic modeling · Information extraction

1 Introduction

Digital libraries are IR systems that work on collections of documents related to specific domains and involve several information retrieval, information extraction (IE) and graph analysis tasks. While the IR tasks in digital libraries pertain to identifying relevant documents to be indexed and facilitating search functionalities, the IE tasks address the extraction of structured data and entities from unstructured documents. The extracted entities are further used for diverse data mining applications such as link analysis, community detection, and author profiling.

© Springer International Publishing Switzerland 2015
P. Braslavski et al. (eds.): RuSSIR 2014, CCIS 505, pp. 3–20, 2015.
DOI: 10.1007/978-3-319-25485-2_1

Scientific digital library portals such as CiteSeer^x [39] and ArnetMiner [54] are large-scale IR systems that work on collections of research literature related to specific disciplines. These systems are non-commercial, **open-access** systems that employ automated techniques to identify and index freely-available Web documents related to scientific research and researchers. The metadata extracted from these document collections is used to construct author-document, citation and co-authorship graphs that facilitate different search applications. Moreover, scientometric and bibliometric measures are estimated based on these extracted networks [1]. Needless to say, the satisfaction of users of such DLs and the accuracy of the extracted networks and computed measures depend *crucially* on acquisition and error-free "parsing" of the relevant documents and webpages.

- What URLs need to be examined on the Web to obtain the research-related content for indexing?
- Given a webpage containing links to research-related and other content, how can we filter out the "noise"?
- Given a crawled document, can we accurately determine if it is a research article and, if so, automatically extract the titles, authors, and its keywords?
- Can we generate author profiles based on the documents associated with a researcher?
- How can we disambiguate researchers having the same name while building citation and co-authorship networks?
- What kind of communities, temporal and topical trends do document and co-authorship networks exhibit?

In this tutorial, we draw on our research related to various modules in CiteSeer^{x1} and show that ML techniques can be employed to answer several of the above questions with reasonable accuracy. CiteSeer^x is an open-source digital library portal for scientific and academic papers for Computer Science and related areas. CiteSeer^x is widely considered the first search engine for academic paper search and a predecessor to Google Scholar[2] and Microsoft Academic Search[3]. With the objective of rapid dissemination of scientific scholarly knowledge, CiteSeer^x currently indexes over a million scholarly documents and provides various functionalities such as automatic citation indexing, author disambiguation, reference linking, and metadata extraction over these documents.

We consider the tasks - **classification**, **metadata extraction**, and **content analysis** in this tutorial. Over the last two decades, these topics have received considerable interest in the ML community due to their applicability in diverse domains including the Web, Biology, Politics, and Law [13,31,40,41]. Consequently, efficient, state-of-the-art ML algorithms are now available for solving these tasks. Applying ML algorithms within scientific digital libraries pose significant challenges. Several issues pertaining to scalability, feature design, noise and multiple modalities in the input data need to be dealt with for applying ML algorithms for digital library tasks.

---

[1] http://citeseerx.ist.psu.edu/.

[2] http://scholar.google.com/.

[3] http://academic.research.microsoft.com/.

Organization: In each section of this tutorial, we present an overview and challenges related to one of the tasks–classification, metadata extraction, and content analysis in the context of digital libraries. We describe the commonly-used ML techniques for these tasks in CiteSeer^x and other comparable systems. This tutorial provides a reference summary of the material presented at the 8th Russian Summer School in Information Retrieval (RuSSIR 2014)[4].

Section 2 describes classification models for identifying researcher homepages and scientific documents in Computer Science. In Sect. 3, we describe the extraction of researcher information from homepages and the extraction of keyphrases from research papers. Finally, in Sect. 4, we present a summary on the usage of topic modeling tools for content analysis and ranking tasks in digital libraries. We focus on discussing semi-supervised and unsupervised ML techniques with the objective of reducing requirements for human-labeled data for learning accurate models.

2 Identifying Research Documents

Classification modules comprise core components in digital libraries. Given that digital library portals provide domain-specific search functionalities on specialized document collections, *how can we ensure that only relevant documents are indexed in the digital libary collection?* Publishers such as ACM DL[5] and PubMed[6], depend on manually-provided information and "trusted" sources to obtain and maintain documents relevant to their respective domains. In contrast, Web crawlers are employed for obtaining publicly-available documents from the Web that are relevant to the domain in open-access systems such as CiteSeer^x.

A Web crawler is a special software that systematically pulls content from the World Wide Web for the purpose of indexing it locally [6]. Given the infeasibility of examining the entire content on the ever-changing Web, *how do open-access IR systems ensure that their indexed collections are relevant and up-to-date?* Periodic and focused crawling of websites where relevant documents are likely to be found is employed for this purpose. A focused crawler aims at minimizing the use of network bandwidth and hardware by selectively crawling only pages relevant to a (specified) set of topics. A key component in such a crawler is a classification module that identifies whether a webpage being accessed during the crawl process is potentially useful to the collection [7].

For a digital library portal such as CiteSeer^x, the size and quality of the indexed collection depends on the accurate identification of various **research-related documents** during periodic crawls of "whitelist academic URLs" [39,55,57]. The relevant documents for CiteSeer^x include scientific publications and researcher-related webpages such as professional homepages. Given the URL of a website where such documents are typically hosted, *how can we automatically identify researcher homepages from "irrelevant" pages such as course pages,*

---

[4] http://romip.ru/russir2014/.
[5] http://dl.acm.org/.
[6] http://www.ncbi.nlm.nih.gov/pubmed.

Table 1. Overlap in the top URL and content features based on information gain between training and crawl datasets

URL		Content	
Training	Crawl	Training	Crawl
TILDENODICT	**ALPHANUMBER**	gmt	**university**
TILDENODICT_SEQEND	TILDENODICT	server	computer
ALPHANUMBER	ALPHANUMBER_ALPHANUMBER	type	science
NONDICTWORD	HYPHENATEDWORD	html	department
Courses	**ALPHANUMBER_SEQEND**	content	numImages
ALPHANUMBER_SEQEND	**TILDENODICT_SEQEND**	text	numLinks
users_NONDICTWORD	QMARK	date	cs
users	NUMBER	professor	box
NONDICTWORD_SEQEND	Courses	university	ri
Homes	NUMBER_SEQEND	research	providence

seminar postings, and other academic webpages. Similarly, given a crawled document, can we automatically identify whether it is a research article or non-research article? We present some studies on CiteSeer^x that address these questions using novel, problem-specific features.

The identification of researcher homepages in the context of the ever-changing Web was addressed in [17]. In this work, the authors raised the concern of training classifiers when labeled datasets do not provide sufficient coverage of "negative" or "irrelevant" documents. For example, although WebKB[7] is a well-known labeled dataset used for training researcher homepage classifiers, due to its outdated nature[8], content-based classifiers trained on WebKB were found to be inaccurate for classifying content on the current-day websites. This low-performance was attributed to the presence of new types of webpages corresponding to jobs, code, seminars, calendars, and lecture material available on current-day academic websites [48]. The newer types of webpages are different from the types covered in the WebKB labeled dataset, which contains faculty and student homepages, course, staff, and project pages [17].

Various features based on surface-patterns and term presence in the WordNet[9] dictionaries were designed from the URL strings and shown to be more consistent across datasets for discriminating non-homepages [17]. This aspect is illustrated in Table 1 from [17]. The table highlights the overlap in the top URL and content features based on Information Gain between the training and crawled datasets. The low overlap in content features in the two columns illustrates why these features are not discriminative in identifying researcher homepages in the newer crawls when a classifier trained on WebKB is used.

In [17], techniques for improving content-based classification of researcher homepages by bootstrapping with URL features are discussed. Multiview

---

[7] http://www.cs.cmu.edu/afs/cs/project/theo20/www/data/.

[8] The WebKB dataset was created in 1997.

learning and particularly **co-training** that works with independent views of the learning instances is used for researcher homepage classification by treating URL and content-based features as independent views of the data. The algorithm is shown via a schematic diagram in Fig. 1. The co-training procedure starts by training two classifiers on independent feature views of a small number of labeled classification instances. In every subsequent iteration, predictions made on unlabeled instances by the two classifiers are used to expand the training dataset for the next round of training. Co-training was shown to significantly improve content-based researcher homepage classification when compared to other semi-supervised algorithms that use a single view of webpages (URL+content features together) (see [17] for more details).

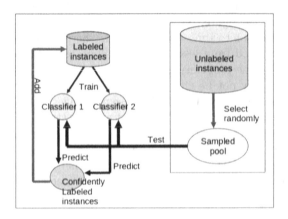

Fig. 1. Illustration of co-training. Credit: http://web.cs.gc.cuny.edu/~zhengchen/papers/naacl09-bootstrap-slides.ppt

Although text and bag-of-words (BoW) representations are common in text classification problems [25,45], as illustrated in the researcher homepage identification problem in CiteSeer^x, it is often beneficial in digital libraries to design features based on the particular problem and domain rather than an "off-the-shelf" application of existing techniques. We describe another classification task in CiteSeer^x where improvements beyond BoW are obtained using simple structural features specific to research documents.

In CiteSeer^x, it is desirable to accurately identify research articles from a set of crawled documents and index these articles in the library for fast search and retrieval of information. A rule-based system that classifies documents as research articles if they contain any of the words `references` or `bibliography` will mistakenly classify documents such as curriculum vita or slides as research articles whenever they contain the word `references` in them, and will fail to identify research articles that do not contain any of the two words. On the other hand, the commonly used "bag of words" representation for document classification can result in prohibitively high-dimensional input spaces. Machine learning algorithms applied to these input spaces may be intractable due to the large number of dimensions. In addition, the "bag of words" may not capture the specifics of research articles, e.g., due to the diversity of the topics covered in CiteSeer^x. As an example, an article in Human Computer Interaction may have a different vocabulary space compared to a paper in Information Retrieval, but some essential terms may persist across the papers, e.g., "references" or "abstract". The number of tokens in a document could be also very informative, i.e., the number of tokens in a research article is generally much higher than in a set of slides, but much smaller than in a PhD thesis. However, these aspects are not captured by the "bag of words" representation.

The number of crawled documents in CiteSeer^x are in the order of millions. Figure 2 shows the increase in both the number of crawled documents as well as the number of research articles indexed by CiteSeer^x between 2008 and 2012. As can be seen from the figure, the number of crawled documents has increased from less than two million to almost eight million, whereas the number of indexed documents has increased from less than one million to more than two million. Due to this scale and the problems described in the previous paragraph, "bag of words" approaches may not be efficient for run-time handling of research article identification in CiteSeer^x. To handle these challenges, novel features, called structural features, extracted from the content and the structure of crawled documents were proposed in [5]. Some of these features include keywords such as "abstract", "references", "bibliography", "introduction", $n-$gram features such as "this paper", "this report", "this manual", and other features such as "number of pages", "number of words in the document", "percentage of space" and "number of lines per page".

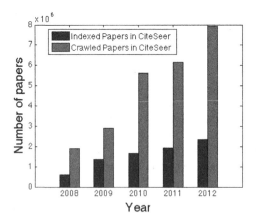

Fig. 2. The growth in the number of crawled documents as well as in the number of research papers indexed by CiteSeer^x between 2008 and 2012.

3 Metadata Extraction

Digital libraries often work with multiple types of documents. For instance, in CiteSeer^x, both research publications and technical reports (usually in PDF format) are crawled along with researcher homepages (HTML). Once again, in contrast with systems such as ACM DL and PubMed that work on human-provided clean metadata, various supervised, semi-supervised and unsupervised techniques are employed in automated systems to extract metadata and other information from pdfs and webpages. For example, classification models trained using Support Vector Machines [4] are used to extract the title and author information from the header of a research publication [24]. Similarly, sequential modeling is employed using Conditional Random Fields (CRFs) [38] in the ParsCit tool for extracting citations and the structure from a scientific document [9]. CRFs are also employed to extract researcher metadata such as email, university affiliations, and job positions from their homepages in ArnetMiner [54].

Although supervised learning models that are trained on human-annotated data are popular for learning metadata parsers, sometimes, simple rules and heuristics can be employed for the same. For instance, based on the heuristic that the first 'person' name in a researcher homepage corresponds to the researcher, the Named Entity Recognition tool from Stanford[10] was used directly to extract researcher names from homepages [18]. Similarly, regex patterns are effective for extracting phone and fax numbers from webpages [52,54]. In general, accurate rules are desirable in digital libraries since supervised learning techniques require large amounts of labeled training data that are tedious to annotate for information/metadata extraction tasks [52]. In this section, we describe some weakly-supervised and unsupervised techniques that can be used to extract certain types of information from scientific documents.

Consider the task of extracting metadata fields: *employment position, university, department affiliations* and *contact information* such as email, phone and fax from a researcher homepage. The corresponding sequence labeling problem (also known as tagging problem or annotation problem) involves predicting for each token from the content of a homepage, a tag/label from the set: {AFFL, EMAIL, FAX, PHN, POS, UNIV, O} where these labels correspond to "affiliation", "email id", "fax number", "phone number", "employment position", "university" and "other" fields, respectively. An example is illustrated in Table 2.

Previous research on this task showed that tagging or sequence labeling approaches out-perform classification approaches due to dependencies among the tags [54,59]. For instance, it is common to find employment position information followed by the affiliation information on a homepage (e.g., "professor" in the "Computer Science department" at "Stanford"). Such dependencies are captured via sequential models rather than classification techniques that make predictions for a given token position independent of the predictions for neighboring tokens.

---

[10] http://nlp.stanford.edu/ner/index.shtml.

Table 2. Example illustrating homepage tagging.

I	am	a	Student	at	Penn	State	and	Work
O	O	O	POS	O	UNIV	UNIV	O	O
with	**Professor**	**Xxxxx**	**Yyyyy**	**on**	**Finite**	**State**	**Automata**	...
O	O	O	O	O	O	O	O	...

Consider sample cue words show in Table 3 that typically surround researcher metadata on their homepages. *Can these cue words provide "weak supervision" while learning annotation models without having to train on fully-annotated examples?* Mann, Druck and McCallum [12] proposed **feature labeling** to answer this question. Rather than fully-annotated instances, "weak supervision" provided via (feature, label) affinities were employed by them to train discriminative classification and tagging models [12,44].

Consider the example in Table 2. Even without annotating the entire snippet, from domain knowledge, one can expect the correct label for the token "student" to be "POS", "most" of the time. This hint can be imposed as a soft preference or a constraint by specifying the (feature, label) distribution. For example, the labeled feature "student POS:0.8, O:0.2", indicates a preference for marking the token "student" with the label "POS" 80 % of the time. The probability for the "O" tag is to capture scenarios when the token does not indicate a position on the webpage. For example, the homepage could belong to a researcher who mentions a list of his current "students" as opposed to a student's homepage where the position information is indicated as "graduate student".

Table 3. Sample cue words for different metadata fields.

AFFL: center, centre, college, department, dept, dipartimento, laboratory
UNIV: universiteit, universitat, university, univ
PHN: cell, ext, extn, homephone, mobile, numbers, ph, phonefax, phone
FAX: ext, extn, facsimile, fax, faxno, faxnumber, telefax, tel/fax
EMAIL: contact, email, firstname, lastname, gmail, mail, mailbox, mailto
POS: president, prof, professor, gradstudent, researcher, scholar, scientist,

Generalized Expectation (GE) and Posterior Regularization (PR) are two frameworks studied previously for imposing preferences expressed as labeled features while learning discriminative models [15,44]. Using the same amount of labeled data, researcher metadata extraction accuracy was improved by $2 - 8\%$ by adding weak supervision via various term and layout-specific labeled features in [21]. Labeled features effectively reduce training data requirements while learning researcher metadata extraction from their homepages.

In addition to the extraction of structured metadata such as researcher information, other types of information is desirable from research papers in digital libraries. For example, the "concepts" in such papers are not always provided directly with these papers. However, accurate extraction of such concepts (or keyphrases) from research papers can allow for *efficient* processing of more information in less time for top-level data mining applications on research document collections such as topic tracking, information filtering, and search.

Keyphrase extraction is defined as the problem of automatically extracting descriptive phrases or concepts from a document. *Keyphrases* act as a concise summary of a document and have been successfully used in several data mining, machine learning and information retrieval applications such as query formulation, document clustering, recommendation, and summarization [23,32,50,58].

Keyphrase extraction was previously studied using both supervised and unsupervised techniques for different types of documents including scientific abstracts, newswire documents, meeting transcripts, and webpages [14,30,42,46,47]. Based on recent experiments in [3,36,37], the PageRank family of methods and *tf-idf* based scoring can be considered the state-of-the-art for unsupervised keyphrase extraction. We describe CiteTextRank, a fully unsupervised graph-based algorithm that incorporates evidence from multiple sources (citation contexts as well as document content) in a flexible manner to score keywords for keyphrase extraction in CiteSeer^x [16].

Let T represent the types of available contexts for a document, d. These types include the *global* context of d, $\mathcal{N}_d^{Ctd}$, the set of *cited* contexts for d, and $\mathcal{N}_d^{Ctg}$, the set of *citing* contexts for d. The global context of d refers to the document's content whereas cited and citing contexts refer to the short text segments around citations to the document d and made by d in the overall document network. An undirected graph, $G = (V, E)$ for d is constructed as follows:

1. For each unique candidate word extracted from all available contexts of d, add a vertex in G.
2. Add an undirected edge between two vertices v_i and v_j if the words corresponding to these vertices occur within a window of w contiguous tokens in any of the contexts.
3. The weight w_{ij} of an edge $(v_i, v_j) \in E$ is given as

$$w_{ij} = w_{ji} = \sum_{t \in T} \sum_{c \in C_t} \lambda_t \cdot \mathrm{cossim}(c, d) \cdot \#_c(v_i, v_j) \tag{1}$$

where $\mathrm{cossim}(c, d)$ is the cosine similarity between the *tf-idf* vectors of any context c of d and d [45]; $\#_c(v_i, v_j)$ is the co-occurrence frequency of words corresponding to v_i and v_j in context c; C_t is the set of contexts of type $t \in T$; and λ_t is the weight for contexts of type t.

The vertices in G (and the corresponding candidate words) are scored using the PageRank algorithm [49]. That is, the score s for vertex v_i is obtained by recursively computing the equation:

$$s(v_i) = (1 - \alpha) + \alpha \sum_{v_j \in Adj(v_i)} \frac{w_{ji}}{\sum_{v_k \in Adj(v_j)} w_{jk}} s(v_j) \tag{2}$$

where α is the damping factor typically set to 0.85 [26].

Unlike simple graph edges with fixed weights, notice that the above equations correspond to *parameterized* edge weights. The notion of "importance" of contexts of a certain type is incorporated using the λ_t parameters. For instance, one might assign higher importance to citation contexts over global contexts, or cited over citing contexts. One way to visualize the edges is to imagine the two vertices in the underlying graph to be connected using multiple edges of different types. For example, in Fig. 3, the two edges between "logic" and "programming" could correspond to *cited* and *global* contexts respectively.

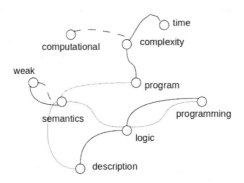

Fig. 3. A small word graph shown in [16]. The edges added due to different context types are shown using different colors/line-styles (Color figure online)

We refer the reader to [16] for more details. It was shown in this work that including information from the interlinked document network available in CiteSeer^x provides statistically significant improvements over the existing state-of-the-art models for keyphrase extraction [16].

4   Content Analysis Using Topic Models

So far, we discussed "document-level" identification and extraction tasks in digital libraries. *How can we obtain a "macro view" of a given document collection without analyzing each document?* Clustering along with visualization and ontology extraction techniques are often employed in IR systems for obtaining aggregate views of the underlying document collections [45]. Most clustering techniques represent documents using bag-of-words techniques. For example, vector-space models use vectors in high-dimensional term spaces to represent documents [51].

In contrast to the vector view, probabilistic modeling expresses each document using multinomial distributions on terms where each document is assumed to belong to one of the latent topics in the collection [6,43]. More recently, however, documents are being modeled as "topical mixtures" where a document can potentially cover multiple topics. Given a document collection, the latent concepts or "topics" can be extracted by applying techniques from Linear Algebra on the underlying term-document matrices [10, 45].

Unsupervised models such as Latent Semantic Indexing (LSI) and their probabilistic counterparts such as Latent Dirichlet Allocation (LDA) and probabilistic Latent Semantic Indexing (pLSI) extract latent topics or concepts in a document collection and estimate probability distributions on terms in the vocabulary for each topic [2, 29]. In these models, each document can be associated with a vector in a low-dimension space corresponding to the topics in the collection. Previous studies show the effectiveness of LDA and LSI models in analyzing text corpora in terms of its topics for a multitude of applications (for example, [8, 33, 35, 56]). We discuss the application of topic modeling including LDA and its extensions in a few tasks related to digital libraries.

We refer the reader to [2, 22, 28] for details on the document generation process and parameter estimation in LDA. Here, we describe the output from an LDA run for gaining the intuition behind topic models. Given a collection of documents and the number of topics, as part of parameter estimation, LDA outputs a topic-term association matrix, ϕ, of size $K \times V$, where K is the number of topics and V, the size of the vocabulary. The entries of ϕ correspond to predictive distributions of words/terms given topics. That is, $\phi_{w,i}$ is the probability of a word w given the topic i. These probabilities can be used to express a document, d as a mixture of topics, θ_d. This K-component vector captures the proportion of each topic in the given document. The terms with high probabilities for a given topic in ϕ, upon manual examination, often indicate the underlying concept captured by a topic in the given corpus.

LDA was used to understand the content of an average Computer Science researcher homepage in CiteSeer^x. Table 4 shows the top words of topics indicative of homepages obtained with LDA on a dataset of researcher homepages obtained from DBLP [18]. Notice that the terms in these topics capture information related to contact information, teaching and professional activities of a researcher. In addition, as illustrated in the topics shown in Table 5 obtained in the same run of LDA, it seems typical for Computer Science researchers to mention information related to their research projects and publications on their homepages.

The topic-term probabilities estimated using LDA were used to identify researcher homepages among other types of webpages [18]. In addition, the topics corresponding to subject areas (Fig. 5) were used to rank fixed-length text segments in a homepage to extract text segments corresponding to research descriptions. Let t be a topic related to a subject area and w, a word inside a text segment, s. The score for s with respect to a topic t is given by

$$\text{score}(s,t) = \sum_{w \in s} \phi_{w,t}$$

The research description segment is extracted using

$$p = argmax_{t \in ST, s \in S}\, \text{score}(s,t)$$

Table 4. Top words of topics related to homepages

Talk	Page	Students	Member
Slides	Home	Graduate	Program
Invited	Publications	Faculty	Committee
Part	Links	Research	Chair
Talks	Contact	cse	Teaching
Tutorial	Personal	Student	Board
Seminar	List	Undergraduate	Editor
Summer	Updated	College	Courses
Book	Fax	Current	State
Introduction	Email	ph	Activities

where S is all possible segments in the homepage with a given size sz and ST is the set of all topics indicating subject areas. Anecdotal examples of research descriptions extracted using this method from [18] are shown in Fig. 4.

4.1 Improving Ranking Tasks Using Topic Models

The insights from the topics extracted by LDA models can be used to improve diverse ranking and recommendation tasks in digital libraries. For example, the terms identified for homepage topics were combined effectively with other features based on URLs and HTML structure to train a ranking function for ranking homepages in response to researcher name queries [19]. Citation links were incorporated into extended LDA models for identifying author interests and influence [35], citation recommendation [34] and for identifying topical trends over time [27].

Table 5. Top words from topics related to subject areas

Data	Multimedia	Systems	Design
Database	Content	Distributed	Circuits
Databases	Presentation	Computing	Systems
Information	Document	Peer	Digital
Management	Media	Operating	Signal
Query	Data	Grid	vlsi
Systems	Documents	Storage	ieee
xml	Based	Middleware	Hardware
acm	Hypermedia	System	fpga
vldb	Video	Scale	Implementation

http://yann.lecun.com/
Note: the best way to reach me is by email or through Hong (I don't check my voicemail very often).
My main research interests are Machine Learning, Computer Vision, Mobile Robotics, and Computational Neuroscience. I am also interested in Data Compression, Digital Libraries, the Physics of Computation, and all the applications of machine learning (Vision, Speech, Language, Document understanding, Data Mining, Bioinformatics).
http://www.cs.colostate.edu/~whitley/
From 1997 to 2002 Prof. Whitley served as Editor-in-Chief for the journal Evolutionary Computation published by MI Press. In 2005 ISGEC became a Special Interest Group (Sigevo) of ACM.
In 2007 Prof. Whitley was elected Chair of Sigevo. Research interests Genetic Algorithms,Neural Networks, Local Search, Elementary Landscapes, Scheduling Applications, Theoretical Foundations of Genetic Algorithms. Publications and Biographical Information
http://domino.research.ibm.com/comm/research_people.nsf/pages/rshankar.index.html
PhD in Computer Science from the University of São Paulo (USP). Disciplinas 2010-1 Compiladores | Programação Research interests Machine learning (especially unsupervised learning, online learning), one-class classification, novelty detection, concept drift, natural computing and bio-inspired computing (especially evolutionary computation, genetic programming, genetic algorithms and artificial neural networks),

Fig. 4. Sample research description segments extracted from homepages

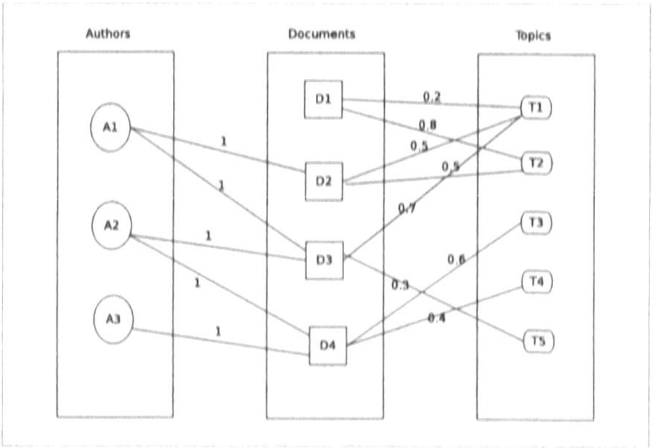

Fig. 5. An example author-document-topic (ADT) graph

More commonly, authors in digital libraries can be represented in terms of their term distributions or topical profiles obtained with LDA [11,22,53]. We describe a graph-based model for scoring authors for expert ranking and similar expert search using the output from LDA [20]. Let T be the set of all topics for a given collection of documents. Intuitively, an expert on a topic, $t \in T$ would have authored documents related to t and other closely-related topics. Similarly, if an author, a has expertise on a topic $t \in T$, authors similar to a could be expected to write about t and topics related to t.

The associations between documents and their authors and documents and their topics can be represented by a weighted tri-partite graph as follows: Let $G = (V, E)$ represent such a graph where the vertex set, $V = A \cup D \cup T$ is the union of author, A, document, D and topic nodes, T. Edges between A and D reflect the authorship relation between documents and authors whereas edges between D and T reflect the topical association of documents. Weights assigned to the edges in ADT capture the association strength between two nodes. For instance, the edges between document and topic nodes can be assigned weights using the proportion of that topic in the document.

Table 6. Top expert recommendations using ADT models in CiteSeer^x.

Natural Language Processing	Machine Learning	Information Retrieval	Semantic Web
Hermann Ney	Raymond J. Mooney	W. Bruce Croft	Ian Horrocks
Aravind K. Joshi	Vasant Honavar	Douglas W. Oard	Dieter Fensel
Raymond J. Mooney	Manuela Veloso	Hermann Ney	Enrico Motta
Bonnie J. Dorr	Jude Shavlik	Jamie Callan	Amit Sheth
Alex Waibel	David B. Leake	Hector Garcia-molina	Steffen Staab

Table 7. Top "similar expert" recommendations using ADT models in CiteSeer^x

Christopher D. Manning	Tom M. Mitchell	W. Bruce Croft	James Hendler
Aravind K. Joshi	Raymond J. Mooney	Douglas W. Oard	Ian Horrocks
Martha Palmer	Sebastian Thrun	Jamie Callan	Dieter Fensel
Raymond J. Mooney	Peter Stone	Justin Zobel	Amit Sheth
Timothy Baldwin	Jude Shavlik	Norbert Fuhr	Frank Van Harmelen
Bonnie J. Dorr	Vasant Honavar	Maarten De Rijke	Wolfgang Nejdl

An example ADT graph is shown in Fig. 5. We refer the reader to [20] for details regarding scoring author nodes using this graph and show for illustration some of the anecdotal examples included in this paper. The top-5 author recommendations obtained for sample topic and name queries in Computer Science using the ADT graph generated from the CiteSeer^x collection provided in [20] are shown in Tables 6 and 7. As the presented examples illustrate, topic models provide insights into the document collections that be incorporated for learning specific extraction and ranking tasks in digital libraries.

5   Summary and Conclusions

In this tutorial, we summarized some common tasks in digital libraries and presented automated techniques based on our own research experiences in CiteSeer^x, an open-access, digital library portal. In particular, we described a few unsupervised, weakly-supervised, and semi-supervised techniques for performing metadata extraction and classification tasks related to research documents in Computer Science and related areas. Based on the experimental results provided in the referenced papers, we conclude that techniques combining machine learning algorithms and domain-specific insights yield models that perform competitively on several tasks which once involved intense human labor. We hope the presented techniques provide an overview of the challenges for applying machine learning research to specific retrieval and extraction tasks in a large, practical system and possible solutions for addressing the same.

References

1. Hood, W.W., Wilson, C.S.: The literature of bibliometrics, scientometrics, and informetrics. Scientometrics **52**(2), 291–314 (2001)
2. Blei, D.M., Ng, A.Y., Jordan, M.I.: Latent dirichlet allocation. J. Mach. Learn. Res. **3**, 993–1022 (2003)
3. Boudin, F.: A comparison of centrality measures for graph-based keyphrase extraction. In: IJCNLP (2013)
4. Burges, C.J.C.: A tutorial on support vector machines for pattern recognition. Data Min. Knowl. Discov. **2**(2), 121–167 (1998)
5. Caragea, C., Wu, J., Williams, K., Gollapalli, S.D., Khabsa, M., Teregowda, P., Giles, C.L.: Automatic identification of research articles from crawled documents. In: Web-Scale Classification: Classifying Big Data from the Web, Co-Located with WSDM (2014)
6. Chakrabarti, S.: Mining the Web: Discovering Knowledge from Hypertext Data. Morgan-Kauffman, Burlington (2002)
7. Chakrabarti, S., van den Berg, M., Dom, B.: Focused crawling: a new approach to topic-specific web resource discovery. Comput. Netw. **31**(11–16), 1623–1640 (1999)
8. Chen, B., Zhu, L., Kifer, D., Lee, D.: What is an opinion about? exploring political standpoints using opinion scoring model. In: AAAI (2010)
9. Councill, I.G., Giles, C.L., Kan, M.-Y.: Parscit: an open-source crf reference string parsing package. In: LREC (2008)
10. Deerwester, S.C., Dumais, S.T., Landauer, T.K., Furnas, G.W., Harshman, R.A.: Indexing by latent semantic analysis. JASIS **41**(6), 391–407 (1990)
11. Deng, H., King, I., Lyu, M.R.: Formal models for expert finding on dblp bibliography data. In: Proceedings of the 2008 Eighth IEEE International Conference on Data Mining, ICDM 2008, pp. 163–172. IEEE Computer Society, Washington, DC, USA (2008)
12. Druck, G., Mann, G., McCallum, A.: Learning from labeled features using generalized expectation criteria. In: Proceedings of the 31st Annual International ACM SIGIR Conference on Research and Development in Information Retrieval, SIGIR 2008, pp. 595–602. ACM, New York (2008)
13. Firdhous, M.: Automating legal research through data mining. CoRR, abs/1211.1861 (2012)
14. Frank, E., Paynter, G.W., Witten, I.H., Gutwin, C., Nevill-Manning, C.G.: Domain-specific keyphrase extraction. In: IJCAI (1999)
15. Ganchev, K., Graça, J., Gillenwater, J., Taskar, B.: Posterior regularization for structured latent variable models. J. Mach. Learn. Res. **11**, 2001–2049 (2010)
16. Gollapalli, S.D., Caragea, C.: Extracting keyphrases from research papers using citation networks. In: AAAI, pp. 1629–1635 (2014)

17. Gollapalli, S.D., Caragea, C., Mitra, P., Giles, C.L.: Researcher homepage classification using unlabeled data. In: Proceedings of the 22nd International Conference on World Wide Web, WWW 2013, pp. 471–482. International World Wide Web Conferences Steering Committee, Republic and Canton of Geneva, Switzerland (2013)

18. Gollapalli, S.D., Giles, C.L., Mitra, P., Caragea, C.: On identifying academic homepages for digital libraries. In: Proceedings of the 11th Annual International ACM/IEEE Joint Conference on Digital Libraries, JCDL 2011, pp. 123–132. ACM, New York (2011)

19. Gollapalli, S.D., Mitra, P., Giles, C.L.: Learning to rank homepages for researcher-name queries. In: SIGIR Workshop on Entity Oriented Search (2011)

20. Gollapalli, S.D., Mitra, P., Giles, C.L.: Ranking experts using author-document-topic graphs. In: Proceedings of the 13th ACM/IEEE-CS Joint Conference on Digital libraries, JCDL 2013, pp. 87–96, ACM, New York (2011)

21. Gollapalli, S.D., Qi, Y., Mitra, P., Giles, C.L.: Extracting researcher metadata with labeled features. In: SDM, pp. 740–748 (2014)

22. Griffiths, T.L., Steyvers, M.: Finding scientific topics. Proc. Natl. Acad. Sci. U.S.A. **101**(Suppl 1), 5228–5235 (2004)

23. Hammouda, K.M., Matute, D.N., Kamel, M.S.: Corephrase: keyphrase extraction for document clustering. In: Machine Learning and Data Mining in Pattern Recognition (2005)

24. Han, H., Giles, C.L., Manavoglu, E., Zha, H., Zhang, Z., Fox, E.A.: Automatic document metadata extraction using support vector machines. In: Proceedings of the 3rd ACM/IEEE-CS Joint Conference on Digital libraries, JCDL 2003, pp. 37–48. IEEE Computer Society, Washington, DC, USA (2003)

25. Han, J.: Data Mining: Concepts and Techniques. Morgan Kaufmann Publishers Inc., Burlington (2005)

26. Haveliwala, T., Kamvar, S., Klein, D., Manning, C., Golub, G.: Computing pagerank using power extrapolation. Number 2003–45. Stanford (2003)

27. He, Q., Chen, B., Pei, J., Qiu, B., Mitra, P., Giles, C.L.: Detecting topic evolution in scientific literature: how can citations help? In: CIKM, pp. 957–966 (2009)

28. Heinrich, G.: Parameter estimation for text analysis. Technical report (2008)

29. Hofmann, T.: Probabilistic latent semantic indexing. In: Proceedings of the 22nd Annual International ACM SIGIR Conference on Research and Development in Information Retrieval, SIGIR 1999, pp. 50–57. ACM, New York (1999)

30. Hulth, A.: Improved automatic keyword extraction given more linguistic knowledge. In: EMNLP, pp. 216–223 (2003)

31. Jakulin, A., Buntine, W., La Pira, T., Brasher, H.: Analyzing the U.S. senate in 2003: similarities, clusters, and blocs. Polit. Anal. **17**(3), 10 (2009)

32. Jones, S., Staveley, M.S.: Phrasier: a system for interactive document retrieval using keyphrases. In: SIGIR (1999)

33. Kataria, S., Kumar, K.S., Rastogi, R., Sen, P., Sengamedu, S.H.: Entity disambiguation with hierarchical topic models. In: KDD, pp. 1037–1045 (2011)

34. Kataria, S., Mitra, P., Bhatia, S.: Utilizing context in generative bayesian models for linked corpus. In: AAAI (2010)

35. Kataria, S., Mitra, P., Caragea, C., Giles, C.L.: Context sensitive topic models for author influence in document networks. In: IJCAI, pp. 2274–2280 (2011)

36. Kim, S.N., Kan, M.-Y.: Re-examining automatic keyphrase extraction approaches in scientific articles. In: Proceedings of the Workshop on Multiword Expressions: Identification, Interpretation, Disambiguation and Applications, MWE 2009 (2009)

37. Kim, S.N., Medelyan, O., Kan, M.-Y., Baldwin, T.: Automatic keyphrase extraction from scientific articles. Lang. Resour. Eval. **47**(3), 723–742 (2013)
38. Lafferty, J.D., McCallum, A., Pereira, F.C.N.: Conditional random fields: probabilistic models for segmenting and labeling sequence data. In: Proceedings of the Eighteenth International Conference on Machine Learning, ICML 2001, pp. 282–289, Morgan Kaufmann Publishers Inc., San Francisco (2001)
39. Li, H., Councill, I.G., Bolelli, L., Zhou, D., Song, Y., Lee, W.-C., Sivasubramaniam, A., Giles, C.L.: Citeseerx: a scalable autonomous scientific digital library. In: Proceedings of the 1st International Conference on Scalable Information Systems, InfoScale 2006. ACM, New York (2006)
40. Li, X., Ng, S.-K., Wang, J.T.L.: Biological Data Mining and Its Applications in Healthcare, 1st edn. World Scientific Publishing Co., Inc., Singapore (2013)
41. Liu, B.: Web Data Mining: Exploring Hyperlinks, Contents, and Usage Data (Data-Centric Systems and Applications). Springer-Verlag New York Inc., New York (2006)
42. Liu, F., Pennell, D., Liu, F., Liu, Y.: Unsupervised approaches for automatic keyword extraction using meeting transcripts. In: Proceedings of NAACL 2009, pp. 620–628 (2009)
43. Liu, X., Croft, W.B.: Statistical language modeling for information retrieval. ARIST **39**(1), 1–31 (2005)
44. Mann, G.S., McCallum, A.: Generalized expectation criteria for semi-supervised learning with weakly labeled data. J. Mach. Learn. Res. **11**, 955–984 (2010)
45. Manning, C.D., Raghavan, P., Schütze, H.: Introduction to Information Retrieval. Cambridge University Press, New York (2008)
46. Marujo, L., Ribeiro, R., de Matos, D.M., Neto, J.P., Gershman, A., Carbonell, J.G.: Key phrase extraction of lightly filtered broadcast news. CoRR (2013)
47. Nguyen, T.D., Kan, M.-Y.: Keyphrase extraction in scientific publications. In: Goh, D.H.-L., Cao, T.H., Sølvberg, I.T., Rasmussen, E. (eds.) ICADL 2007. LNCS, vol. 4822, pp. 317–326. Springer, Heidelberg (2007)
48. Ortega-Priego, J.-L., Aguillo, I.F., Prieto-Valverde, J.A.: Longitudinal study of contents and elements in the scientific web environment. J. Inf. Sci. **32**(4), 344–351 (2006)
49. Page, L., Brin, S., Motwani, R., Winograd, T.: The pagerank citation ranking: Bringing order to the web. Technical report (1999)
50. Pudota, N., Dattolo, A., Baruzzo, A., Ferrara, F., Tasso, C.: Automatic keyphrase extraction and ontology mining for content-based tag recommendation. Int. J. Intell. Syst. **25**(12), 1158–1186 (2010)
51. Salton, G., McGill, M.J.: Introduction to Modern Information Retrieval. McGraw-Hill Inc., New York (1986)
52. Sarawagi, S.: Information extraction. Found. Trends Databases **1**(3), 261–377 (2008)
53. Tang, J., Jin, R., Zhang, J.: A topic modeling approach and its integration into the random walk framework for academic search. In: Proceedings of the 2008 Eighth IEEE International Conference on Data Mining, ICDM 2008, pp. 1055–1060. IEEE Computer Society, Washington, DC, USA (2008)
54. Tang, J., Zhang, J., Yao, L., Li, J., Zhang, L., Su, Z.: Arnetminer: extraction and mining of academic social networks. In: Proceedings of the 14th ACM SIGKDD International Conference on Knowledge Discovery nd Data Mining, KDD 2008, pp. 990–998. ACM, New York (2008)

55. Teregowda, P.B., Councill, I.G., Fernández, R.J.P., Khabsa, M., Zheng, S., Giles, C.L.: Seersuite: developing a scalable and reliable application framework for building digital libraries by crawling the web. In: Proceedings of the 2010 USENIX Conference on Web Application Development WebApps 2010 (2010)

56. Tuarob, S., Pouchard, L.C., Giles, C.L.: Automatic tag recommendation for metadata annotation using probabilistic topic modeling. In: Proceedings of the 13th ACM/IEEE-CS Joint Conference on Digital Libraries, JCDL 2013, pp. 239–248. ACM (2013)

57. Wu, J., Williams, K., Chen, H.-H., Khabsa, M., Caragea, C., Ororbia, A., Jordan, D., Giles, C.L.: Citeseerx: Ai in a digital library search engine. In: IAAI (2014)

58. Zha, H.: Generic summarization and keyphrase extraction using mutual reinforcement principle and sentence clustering. In: SIGIR (2002)

59. Zheng, S., Zhou, D., Li, J., Giles, C.L.: Extracting author meta-data from web using visual features. In: Proceedings of the Seventh IEEE International Conference on Data Mining Workshops, ICDMW 2007, pp. 33–40. IEEE Computer Society, Washington, DC, USA (2007)

Online Experimentation
for Information Retrieval

Katja Hofmann[✉]

Microsoft Research, Cambridge, UK
katja.hofmann@microsoft.com

Abstract. Online experimentation for information retrieval (IR) focuses on insights that can be gained from user interactions with IR systems, such as web search engines. The most common form of online experimentation, A/B testing, is widely used in practice, and has helped sustain continuous improvement of the current generation of these systems.

As online experimentation is taking a more and more central role in IR research and practice, new techniques are being developed to address, e.g., questions regarding the scale and fidelity of experiments in online settings. This paper gives an overview of the currently available tools. This includes techniques that are already in wide use, such as A/B testing and interleaved comparisons, as well as techniques that have been developed more recently, such as bandit approaches for online learning to rank.

This paper summarizes and connects the wide range of techniques and insights that have been developed in this field to date. It concludes with an outlook on open questions and directions for ongoing and future research.

Keywords: Online evaluation · A/B testing · Contextual bandits · Dueling bandits · Interleaved comparison · Online learning to rank · Counterfactual analysis · Experiment design

1 Introduction

Online experimentation for information retrieval (IR) refers to experiments that rely on natural user interactions. For example, an *online evaluation* experiment might compare alternative search interfaces, or alternative methods for ranking search results (often referred to as *rankers*). Such controlled experiments allow researchers or system developers to gain a better understanding of how to support the searchers' goals, or to test models of information seeking behavior. Extending the controlled experiment scenario, *online learning* approaches can automate the experimentation process to efficiently search a large space of IR solutions.

Many examples and success stories of online evaluation have been published in previous years. For example, Kohavi et al. [44] show how AB testing was used to identify a search widget for MSN Real Estate that would maximize revenue.

© Springer International Publishing Switzerland 2015
P. Braslavski et al. (eds.): RuSSIR 2014, CCIS 505, pp. 21–41, 2015.
DOI: 10.1007/978-3-319-25485-2_2

The result of the experiment described there led to a 10 % increase in revenue, illustrating that online experiments can lead to high real-world impact that often cannot be predicted even by domain experts.

The techniques that have been developed for online experimentation for IR complement more traditional experimentation techniques. Test-collection based IR experimentation [63,75] can efficiently compare search solutions at various levels of abstraction, but require expensive manual annotations. For example, they allow experimenters to focus on an isolated concept like topical relevance, while abstracting from individual differences between searchers. In contrast, online experimentation techniques have been developed to evaluate and tune systems to directly optimize systems' online performance. As a result, they reflect expectations and behavior of real users and can adapt to their preferences.

While the initial focus of online experimentation was primarily on improving the performance of a given system, they are not only applicable to system development, but can also provide new insights from a research perspective. Online experimentation is particularly closely related to research in interactive IR [38]. The focus of interactive IR research has led to valuable insights in experimental design, and has been particularly informative in testing hypotheses and developing theory, e.g., of search behavior. This more theoretical view can benefit online experimentation, showing a path beyond the optimization of individual system performance. Correspondingly, techniques developed for online experimentation can add to the tool set available for interactive IR research, in particular where unobtrusive measurement is required for naturalistic studies.

The focus of online experimentation on studying natural user interactions in realistic settings results in a unique set of challenges. First, exploring solutions of unknown performance creates the risk of hurting the user experience. At the same time, feedback on these solutions can only be obtained by trying them out. This results in a trade-off between exploration of new solutions, and exploitation of good solutions known at a given point in time. Second, in a natural setting we often cannot control many sources of variance (e.g., search goal) that would be controlled in a more traditional experiment, such as a lab study. This can result in high variance, but this problem is typically alleviated by large sample sizes that can be collected, at least in frequently-used systems. Finally, online experiments in natural settings can offer a very narrow windows of observation. Instead of e.g., recordings and follow-up interviews that may be collected in a lab study, and that can help interpret results, we now have to rely exclusively on behavior traces (e.g., clicks, page views) that result from users' natural interactions. This limited bandwidth of observation can be particularly problematic when unobserved confounding variables affect measured outcomes (e.g., effects of search task or user characteristics). Therefore it is particularly important to carefully design the experiment, and especially the measurements designed to evaluate or compare solutions.

In the remainder of this overview paper, we outline the techniques that have been developed to address these challenges. First, we further motivate the need for controlled experiments, and introduce online evaluation using the example of A/B testing (Sect. 2). The next two sections focus on measurement, first in the form of estimating online metrics from exploration data (Sect. 3), then in the form of paired comparisons enabled by interleaved comparisons (Sect. 5). After discussing questions related to online evaluation, we turn to the question of how these evaluation techniques can be used to automatically optimize system performance, for example when many system configurations are feasible. We turn to online learning, where we first introduce a common problem formulation – bandits (Sect. 6). Finally, we focus on learning from relative feedback (Sect. 7). Sections 6 and 7 build on the earlier sections, in that many of the learning approaches utilize the previously introduced online evaluation methods to infer feedback for learning.

2   Controlled Experiments

This section outlines the role that controlled experiments play in identifying causal relationships. We establish the connection between these concepts and IR, and discuss examples of controlled experiments, from small-scale lab studies to web-scale experiments.

Scientific discovery can take many forms. Following the three-fold distinction of Babbie [5], exploratory studies are valuable for identifying phenomena of interest and formulating hypotheses. Descriptive studies describe phenomena and their observed relationships. Finally, explanatory studies aim to uncover *causal relationships*, that explain the mechanisms that lead to the observed phenomena and relationships.

Identifying causal relationships is particularly valuable, because they are the most robust [54,55]. By explaining the mechanisms of why events happen, they allow us to make predictions under changing conditions. These insights are crucial for predicting the effects of actions. In everyday life, knowing causal relationships and consequences of our actions lets us make informed decisions on how to achieve our goals. For developers of an interactive system, identifying causal relationships enables data-driven decisions on how the system should interact with the user.

In our everyday life, we are used to thinking in terms of cause and effect, and many causal relationships are obvious to us. Following an example of Babbie [5], when we observe a correlation between ice cream consumption and death by drowning in lakes, we would not conclude that one causes the other. Rather, we know that there is a common cause, temperature or season, that affects both.

In many cases, causal structures are far less obvious, especially when systems are complex, and potential causes cannot be observed. In particular, observational data alone is not enough to draw any conclusions on causal relationships [55]. Observing a correlation (also called association) between events could result from infinitely many possible causal relationships. This is illustrated in the path diagram in Fig. 1. Returning to the example above, observing a correlation between ice cream consumption and deaths by drowning does not, by itself, allow us to infer the causal structure that explains these events. Only with

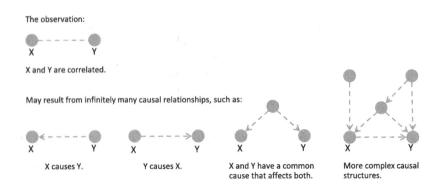

Fig. 1. Path diagram of an example observation (X and Y are correlated), and possible causal structures that may explain the observation. Possible correlation between two variables is denoted by bi-directional arrows. Hypothesized causal relations are denoted by directed arrows from cause to effect.

additional information can we reason that the common cause explanation is the most likely. Conversely, the absence of correlation in observational data does not exclude the possibility of a causal relationship.

For a concrete example from information retrieval, let us consider two studies of searchers' click behavior in web search. Granka et al. [22] report on an observational study, which measures the time searchers spend looking at search result summaries (document titles and snippets) per rank, and the number of clicks per rank. They find that searchers spend more time examining higher-ranked documents, and that they click on higher-ranked documents more often than on lower-ranked documents. The findings describe searcher behavior, but note that it does not allow us to draw conclusions about causal relationships. For example, it is possible that examination and clicks are caused by document rank, or by some other factor, such as the attractiveness of the snippets (e.g., if more attractive snippets are ranked higher). In IR, we are often interested in document relevance, but again, observational data alone does not allow us to draw conclusions on whether relevance may be causally related to the observed behavior.

The most reliable method for identifying causal relationships are controlled experiments [55]. To explain observed correlations between click behavior, search result rank, and document relevance, Guan and Cutrell [23] conducted a controlled experiment in a lab study or search behavior. They manipulated the search result pages shown to study participants to include a single relevant document, and randomly selected the rank at which this target document was shown. They found that, when the target was ranked lower in the result list, participants were often not able to find it, and were likely to click on less relevant higher-ranked documents. The results suggest that both rank and document relevance affect searchers' click behavior. Based on this insight, numerous click models have been proposed that model document relevance using causal assumptions about rank, and observed user interactions (e.g., [14, 18]).

Conducting a controlled experiment means that, instead of observing naturally occurring values of all variables, we specifically set the value of the hypothesized causal variable (e.g., X in Fig. 1). In doing so, we break the causal chain that may carry associations between this variable and the hypothesized effect (e.g., Y). The key to a successful intervention is that the assignment of values to X is done at random. This ensures that the decision to assign the chosen value is independent of any other variables that could carry information between X and Y. Any remaining relationship between X and Y then has to reflect the strength of the causal relationship between the two variables.

Designing controlled experiments may be difficult or impossible in some settings, e.g., when studying the economy of a country. In some of these cases, it may be possible to infer causal structure from initial causal assumptions in combination (i.e., non-experimental data that was observed outside of a controlled experiment) [55]. In the present discussion we focus on settings where controlled experiments are possible. Combining the insights that result from controlled experiments with causal inference mechanisms can result in an even more powerful discovery process.

Luckily, many interactive systems are well suited for experimental control. For example, if we are interested in whether a redesign of a website affects user engagement, we can conduct a controlled experiment (also called "A/B test" or "bucket test" in the context of web-scale studies) [43–45,72]. We can do this by deciding for each new incoming user at random whether they are directed to the original version (often called control) or the redesigned version (often called treatment). We measure the target quantity we are interested in (e.g., number of click, or time spent on the page). When comparing the measurements for control and treatment group, statistical significance testing is used to establish whether any observed differences are likely to result from random noise [11]. If statistically significant differences are observed, these can be attributed solely to the differences between the two versions of the website, because our random assignment to control and treatment group blocked any other possible causal effects. Once a causal relationship has been established, it can guide decisions on how to change the current system.

Controlled experiments can be conducted in interactive systems of any scale, ranging from small-scale laboratory experiments to crowdsourced experiments with hundreds or thousands of crowdworkers, to experiments with millions of users for large web systems. Each of these affords a different level of control. In lab studies, where it is typically only feasible to work with a small number of participants, it is often useful to work with a complex experimental design that investigate several variables at once (see [38] for a detailed discussion of experiment design, especially in interactive IR studies). For example, in an eye-tracking study of user interactions with query auto-completion (QAC), we investigated the effect of QAC quality on 10 variables that captured user behavior, while controlling for effects of search task and user differences [31]. Crowdsourcing environments can afford an interesting balance between scale and control. For example, Kazai et al. [37] studied the effect of quality-control mechanisms and page ordering on the quality of annotations in a relevance labeling task with hundreds of workers. At web scale, experimental designs are often less complex, to keep results interpretable. For example, Bendersky et al. [7] examined the effects of retrieval-based video recommendation strategies on users' viewing behavior in a month-long experiment that involved millions of users.

The methodology for running controlled experiments on the web, in the form of A/B tests, has been refined over recent years. Kohavi et al. [43] gives a detailed account of the key aspects that need to be considered, such as deciding on a metric, implementing the split in control and treatment group, and estimating the required sample size. An important challenge is the question bandwidth available for running experiments. Often, online experiments run for several days or weeks, and running only one experiment at a time would lead to very slow progress (especially considering that typically few of the tested changes improve system performance or lead to significant insights) [45]. This can be improved by running several mutually independent experiments in parallel [43,72]. Improving the sensitivity of online controlled experiments is an area of active research [19].

This section motivated the need for controlled experiments as a basis for evaluating interactive systems. The most prominent methodology for running these experiments online, A/B testing, was briefly introduced. In the next section, we discuss extensions of the controlled experiment setup that allow the use of exploration for large-scale offline evaluation.

3   Offline Estimates of Online Performance

In the previous section we discussed how controlled experiments allow us to assess effects of system changes metrics on their users. In this section, we introduce methods that allow system comparisons using so-called exploration data. Recall that the key requirement for controlled experiments is that the assignment of users to control and treatment conditions has to be independent of any other factors that may carry information between the hypothesized cause and effect, and that randomization is a reliable way of blocking any such interactions. The same principle is exploited in the methods introduced here.

The first approach we discussed is called exploration scavenging [48].[1] The question raised there is whether a data set that was collected under an arbitrary data collection policy[2] π_D can be used to evaluate alternative policies π_A (i.e., other configurations of the system). The problem is formulated as the task of obtaining an unbiased and consistent estimate of the online performance of π_A

---

[1] The terminology comes from the area of reinforcement learning, a type of machine learning in which an intelligent agent (e.g., an interactive IR system) learns from interactions with its environment (e.g., users) by trying out actions and observing rewards. This is a natural model for learning in interactive IR, and is discussed in more detail in Sect. 6.

[2] A policy defines a distribution over system actions, often conditioned on additional information, such as the history of previous interactions, or information about context, such as a query posed by the user.

in terms of a target metric (e.g., clickthrough rate – CTR) offline, i.e., without running an actual experiment and using previously collected data. Langford et al. [48] show that this is not the case generally, but that there are cases where exploration data can be used to obtain unbiased estimates of the online performance of alternative policies. In particular, this is the case when π_D selects actions independently of the information used by π_A, and when it selects all actions that are available to π_A sufficiently often.

Li et al. [49,50] propose and analyze a specific exploration scheme, and show that it allows very accurate prediction of online performance. Their approach relies on an exploration policy that selects actions during data collection uniformly at random. They show that this data can be used to obtain unbiased offline estimates of online performance. The method uses rejection sampling, where an observed sample is accepted to contribute to the estimate if it matches the action that would have been selected by the system under evaluation, and rejected otherwise. The effectiveness of the method is demonstrated in the context of a news recommendation application, where the task is to learn how to select the news article to display the most prominently to maximize user engagement (in terms of CTR). Recently, Li et al. [51] demonstrated an application of unbiased estimation from exploration data to optimize components of a commercial search engines (here: speller) in a large parameter space. They also propose non-uniform sampling during exploration, and show that very accurate estimates of online performance can be obtained.

The key benefit of the proposed offline evaluation techniques is that they allow infinitely many system comparisons once a set of exploration data has been collected. When exploration was done uniformly at random, meaning that it is independent of any information that the system might use, we can evaluate any system, including those where decisions are based on user attributes, or interaction history. This creates a powerful set up for testing effects of these factors on user interactions, and can be used to optimize system performance directly in terms of online metrics.

The methods proposed in [48–51] are powerful when the number of actions available to a system are relatively small compared to the sample size, e.g., when selecting from a small set of news articles, or from a small set of ads to be placed on the result page for a relatively frequent query. The amount of data required to obtain accurate (low-variance) estimates grows linearly in the number of available actions, making the approaches infeasible for large action spaces. An alternative is proposed in [8], where exploration is not in terms of the set of available actions, but instead in terms of the parameter space of the system.

Bottou et al. [8] propose a counterfactual reasoning approach. In it, exploration is achieved by changing system decisions from being deterministic to following some distribution (e.g., instead of using a fixed setting for a given parameter, use a continuous distribution over that parameter, from which the actual value for each impression is sampled during data collection). Given exploration data collected this way, counterfactual reasoning can be used to answer "what if" questions of the form *What would have happened if we had used a different system configuration?*. Answers are obtained using importance sampling, where observed samples are reweighted by the ratio of their probability during data collection, and in the system under evaluation. Finally, Bottou et al. [8] propose a learning approach that utilizes counterfactual reasoning to compute the direction of parameter updates.

The counterfactual reasoning approach extends the principle of controlled experiments to settings where it is impossible to split a user population into control and treatment. The experimental unit is the impression, and every user experiences various parameter settings at various times. This allows the method to be applied to complex system, which Bottou et al. [8] demonstrate in the example of an online advertising marketplace.

The methods for offline evaluation that have been discussed in this section are inspired by methods from reinforcement learning, where off-policy evaluation is used to greatly speed up learning [56]. They enable methods for learning in interactive IR that directly maximize online performance (Sect. 6), but can naturally be used for system evaluation, or experimentation to test IR models that go beyond optimizing the performance of a specific application. In addition to the possibilities these methods open up for individual researchers or teams, the idea of using exploration data for unbiased evaluation may open up a path to sharing (annonymized) data in a future form of test-collection based IR evaluation.

So far, we have assumed that our goal is to evaluate systems in terms of an arbitrary online evaluation metric. Insights into what to measure, and proposed online evaluation metrics, are discussed in the next section (Sect. 4). The idea of using exploration data for offline evaluation is extended to within-subject experiments in the context of interleaved comparisons, discussed in Sect. 5.

4 Online Metrics

While implicit feedback has been used for IR evaluation for a long time [39], ubiquitous access to the web and web search engines have emphasized the need for reliable and interpretable online metrics. Consequently, research efforts in this direction have dramatically intensified in recent years and much progress has been made. Because this is such a large and active research area, the overview here only mentions a number of selected approaches that illustrate certain trends and developments. More detailed considerations regarding measurement in interactive experiments can be found in [40].

The specific choice of target metric very much depends on the specific application, and the goals of the experiment. For commercial applications, revenue, the number of purchases, or the value of purchases per buyer [43]. Similar metrics can be considered for recommendation systems and advertising platforms. As a general measure of user engagement, Dupret and Lalmas [21] recently proposed modeling absence time, i.e., how long a user waits before returning to a website. Crucially, the target metric should be decided on before the experiment is run. Other considerations include the variance of the metric and the expected difference between systems, as these affect the size of the sample required to detect statistically significant differences between system (the power of a controlled experiment) [43].

In search, it is notoriously difficult to identify a single reliable online metrics. For example, changes in the number of clicks per query might mean that the user needs more clicks to find what they are looking for, or that several highly relevant pages are shown and keep the user engaged. An increase in abandonment rate could indicate that searchers give up in frustration, or that they can easily find the answer to their question directly on the search result page. Especially user clicks were shown to be affected by biases, e.g., due to result presentation [27,82], and to vary substantially across search tasks and users [64]. Consequently, these and similar absolute evaluation metrics have been found to exhibit high variance, and caution has to be taken in interpreting their results [61].

Many recently proposed metrics take a more long-term or holistic view on measuring search engine quality. Joachims et al. [41] used per-query type models of dwell time to capture user satisfaction with search results, and personalized models of user satisfaction are explored in [24]. Song et al. [69] analyzed the long-term behavior of metrics, and showed that users may initially compensate for changes in search engine quality. Absence time as a measure of search effectiveness was considered in [12].

The interpretation of user signals as relative feedback has been proposed as an alternative to high-variance absolute metrics. Joachims et al. [35] show that the interpretation of clicks as relative preferences between documents, using so-called click-skip heuristics, can lead to accurate relative judgments. A proposed aggregation of these rules into a result page-level metric is PSkip [76]. Based on the construction of a controlled experiment, FairPairs infers the relative preference between documents from their relative CTR [59].

A difficulty with per-document relative metrics can be the large amount of exploration required for obtaining accurate estimates. This problem is avoided by interleaved comparison methods, which aggregate interactions with documents into a ranking level comparison. This can be thought of in similar terms as the exploration strategies over specific actions as opposed to exploration in terms of system parameters discussed in the previous section. Interleaved comparison methods are discussed in the next section.

5   Interleaved Comparisons

Interleaved comparisons have been developed to provide unbiased, relative comparisons of ranked lists [61]. In comparison to A/B tests, which run controlled experiments between users (each user is either in the control or in the treatment condition), interleaving experiments can be thought of as a within-subject experiment, where each user is presented with results that combine two competing rankings. To avoid introducing bias in such a setting, the interleaved (combined) result lists presented to users need to be constructed in a way that is fair to both rankers in expectation, and it has to be ensured that users cannot distinguish between the results contributed by either ranker.

The most widely-known interleaved comparison methods is Team-Draft interleaving [15,57,61]. We briefly describe the general principle of interleaving using this method. In Team-Draft interleaving, interleaved result lists are constructed in a way that is designed to ensure that each original ranker contributes the same number of its documents at a given rank to a given rank of the target list in expectation over impressions. This is done as follows. To fill the first two ranks in the interleaved list, a coin-flip determines which rankers first contributes a document. This ranker deterministically choses its highest document that is not yet part of the interleaved list. Then the competing ranker contributes its highest-ranked document. The process continues until a result list of the desired length has been constructed. During interleaving, the system keeps track of which ranker contributed which document. The constructed interleaved result list is then shown to the user, and user clicks (or, potentially, other interactions) on the presented documents are recorded. The observed clicks are then interpreted as preference indications for one of the rankers. Only the clicks on results contributed by a ranker are counted in its favor. Aggregating over multiple impressions results in an estimate of how much a ranker would be preferred over its competitor.

Team-Draft interleaving constructs a controlled within-subject experiment to compare between two rankers. This setup is extended by Probabilistic Interleave [26,30]. Probabilistic Interleave is based on the idea of generating interleaved result lists from probability distributions over documents. These distributions are based on the rankings to be compared, in order to maintain the fairness of the interleaving. Interleaving outcomes can be computed by marginalizing over possible ways in which the observed interleaved result list could be generated. The result is a highly sensitive comparison method in which the magnitude of assigned click weights reflects the magnitude of ranking differences between the original rankings.

Crucially, probabilistic interleave defines an exploration policy, similar to those discussed in Sect. 3. This means that the collected data can be used to obtain unbiased estimates of online metrics, in this case of online interleaved comparison outcomes [28]. This results in a flexible online/offline evaluation setup, where interleaved comparisons can be performed online, and the observed data can be used as exploration data for further comparisons. Conversely, exploration data that was not collected using interleaving, but covers the same action space, can be used to obtain interleaved comparison outcomes for rankers in that same space. In Sect. 7 we show how the resulting method can be used to learn very efficiently from interleaving feedback.

Several extensions of interleaving have been devised. Optimized interleaving [58] considers the construction of a distribution over interleaved lists as a constrained optimization problem designed to obtain accurate comparisons between known rankers from as few samples as possible. Vertical-aware interleaving shows that the interleaved comparison approach can be extended to settings where the linear ranking assumption is violated, e.g., in the presence of vertical search results [16,17]. Most recently, multileave was proposed to efficiently compare sets of rankers without having to perform all pairwise comparisons [67].

Interleaved comparison methods are particularly interesting for online evaluation because they allow within-subject experiments that result in particularly low variance. This provides highly sensitive comparisons with up to two orders of magnitude smaller sample sizes than those that would be required for comparable A/B tests [57].

6   Online Learning for Information Retrieval

Up to now we have focused on the use of controlled experiments for online evaluation. Given one or more systems, online evaluation techniques assess their absolute or relative online performance. However, in an online system, it is often not necessary to accurately determine the online performance of each candidate system. Rather, we are interested in identifying the best performing system as quickly as possible. When we need to chose from a small fixed set of systems, the earlier we know which one performs best, the sooner we can stop exploring the alternatives. In this setting, online learning can avoid over-exploring sub-optimal systems, leading to better online performance while learning. If, instead, we the set of possible systems is infinite (e.g., when a system is identified in terms of the settings of continuous parameters), online learning can allow us to find the best such system efficiently.

Within this paper, we define an online learning system as a system that changes its behavior through interaction with its environment. A natural fit for this task are problem formulations from reinforcement learning [71]. Reinforcement learning is a branch of machine learning where agents (e.g., an IR system) interact with an environment (e.g., users) and learn by trying out actions (e.g., documents, news items, etc.) and observing rewards (e.g., interpret user actions as absolute or relative feedback). The full reinforcement learning problem also specifies states in which the environment can be in, and transitions between states, which may depend on the agent's action. In this paper, we focus on a subset of problems called bandit problems, where system actions do not affect future states. Initial work on taking state transitions into account has been conducted in the context of exploratory search [33] and session search [52].

Bandit problems are a natural fit for many online learning tasks in information retrieval, where characteristics of incoming users are independent of other users. One mapping to web search is shown in Fig. 2 (analogous mappings hold for other IR tasks, such as news recommendation, ad placement, etc.). Here, the system learns from user interactions, by taking actions (selecting documents or document rankings), and observing user feedback. Interactions are modeled in rounds or discrete timesteps, where in each timestep the agent may observe some context, generates an action, and observes and applies feedback. A crucial difference to learning in a supervised setting is that only feedback for selected actions is observed. The task of the learner is to optimize online performance, i.e., performance while learning. These two characteristics result in the exploration-exploitation challenge, because actions with unknown performance have to be explored to learn better solutions. An important benefit of reducing IR problems

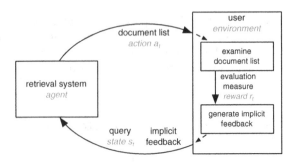

Fig. 2. Example formulation of search as a contextual bandit problem, with information retrieval terminology shown in black, and reinforcement learning terminology shown in green (Color figure online).

to bandit approaches is that the rich body of work on bandit approaches can be leveraged. At the same time, IR poses some unique challenges that further drive development in bandit research, such as approaches that work with relative feedback (discussed in Sect. 7).

Many types of bandit approaches have been developed. Here, we divide these approaches in terms of how they interpret feedback for learning. In this section, we focus on approaches with absolute feedback. We outline work in three areas, and show how the developed approaches relate to IR applications. We start from the non-contextual K-armed bandit setting, where the payoffs of available actions are independent. This is extended to the contextual setting, where context provides additional information on when an action may have high reward. Finally, we consider extensions to large or infinite action spaces. A detailed survey of bandit approaches and their analysis can be found in [9].

In the classic *K-armed bandit* setting, the learner has to select from a finite set of available actions. A simple approach that often works surprisingly well in practice is ϵ-greedy [77]. At each timestep, it explores with probability ϵ by randomly selecting an available action, and exploits the empirically best action with probability $1 - \epsilon$. Convergence guarantees are known for appropriate choice of ϵ. Another popular type of approach is UCB (upper confidence bound) [3]. It maintains estimates of the expected payoff for each available action, constructs confidence intervals around these estimates, and at each timestep selects the action with the highest upper confidence bound. Convergence guarantees exist for the stochastic setting, where payoff for each action is assumed to be independently sampled from a stationary distribution. An approach that does not rely on stochastic feedback is EXP [4]. Approaches of this type maintain a distribution over actions' expected payoff and sample from this distribution. Because of its stochastic nature, EXP has performance guarantees even in adversarial settings, where payoffs are selected by an adversary that competes with the learner. Recently, approaches based on Thompson Sampling have been shown to achieve good empirical performance [13]. This finding triggered much theoretical work to better understand properties of this approach analytically [1,36,62].

The approach works by maintaining a distribution over expected payoffs for all arms, and at each timestep sampling from this distribution and acting optimally according to the drawn sample.

A pioneering approach that applied bandit approaches to IR was proposed by [60]. The authors formulated the task of learning diverse rankings with bandit feedback. Assuming a user population with diverse information needs, the task is to learn rankings that satisfy as many users as possible (i.e., show at least one relevant document per intent). This problem was later generalized to the submodular bandit problem, where a set of items has to be selected to optimize submodular utility functions [46, 70, 78].

An extension of the classic K-armed bandit problem that is particularly relevant to IR problems is the *contextual bandit problem* (also known as bandits with side-information, associative bandits, and bandits with expert advice) [49]. Here, the learner is given additional information in each round, that can help identify the action to select. In an IR setting, this context information can consist of, for example, a user profile or history, a query, a website on which an ad must be placed, etc. Naively, K-armed bandit approaches can be applied to this setting by learning a separate bandit for each context. However, this approach results in a large increase in the amount of required exploration (all actions have to be explored sufficiently often in each context), and consequently a reduction in online performance. However, extensions to K-armed bandit approaches have been developed that efficiently generalize over contexts. For example, EXP4 generalizes over actions by transforming the exploration problem to exploration in some contextual policy space [4]. Langford and Zhang [47] extend ideas from ϵ-greedy to continuous contexts. LinUCB extends UCB to the contextual bandit setting by generalizing it to a linear reward model [49], and similar approaches are explored for Thompson Sampling [62]. A linear approach with submodular utility functions is proposed in [78].

Much of the work on contextual bandit approaches was informed by, and empirically validated on, IR problems such as news recommendation [49, 78], ad placement [48], vertical selection [20, 32], comment recommendation [53], ad format selection [73], and, most recently, spell correction in search queries [51]. These problems can be accurately modeled as contextual bandit problems with small action sets and high-dimensional context information, with absolute reward metrics such as clickthrough rate.

An orthogonal extension of bandit approaches is to consider large or infinite action spaces. In settings, where the number of actions is large, as is the case when searching large document collections, exploring all available actions is prohibitive. Approaches that tackle this problem exploit information about the similarity of actions. This information can be provided explicitly, often in the form of a tree structure [42]. Extensions of this work generalize to cases where properties of the underlying space are unknown and feedback stochastic [74]. Slivkins et al. [68] extend this approach to the ranked bandit setting, to learn diverse subsets of large or infinite action spaces.

In this section we discussed online learning approaches for IR. We focused on bandit approaches for learning in the K-armed and contextual setting, and briefly outlined approaches for learning in settings with large or infinite action spaces. The approaches described so far learn effectively in settings where reliable absolute feedback, such as clickthrough, can be observed. In the next section, we discuss bandit approaches that learn from relative feedback.

7   Online Learning from Relative Feedback

In many interactive IR systems, absolute reward may not accurately reflect user satisfaction or other target quantities, because they are too context dependent and noisy (cf., Sect. 4). For these settings, relative feedback methods have been developed, and have been shown to be substantially more robust (cf., Sect. 5). Naturally, we would like to use these relative metrics as feedback for online learning. While classic approaches (such as the bandit approaches discussed in Sect. 6) focused on absolute feedback settings, the first relative approaches have been proposed recently. We give a brief overview of these in this section. A thorough review of relative bandit algorithms (also called preference-based multiarmed bandits) was recently published by Busa-Fekete and Hüllermeier [10].

Supervised learning approaches for learning from relative feedback go back to at least [34]. In IR, this approach has been very successfully applied to supervised learning to rank problems, where expert relevance labels can be interpreted as relative feedback. However, supervised approaches do not address the exploration-exploitation challenge, and directly applying supervised approaches to interaction data is very susceptible to noise and bias. If applied to learn in a batch setting from exploration data, very high levels of exploration would be required to combat bias [25]. Dueling bandit approaches naturally address the exploration-exploitation challenge, while working with relative feedback.

The K-armed dueling bandit problem was first formulated by Yue et al. [81,83], and was directly motivated by the need to learn from relative feedback in IR settings. It generalizes multiarmed bandit problems to settings where absolute performance cannot be quantified, but comparisons between two arms can be made. These comparisons can be stochastic, such that the a better arm i has a probability of winning a comparison against a worse arm j of $p_{ij} > 0.5$. They propose an approach to solving this problem, called Interleaved Filter (IF), which works in rounds in which it eliminates an arm when it is proven to have low performance. Since then, new dueling bandit approaches have been developed that substantially improve over both the empirical performance of IF, and over its theoretical guarantees. For example, Beat-the-Mean compares each arm to the sampled mean of all arms [80].

The main challenge addressed by dueling bandit approaches is to select the arms to compete in each round such that the competition quickly focuses on the best arms, to avoid excessive exploration of bad arms. Zoghi et al. [85] employs a relative UCB-style approach, such that it always selects the arm with the highest confidence bound as one competitor, and plays it against the arm that has the best chance of beating it. A strategy based on Thompson Sampling is proposed in [84]. Another very recent approach is by Ailon et al. [2], who provide several reductions from dueling bandits to classic cardinal bandits.

In addition to the K-armed dueling bandit problem, Yue and Joachims [79] proposed a contextual problem formulation, resulting in a generic dueling bandit formulation. In this formulation, the learner has to optimize a linear function in d dimensions, using only relative feedback about the relative performance of two such solutions. A stochastic gradient descent approach to solving this problem is the Dueling Bandits Gradient Descent (DBGD) [79]. Briefly, it it learns by interacting with the environment in rounds, and observing relative feedback as follows. At all times, the learner maintains a "current best" solution w_t (a solution is a weight vector for linear weighted combination of context features). In each round, it generates a "challenger" w_t', by randomly sampling from a unit sphere around w_t. Then, w_t and w_t' are compared (e.g., when learning rankings, this could be done using interleaving). If w_t' wins the comparison, w_t is updated by a learning step in the direction of w_t'. If the solution space is convex, this approach is guaranteed to converge [79].

DBGD can be directly applied to e.g., online learning to rank settings, and was empirically validated on such a task [79]. However, it is more generally applicable, as it makes no specific assumptions about how solutions are compared, as long as assumptions of the algorithm regarding their stochastic characteristics hold. Hofmann et al. [29] demonstrated that, by taking structure into account, substantially better online and offline performance can be achieved in specific applications. For the task of online learning to rank from interleaved comparisons, an approach called Candidate Pre-Selection (CPS) was proposed. It leverages the exploration that is a side-effect of probabilistic interleave (see Sect. 5), to evaluate new ranker candidates. In comparison to DBGD, which explores uniformly around the current best solution, CPS uses offline estimates derived from exploration data to focus on the most promising candidates. The resulting approach learns significantly faster than the structure oblivious approach, and is particularly robust to noisy feedback [29]. Schuth et al. [66] that dueling bandit approaches can be successfully applied to learning the parameters of non-linear ranking functions.

In this and the previous section, we have provided a summary of the many online learning approaches that can enable interactive retrieval systems to learn directly from interactions with their users. Interestingly, the unique challenges posed by IR applications have motivated many recent advances in e.g., contextual and dueling bandit approaches. As these make their way into more and more interactive IR systems, we can expect to discover and solve new challenges.

8   Conclusion

In this paper we have presented an overview of techniques for online experimentation for IR. With the increase in web-scale IR systems, controlled experiments have been adapted to deal with the challenges of scale and complexity that these systems present. As well as moving insights into experimentation methodology into practical settings, new methods for measurement an learning have been developed that can in turn benefit IR research.

The basis of online experimentation for IR naturally are controlled experiments. In Sect. 2 we motivated why the causal relations they let us infer are crucial for systems that learn how to act, or interact, with their users. After motivating the need for controlled experimentation, we introduced the most well-known technique, A/B testing. A/B testing is the technique that is the most general, but limitations in terms of the scalability of comparisons. This gap is filled by methods for estimating online performance from exploration data, as discussed in Sect. 3. In Sect. 4 we briefly discussed recent trends in measuring online performance of IR systems. Section 5 concluded our discussion of online evaluation, by introducing interleaved comparison methods, which allow within-subject controlled experiments.

The first sections of this paper focused on online learning for IR. Online learning goes one step beyond the previously discussed online evaluation approaches where the comparisons to be performed were selected manually. Online learning using bandit approaches in particular can automatically select the required evaluations or comparisons in order to optimize online performance. A key challenge addressed by bandit approaches is the trade-off between exploring new solutions to obtain accurate performance estimates, and exploiting solutions with known high performance. The resulting approaches are especially valuable for online learning in IR systems, as they achieve high online performance while learning.

Following on from this overview article, many of the presented approaches can be tried out in the experimental framework lerot [65]. Using simulations of online interactions, online evaluation and learning approaches can be compared and developed further.

Online evaluation and learning have only recently been introduced to the IR community, and form a growing area of research within this community. Many open questions remain to be addressed. From the perspective of deploying online evaluation and learning approaches, we need to better understand the impact of exploration on the user experience. While exploration allows learning, and therefore improves system performance in the long run, it is not yet well-understood how users are affected in the short run, and how potential risks can be mitigated. Particularly valuable are exploration schemes that limit the risk for individual users. On the other hand, we need to better understand how to effectively explore in applications with large action spaces. The more we know about the solution space of a given IR problem, the more effectively we can design exploration schemes that use this structure to quickly focus on the most promising areas of the solution space.

Online experimentation has been embraced by owners of large web properties, and is a key part of the development process in these companies. In the research community, online experimentation seems to see somewhat slower adoption. Is one reason the difficulty in obtaining data or running experiments in an online setting? Exploration data may be a key to enabling wider participation in online experimentation. Another promising initiative is the CLEF living labs initiative, which brings together IR researchers and search engine operators, by providing a shared platform for online experimentation [6].

What questions can we study using online experimentation for IR? The methods presented in this article build on and complement the traditional toolset of IR experimentation. Online experimentation can expand IR research from small-scale and short-term lab studies to a wide range of naturalistic experiments. This will allow us to gain new insights into information seeking behavior, and into how retrieval systems can best address these.

Online learning for IR can transform the way in which IR systems are currently developed. By learning directly from user interactions, they can quickly adapt to changing user behavior and expectations. We will move away from developing systems for which behavior is completely specified before deployment, and will move towards defining a space of possible solutions. Online evaluation and online learning to rank will drive this development, towards IR systems that learn directly from their users.

References

1. Agrawal, S., Goyal, N.: Analysis of thompson sampling for the multi-armed bandit problem. In: COLT 2012 (2012)
2. Ailon, N., Karnin, Z., Joachims, T.: Reducing dueling bandits to cardinal bandits. In: ICML 2014 (2014)
3. Auer, P., Cesa-Bianchi, N., Fischer, P.: Finite-time analysis of the multiarmed bandit problem. Mach. Learn. **47**(2–3), 235–256 (2002)
4. Auer, P., Cesa-Bianchi, N., Freund, Y., Schapire, R.E.: The nonstochastic multi-armed bandit problem. SIAM J. Comput. **32**(1), 48–77 (2002)
5. Babbie, E.R.: The Practice of Social Research, 13th edn. Cengage Learning, Boston (2012)
6. Balog, K., Kelly, L., Schuth, A.: Head first: Living labs for ad-hoc search evaluation. In: CIKM 2014 (2014)
7. Bendersky, M., Garcia-Pueyo, L., Harmsen, J., Josifovski, V., Lepikhin, D.: Up next: Retrieval methods for large scale related video suggestion. In: KDD 2014 (2014)
8. Bottou, L., Chickering, J., Portugaly, E., Ray, D., Simard, P., Snelson, E.: Counterfactual reasoning and learning systems: The example of computational advertising. J. Mach. Learn. Res. **14**(1), 3207–3260 (2013)
9. Bubeck, S., Cesa-Bianchi, N.: Regret analysis of stochastic and nonstochastic multi-armed bandit problems. Found. Trends Mach. Learn. **5**(1), 1–122 (2012)
10. Busa-Fekete, R., Hüllermeier, E.: A survey of preference-based online learning with bandit algorithms. In: Auer, P., Clark, A., Zeugmann, T., Zilles, S. (eds.) ALT 2014. LNCS, vol. 8776, pp. 18–39. Springer, Heidelberg (2014)
11. Carterette, B.: Statistical significance testing in information retrieval: Theory and practice. In: ICTIR 2013 (2013)
12. Chakraborty, S., Radlinski, F., Shokouhi, M., Baecke, P.: On correlation of absence time and search effectiveness. In: SIGIR 2014, pp. 1163–1166 (2014)
13. Chapelle, O., Li, L.: An empirical evaluation of thompson sampling. In: NIPS 2011, pp. 2249–2257 (2011)

14. Chapelle, O., Zhang, Y.: A dynamic bayesian network click model for web search ranking. In: WWW 2009, pp. 1–10 (2009)
15. Chapelle, O., Joachims, T., Radlinski, F., Yue, Y.: Large-scale validation and analysis of interleaved search evaluation. ACM Trans. Inf. Syst. **30**(1), 6:1–6:41 (2012)
16. Chuklin, A., Schuth, A., Hofmann, K., Serdyukov, P., de Rijke, M.: Evaluating aggregated search using interleaving. In: CIKM 2013 (2013)
17. Chuklin, A., Schuth, A., Zhou, K., de Rijke, M.: A comparative analysis of inter-leaving methods for aggregated search. ACM Trans. Inf. Syst. (2014)
18. Craswell, N., Zoeter, O., Taylor, M., Ramsey, B.: An experimental comparison of click position-bias models. In: WSDM 2008, pp. 87–94 (2008)
19. Deng, A., Xu, Y., Kohavi, R., Walker, T.: Improving the sensitivity of online controlled experiments by utilizing pre-experiment data. In: WSDM 2013, pp. 123–132 (2013)
20. Diaz, F.: Adaptation of offline vertical selection predictions in the presence of user feedback. In: SIGIR 2009, pp. 323–330 (2009)
21. Dupret, G., Lalmas, M.: Absence time and user engagement. In: WSDM 2013, p. 173. ACM Press, New York, February 2013
22. Granka, L.A., Joachims, T., Gay, G.: Eye-tracking analysis of user behavior in www search. In: SIGIR 2004, pp. 478–479 (2004)
23. Guan, Z., Cutrell, E.: An eye tracking study of the effect of target rank on web search. In: CHI 2007, pp. 417–420 (2007)
24. Hassan, A., White, R.W.: Personalized models of search satisfaction. In: CIKM 2013, pp. 2009–2018 (2013)
25. Hofmann, K., Whiteson, S., de Rijke, M.: Balancing exploration and exploitation in learning to rank online. In: Clough, P., Foley, C., Gurrin, C., Jones, G.J.F., Kraaij, W., Lee, H., Mudoch, V. (eds.) ECIR 2011. LNCS, vol. 6611, pp. 251–263. Springer, Heidelberg (2011)
26. Hofmann, K., Whiteson, S., de Rijke, M.: A probabilistic method for inferring preferences from clicks. In: CIKM 2011, pp. 249–258 (2011)
27. Hofmann, K., Behr, F., Radlinski, F.: On caption bias in interleaving experiments. In: CIKM 2012, pp. 115–124. ACM Press (2012)
28. Hofmann, K., Whiteson, S., de Rijke, M.: Estimating interleaved comparison out-comes from historical click data. In: CIKM 2012, pp. 1779–1783 (2012)
29. Hofmann, K., Whiteson, S., de Rijke, M.: Balancing exploration and exploitation in listwise and pairwise online learning to rank for information retrieval. Inf. Retrieval J. **16**(1), 63–90 (2013)
30. Hofmann, K., Whiteson, S., de Rijke, M.: Fidelity, soundness, and efficiency of interleaved comparison methods. ACM Trans. Inf. Syst. **31**(4), 1–43 (2013)
31. Hofmann, K., Mitra, B., Radlinski, F., Shokouhi, M.: An eye-tracking study of user interactions with query auto completion. In: CIKM 2014 (2014)
32. Jie, L., Lamkhede, S., Sapra, R., Hsu, E., Song, H., Chang, Y.: A unified search federation system based on online user feedback. In: KDD 2013, pp. 1195–1203 (2013)
33. Jin, X., Sloan, M., Wang, J.: Interactive exploratory search for multi page search results. In: WWW 2013, pp. 655–666 (2013)
34. Joachims, T.: Optimizing search engines using clickthrough data. In: KDD 2002, pp. 133–142 (2002)
35. Joachims, T., Granka, L., Pan, B., Hembrooke, H., Radlinski, F., Gay, G.: Eval-uating the accuracy of implicit feedback from clicks and query reformulations in web search. ACM Trans. Inf. Syst. **25**(2), 1–26 (2007)

36. Kaufmann, E., Korda, N., Munos, R.: Thompson sampling: an asymptotically optimal finite-time analysis. In: Bshouty, N.H., Stoltz, G., Vayatis, N., Zeugmann, T. (eds.) ALT 2012. LNCS, vol. 7568, pp. 199–213. Springer, Heidelberg (2012)
37. Kazai, G., Kamps, J., Koolen, M., Milic-Frayling, N.: Crowdsourcing for book search evaluation: Impact of hit design on comparative system ranking. In: SIGIR 2011, pp. 205–214 (2011)
38. Kelly, D.: Methods for evaluating interactive information retrieval systems with users. Found. Trends Inf. Retrieval **3**(1–2), 1–224 (2009)
39. Kelly, D., Teevan, J.: Implicit feedback for inferring user preference: a bibliography. SIGIR Forum **37**(2), 18–28 (2003)
40. Kelly, D., Gyllstrom, K., Bailey, E.W.: A comparison of query and term suggestion features for interactive searching. In: SIGIR 2009, p. 371. ACM Press, New York, July 2009
41. Kim, Y., Hassan, A., White, R.W., Zitouni, I.: Modeling dwell time to predict click-level satisfaction. In: WSDM 2014, pp. 193–202. ACM, New York (2014)
42. Kleinberg, R., Slivkins, A., Upfal, E.: Multi-armed bandits in metric spaces. In: STOC 2008. ACM Press (2008)
43. Kohavi, R., Longbotham, R., Sommerfield, D., Henne, R.M.: Controlled experiments on the web: survey and practical guide. Data Min. Knowl. Disc. **18**(1), 140–181 (2009)
44. Kohavi, R., Deng, A., Frasca, B., Longbotham, R., Walker, T., Xu, Y.: Trustworthy online controlled experiments: Five puzzling outcomes explained. In: KDD 2012, pp. 786–794. ACM, New York (2012)
45. Kohavi, R., Deng, A., Frasca, B., Walker, T., Xu, Y., Pohlmann, N.: Online controlled experiments at large scale. In: KDD 2013, pp. 1168–1176. ACM, New York (2013)
46. Kohli, P., Salek, M., Stoddard, G.: A fast bandit algorithm for recommendation to users with heterogenous tastes. In: AAAI 2013 (2013)
47. Langford, J., Zhang, T.: The epoch-greedy algorithm for multi-armed bandits with side information. In: NIPS 2008, pp. 817–824 (2008)
48. Langford, J., Strehl, A., Wortman, J.: Exploration scavenging. In: ICML 2008, pp. 528–535 (2008)
49. Li, L., Chu, W., Langford, J., Schapire, R.E.: A contextual-bandit approach to personalized news article recommendation. In: WWW 2010, pp. 661–670 (2010)
50. Li, L., Chu, W., Langford, J., Wang, X.: Unbiased offline evaluation of contextual-bandit-based news article recommendation algorithms. In: WSDM 2011, pp. 297–306 (2011)
51. Li, L., Chen, S., Kleban, J., Gupta, A.: Couterfactual estimation and optimization of click metrics for search engines (2014). arXiv preprint arXiv:1403.1891
52. Luo, J., Zhang, S., Yang, H.: Win-win search: Dual-agent stochastic game in session search. In: SIGIR 2014, pp. 587–596. ACM (2014)
53. Mahajan, D.K., Rastogi, R., Tiwari, C., Mitra, A.: LogUCB: An explore-exploit algorithm for comments recommendation. In: CIKM 2012, pp. 6–15 (2012)
54. Pearl, J.: Causality: Models, Reasoning and Inference, vol. 29. Cambridge University Press, Cambridge (2000)
55. Pearl, J.: An introduction to causal inference. Int. J. Biostatistics 6(2) (2010)
56. Precup, D., Sutton, R.S., Singh, S.P.: Eligibility traces for off-policy policy evaluation. In: ICML 2000, pp. 759–766 (2000)
57. Radlinski, F., Craswell, N.: Comparing the sensitivity of information retrieval metrics. In: SIGIR 2010, pp. 667–674 (2010)

58. Radlinski, F., Craswell, N.: Optimized interleaving for online retrieval evaluation. In: WSDM 2013 (2013)
59. Radlinski, F., Joachims, T.: Minimally invasive randomization for collecting unbiased preferences from clickthrough logs. In: AAAI 2006, p. 1406 (2006)
60. Radlinski, F., Kleinberg, R., Joachims, T.: Learning diverse rankings with multi-armed bandits. In: ICML 2008, pp. 784–791. ACM (2008)
61. Radlinski, F., Kurup, M., Joachims, T.: How does clickthrough data reflect retrieval quality?. In: CIKM 2008, pp. 43–52 (2008)
62. Russo, D., Roy, B.V.: An information-theoretic analysis of thompson sampling. CoRR, abs/1403.5341 (2014). URL http://arxiv.org/abs/1403.5341
63. Sanderson, M.: Test collection based evaluation of information retrieval systems. Found. Trends Inf. Retrieval 4(4), 247–375 (2010)
64. Scholer, F., Shokouhi, M., Billerbeck, B., Turpin, A.: Using clicks as implicit judgments: expectations versus observations. In: Macdonald, C., Ounis, I., Plachouras, V., Ruthven, I., White, R.W. (eds.) ECIR 2008. LNCS, vol. 4956, pp. 28–39. Springer, Heidelberg (2008)
65. Schuth, A., Hofmann, K., Whiteson, S., de Rijke, M.: Lerot: an online learning to rank framework. In: LivingLab 2013, pP. 23–26. ACM (2013)
66. Schuth, A., Sietsma, F., Whiteson, S., de Rijke, M.: Optimizing base rankers using clicks. In: de Rijke, M., Kenter, T., de Vries, A.P., Zhai, C.X., de Jong, F., Radinsky, K., Hofmann, K. (eds.) ECIR 2014. LNCS, vol. 8416, pp. 75–87. Springer, Heidelberg (2014)
67. Schuth, A., Sietsma, F., Whiteson, S., Lefortier, D., de Rijke, M.: Multileaved comparisons for fast online evaluation. In: CIKM 2014 (2014)
68. Slivkins, A., Radlinski, F., Gollapudi, S.: Ranked bandits in metric spaces: learning diverse rankings over large document collections. J. Mach. Learn. Res. 14(1), 399–436 (2013)
69. Song, Y., Shi, X., Fu, X.: Evaluating and predicting user engagement change with degraded search relevance. In: WWW 2013, pp. 1213–1224 (2013)
70. Streeter, M., Golovin, D., Krause, A.: Online learning of assignments. In: NIPS 2009, pp. 1794–1802 (2009)
71. Sutton, R.S., Barto, A.G.: Introduction to Reinforcement Learning. MIT Press, Cambridge (1998)
72. Tang, D., Agarwal, A., O'Brien, D., Meyer, M.: Overlapping experiment infrastructure: More, better, faster experimentation. In: KDD 2010, pp. 17–26 (2010)
73. Tang, L., Rosales, R., Singh, A., Agarwal, D.: Automatic ad format selection via contextual bandits. In: CIKM 2013, pp. 1587–1594 (2013)
74. Valko, M., Carpentier, A., Munos, R.: Stochastic simultaneous optimistic optimization. In: ICML 2013, pp. 19–27 (2013)
75. Voorhees, E.M., Harman, D.K.: TREC: Experiment and Evaluation in Information Retrieval. Digital Libraries and Electronic Publishing. MIT Press, Cambridge (2005)
76. Wang, K., Walker, T., Zheng, Z.: PSkip: estimating relevance ranking quality from web search clickthrough data. In: KDD 2009, pp. 1355–1364 (2009)
77. Watkins, C.J.C.H.: Learning from delayed rewards. Ph.D. thesis, University of Cambridge (1989)
78. Yue, Y., Guestrin, C.: Linear submodular bandits and their application to diversified retrieval. In: Shawe-Taylor, J., Zemel, R., Bartlett, P., Pereira, F., Weinberger, K. (eds.) NIPS 2011, pp. 2483–2491 (2011)
79. Yue, Y., Joachims, T.: Interactively optimizing information retrieval systems as a dueling bandits problem. In: ICML 2009, pp. 1201–1208 (2009)

80. Yue, Y., Joachims, T.: Beat the mean bandit. In: ICML 2011 (2011)
81. Yue, Y., Broder, J., Kleinberg, R., Joachims, T.: The K-armed dueling bandits problem. In: COLT 2009 (2009)
82. Yue, Y., Patel, R., Roehrig, H.: Beyond position bias: examining result attractiveness as a source of presentation bias in clickthrough data. In: WWW 2010, pp. 1011–1018 (2010)
83. Yue, Y., Broder, J., Kleinberg, R., Joachims, T.: The K-armed dueling bandits problem. J. Comput. Syst. Sci. **78**(5), 1538–1556 (2012)
84. Zoghi, M., Whiteson, S.A., de Rijke, M., Munos, R.: Relative confidence sampling for efficient on-line ranker evaluation. In: WSDM 2014, pp. 73–82 (2014)
85. Zoghi, M., Whiteson, S.A., Munos, R., de Rijke, M.: Relative upper confidence bound for the K-armed dueling bandit problem. In: ICML 2014 (2014)

Introduction to Formal Concept Analysis and Its Applications in Information Retrieval and Related Fields

Dmitry I. Ignatov[✉]

National Research University Higher School of Economics, Moscow, Russia
dignatov@hse.ru

Abstract. This paper is a tutorial on Formal Concept Analysis (FCA) and its applications. FCA is an applied branch of Lattice Theory, a mathematical discipline which enables formalisation of concepts as basic units of human thinking and analysing data in the object-attribute form. Originated in early 80s, during the last three decades, it became a popular human-centred tool for knowledge representation and data analysis with numerous applications. Since the tutorial was specially prepared for RuSSIR 2014, the covered FCA topics include Information Retrieval with a focus on visualisation aspects, Machine Learning, Data Mining and Knowledge Discovery, Text Mining and several others.

Keywords: Formal Concept Analysis · Concept lattices · Information retrieval · Machine learning · Data mining · Knowledge discovery · Text mining · Biclustering · Multimodal clustering

1 Introduction

According to [1], "information retrieval (IR) is finding material (usually documents) of an unstructured nature (usually text) that satisfies an information need from within large collections (usually stored on computers)." In the past, only specialized professions such as librarians had to retrieve information on a regular basis. These days, massive amounts of information are available on the Internet and hundreds of millions of people make use of information retrieval systems such as web or email search engines on a daily basis. Formal Concept Analysis (FCA) was introduced in the early 1980 s by Rudolf Wille as a mathematical theory [2,3] and became a popular technique within the IR field. FCA is concerned with the formalisation of concepts and conceptual thinking and has been applied in many disciplines such as software engineering, machine learning, knowledge discovery and ontology construction during the last 20–25 years. Informally, FCA studies how objects can be hierarchically grouped together with their common attributes.

The core contributions of this tutorial from IR perspective are based on our surveys [4–6] and experiences in both fields, FCA and IR. In our surveys we visually represented the literature on FCA and IR using concept lattices as well

© Springer International Publishing Switzerland 2015
P. Braslavski et al. (eds.): RuSSIR 2014, CCIS 505, pp. 42–141, 2015.
DOI: 10.1007/978-3-319-25485-2_3

as several related fields, in which the objects are the scientific papers and the attributes are the relevant terms available in the title, keywords and abstract of the papers. You can see an example of such a visualisation in Fig. 1 for papers published between 2003 and 2009. We developed a toolset with a central FCA component that we used to index the papers with a thesaurus containing terms related to FCA research and to generate the lattices. The tutorial also contains a partial overview of the papers on using FCA in Information Retrieval with a focus on visualisation.

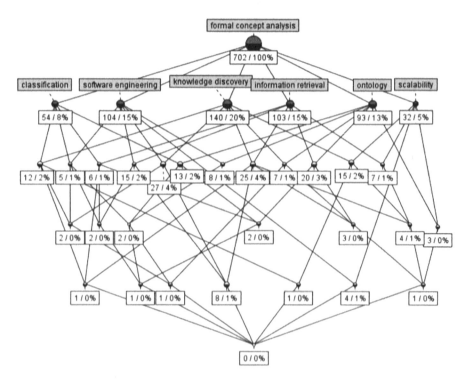

Fig. 1. The lattice diagram representing collection of 702 papers on FCA including 103 papers on FCA and IR (2003–2009).

In 2013 European Conference on Information Retrieval [7] hosted a thematic workshop FCA meets IR which was devoted to two main issues:

- How can FCA support IR activities including but not limited to query analysis, document representation, text classification and clustering, social network mining, access to semantic web data, and ontology engineering?
- How can FCA be extended to address a wider range of IR activities, possibly including new retrieval tasks?

Claudio Carpineto delivered an invited lecture at the workshop – "FCA and IR: The Story So Far". The relevant papers and results presented there are also discussed in the tutorial.

Since the tutorial preparations were guided by the idea to present the content at a solid and comprehensible level accessible even for newcomers, it is a balanced combination of theoretical foundations, practice and relevant applications. Thus we provide intro to FCA, practice with main tools for FCA, discuss FCA in Machine Learning and Data Mining, FCA in Information Retrieval and Text Mining, FCA in Ontology Modeling and other selected applications. Many of the used examples are real-life studies conducted by the course author.

The target audience is Computer Science, Mathematics and Linguistics students, young scientists, university teachers and researchers who want to use FCA models and tools in their IR and data analysis tasks.

The course features five parts. Each part placed in a separate section and contains a short highlight list to ease the navigation within the material. Section 2 contains introduction to FCA and related notions of Lattice and Order Theory. In Sect. 3, we describe selected FCA tools and provide exercises. Section 4 provides an overview of FCA-based methods and applications in Data Mining and Machine Learning, and describes an FCA-based tool for supervised learning, QuDA (Qualitative Data Analysis). Section 5 presents the most relevant part of the course, FCA in Information Retrieval and Text Mining. Penultimate Sect. 6 discusses FCA in Ontology Modeling and gives an example of FCA-based Attribute Exploration technique on building the taxonomy of transportation means. Section 7 concludes the paper and briefly outlines prospects and limitations of FCA-based models and techniques.

2   Introduction to FCA

Even though that many disciplines can be dated back to Aristotles time, more closer prolegomena of FCA can be found, for example, in the Logic of Port Royal (1662)[8], an old philosophical concept logic, where a concept was treated as a pair of its extent and its intent (yet without formal mathematical apparatus).

Being a part of lattice theory, concept lattices are deeply rooted in works of Dedekind, Birkgoff [9] (Galois connections and "polarities"), and Ore [10] (Galois connections), and, later, on Barbut and Monjardet [11] (treillis de Galois, i.e. Galois lattices).

In fact, the underlying structure, Galois connection, has a strong impact in Data Analysis [12–15].

In this section, we mainly reproduce basic definitions from Ganter and Wille's book on Formal Concept Analysis [3]. However, one can find a good introductory material, more focused on partial orders and lattices, in the book of Davey and Priestly [16]. An IR-oriented reader may also find the following books interesting and helpful [15, 17].

There were several good tutorials with notes in the past, for example, a basic one [18] and more theoretical with algorithmic aspects [19].

We also refer the readers to some online materials that might be suitable for self-study purposes[1,2,3].

A short section summary:

- Binary Relations, Partial Orders, Lattices, Line (Hasse) Diagram.
- Galois Connection, Formal Context, Formal Concept, Concept Lattice.
- Concept Lattice drawing. Algorithms for concept lattices generation (naïve, Ganter's algorithm, Close-by-One).
- Attribute Dependencies: implications, functional dependencies. Armstrong Rules. Implication bases (Stem Base, Generator base).
- Many-valued contexts. Concept scaling.

2.1   Binary Relations, Partial Orders, Lattices, Hasse Diagram

The notion of a **set** is fundamental in mathematics. In what follows, we consider only finite sets of objects.

Definition 1. *A **binary relation** R between two sets A and B is a set of all pairs (a, b) with $a \in A$ and $b \in B$., i.e., a subset of their Cartesian product $A \times B$, the set of all such pairs.*

Sometimes it is convenient to write aRb instead $(a, b) \in R$ for brevity. If $A = B$ then $R \subseteq A \times A$ is called a binary relation on the set A.

Definition 2. *A binary relation R on a set A is called a **partial order relation** (or shortly a partial order), if it satisfies the following conditions for all elements $a, b, c \in A$:*

1. *aRa (reflexivity)*
2. *aRb and $a \neq b \implies$ not aRb (antisymmetry)*
3. *aRb and $bRc \implies aRc$ (transitivity)*

We use symbol $\leq$ for partial order, and in case $a \leq b$ and $a \neq b$ we write $a \leq b$. We read $a \leq b$ as "a is less of equal to b". A partially ordered set (or poset) is a pair $(P, \leq)$, where P is a set and $\leq$ is an partial order on P.

Definition 3. *Given a poset $(P, \leq)$, an element a is called a **lower neighbour** of b, if $a \leq b$ and there is no such c fulfilling $a \leq c \leq b$. In this case, b is also an **upper neighbour** of a, and we write $a \prec b$.*

Every finite ordered poset $(P, \leq)$ can be depicted as a **line diagram** (many authors call it Hasse diagram). Elements of P are represented by small circles in the plane. If $a \leq b$, the circle corresponding to a is depicted higher than the circle corresponding to b, and the two circles are connected by a line segment. One can check whether some $a \leq b$ if there is an ascending path from b to a in the diagram.

---

[1] http://www.kbs.uni-hannover.de/~jaeschke/teaching/2012w/fca/.
[2] http://www.upriss.org.uk/fca/fcaintro.html.
[3] http://ddll.inf.tu-dresden.de/web/Introduction_to_Formal_Concept_Analysis_ (WS2014)/en.

Example 1. The poset P is given by its incidence cross-table, where $\times$ in a cell means that the corresponding pair of row and column elements x and y are related as follows $x \le y$.

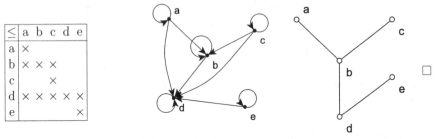

$\le$	a	b	c	d	e
a	$\times$				
b	$\times$	$\times$	$\times$		
c			$\times$		
d	$\times$	$\times$	$\times$	$\times$	$\times$
e					$\times$

The graph of P.

The line diagram of P.

Definition 4. *Let $(P, \le)$ be a poset and A a subset of P. A **lower bound** of A is an element l of P with $l \le A$ for all $a \in A$. An **upper bound** of A is defined dually. If there is a largest element in the set of all lower bounds of A, it is called the **infimum** of A and is denoted by $\inf A$ or $\bigwedge A$. Dually, if there is a smallest element in the set of all upper bounds, it is called **supremum** and denoted by $\sup A$ or $\bigvee A$.*

For $A = \{a, b\}$ we write $x \wedge y$ for $\inf A$ and $x \vee y$ for $\sup A$. Infimum and supremum are also called **meet** and **join**.

Definition 5. *A poset $\mathbf{L} = (L, \le)$ is a **lattice**, if for any two elements a and b in L the supremum $a \vee b$ and the infimum $a \wedge b$ always exist. $\mathbf{L}$ is called a **complete lattice**, if the supremum $\bigvee X$ and the infimum $\bigwedge X$ exist for any subset A of L. For every complete lattice $\mathbf{L}$ there exist its largest element, $\bigvee L$, called the **unit element** of the lattice, denoted by $\mathbf{1}_L$. Dually, the smallest element $\mathbf{0}_L$ is called the **zero element**.*

Example 2. In Fig. 2 there are the line diagrams of the poset P, which is not a lattice, and the lattice L. It is interesting that P has its largest and smallest elements, $\mathbf{1}_P$ and $\mathbf{0}_P$; the pair of its elements, s and t, has its infumum, $s \vee t = \mathbf{0}_P$, but there is no a supremum for it. In fact, p, t does not have a smallest element in the set of all its upper bounds. $\square$

3   Galois Connection, Formal Context, Formal Concept, Concept Lattice

Definition 6. *Let $\varphi : P \to Q$ and $\psi : Q \to P$ be maps between two posets $(P, \le)$ and $(Q, \le)$. Such a pair of maps is called a **Galois connection** between the ordered sets if:*

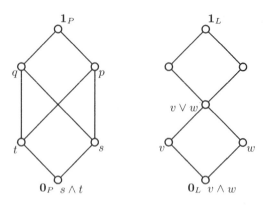

Fig. 2. The line diagrams of the order, which is not a lattice (left), and the order, which is a lattice (right)

1. $p_1 \leq p_2 \Rightarrow \varphi p_1 \geq \varphi p_2$
2. $q_1 \leq q_2 \Rightarrow \psi q_1 \geq \psi q_2$
3. $p \leq \psi \varphi p \Rightarrow q \leq \varphi \psi q$.

Exercise 1. Prove that a pair (φ, ψ) of maps is a Galois connection if and only if $p \leq \psi q \Leftrightarrow q \leq \psi p$.   □

Exercise 2. Prove that for every Galois connection (φ, ψ)

$$\psi = \psi \varphi \psi \text{ and } \varphi = \varphi \psi \varphi.$$

                                                                                                                          □

Definition 7. *A **formal context** $\mathbb{K} = (G, M, I)$ consists of two sets G and M and a relation I between G and M. The elements of G are called the **objects** and the elements of M are called the **attributes** of the context. The notation gIm or $(g, m) \in I$ means that the object g has attribute m.*

Definition 8. *For $A \subseteq G$, let*

$$A' := \{m \in M | (g, m) \in I \text{ for all } g \in A\}$$

and, for $B \subseteq M$, let

$$B' := \{g \in G | (g, m) \in I \text{ for all } m \in B\}.$$

    *These operators are called **derivation operators** or **concept-forming operators** for $\mathbb{K} = (G, M, I)$.*

Proposition 1. *Let (G, M, I) be a formal context, for subsets $A, A_1, A_2 \subseteq G$ and $B \subseteq M$ we have*

1. $A_1 \subseteq A_2$ iff $A_2' \subseteq A_1'$,
2. $A \subseteq A''$,
3. $A = A'''$ (hence, $A'''' = A''$),
4. $(A_1 \cup A_2)' = A_1' \cap A_2'$,
5. $A \subseteq B' \Leftrightarrow B \subseteq A' \Leftrightarrow A \times B \subseteq I$.

Similar properties hold for subsets of attributes.

Exercise 3. Prove properties of operator $(\cdot)'$ from proposition 1.   □

Definition 9. *A **closure operator** on set G is a mapping $\varphi: 2^G \to 2^G$ with the following properties:*

1. $\varphi\varphi X = \varphi X$ *(**idempotency**)*
2. $X \subseteq \varphi X$ *(**extensity**)*
3. $X \subseteq Y \Rightarrow \varphi X \subseteq \varphi Y$ *(**monotonicity**)*

For a closure operator φ the set φX is called **closure** of X.
A subset $X \subseteq G$ is called **closed** if $\varphi X = X$.

Exercise 4. Let (G, M, I) be a context, prove that operators

$$(\cdot)'': 2^G \to 2^G, \ (\cdot)'': 2^M \to 2^M$$

are closure operators.   □

Definition 10. *A **formal concept** of a formal context $\mathbb{K} = (G, M, I)$ is a pair (A, B) with $A \subseteq G$, $B \subseteq M$, $A' = B$ and $B' = A$. The sets A and B are called the **extent** and the **intent** of the formal concept (A, B), respectively. The **subconcept-superconcept relation** is given by $(A_1, B_1) \le (A_2, B_2)$ iff $A_1 \subseteq A_2$ $(B_1 \subseteq B_2)$.*

This definition says that every formal concept has two parts, namely, its extent and intent. This follows an old tradition in the *Logic of Port Royal (1662)*, and is in line with the International Standard ISO 704 that formulates the following definition: "A concept is considered to be a unit of thought constituted of two parts: its extent and its intent."

Definition 11. *The set of all formal concepts of a context $\mathbb{K}$ together with the order relation I forms a complete lattice, called the **concept lattice** of $\mathbb{K}$ and denoted by $\underline{\mathfrak{B}}(\mathbb{K})$.*

Example 3. The context with four geometric figures and four attributes is below.

G \ M	a	b	c	d
1 △	×		×	
2 ◣	×	×		
3 ▭		×	×	
4 ▢		×	×	×

Objects:
1 – equilateral triangle,
2 – rectangle triangle,
3 – rectangle,
4 – square.

Attributes:
a – has 3 vertices,
b – has 4 vertices,   □
c – has a direct angle,
d – equilateral.

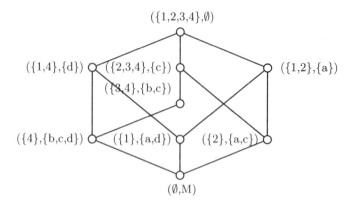

Fig. 3. The line diagram of the concept lattice for the context of geometric figures

Definition 12. *For every two formal concepts* (A_1, B_1) *and* (A_2, B_2) *of a certain formal context their* **greatest common subconcept** *is defined as follows:*

$$(A_1, B_1) \wedge (A_2, B_2) = (A_1 \cap A_2, (B_1 \cup B_2)'').$$

The **least common superconcept** *of* (A_1, B_1) *and* (A_2, B_2) *is given as*

$$(A_1, B_1) \vee (A_2, B_2) = ((A_1 \cup A_2)'', B_1 \cap B_2).$$

We say supremum instead of "least common superconcept", and instead of "greatest common subconcept" we use the term infimum.

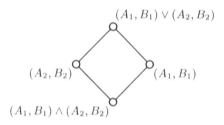

Fig. 4. Supremum and infimum of two concepts

It is possible to define supremum and infumum operations for an arbitrary set of concepts of a certain context. This is done in the first part of Theorem 1.

Definition 13. *A subset* $X \subseteq L$ *of lattice* $(L, \leq)$ *is called* **supremum-dense** *if any lattice element* $v \in L$ *can be represented as*

$$v = \bigvee \{x \in X \mid x \leq v\}.$$

Dually for **infimum-dense** *subsets.*

The Basic Theorem of Formal Concept Analysis below defines not only supre-mum and infimum of arbitrary sets of concepts; it also answer the question whether concept lattices have any special properties. In fact, the answer is "no" since every concept lattice is (isomorphic to some) complete lattice. That is one can compose a formal context with objects G, attributes M and binary relation $I \subset G \times M$ such that the original complete lattice is isomorphic $\underline{\mathfrak{B}}(G, M, I)$. Even though the theorem does not reply how such a context can be built, but rather describes all possibilities to do this.

Theorem 1. *Basic Theorem of Formal Concept Analysis* (*[Wille 1982]*, *[Ganter, Wille 1996]*)

Concept lattice $\underline{\mathfrak{B}}(G, M, I)$ is a complete lattice. For arbitrary sets of formal concepts
$$\{(A_j, B_j) \mid j \in J\} \subseteq \underline{\mathfrak{B}}(G, M, I)$$
their infimum and supremum are given in the following way:

$$\bigwedge_{j \in J}(A_j, B_j) = (\bigcap_{j \in J} A_j, (\bigcup_{j \in J} B_j)''),$$

$$\bigvee_{j \in J}(A_j, B_j) = ((\bigcup_{j \in J} A_j)'', \bigcap_{j \in J} B_j).$$

A complete lattice L is isomorphic to a lattice $\underline{\mathfrak{B}}(G, M, I)$ iff there are mappings $\gamma \colon G \to V$ and $\mu \colon M \to V$ such that $\gamma(G)$ is supremum-dense in $\mathbf{L}$, $\mu(M)$ is infimum-dense in $\mathbf{L}$, and $gIm \Leftrightarrow \gamma g \leq \mu m$ for all $g \in G$ and all $m \in M$. In particular, $\mathbf{L}$ is isomorphic to $\underline{\mathfrak{B}}(L, L, \leq)$.

An interested reader may refer to Ganter&Wille's book on FCA [3] for further detailed and examples.

3.1   Concept Lattice Drawing and Algorithms for Concept Lattices Generation

One can obtain the whole set of concepts of a particular context $\mathbb{K}$ simply by definition, i.e. it is enough to enumerate all subsets of objects $A \subseteq G$ (or attributes $B \subseteq M$) and apply derivation operators to them. For example, for the context from Example 3 and empty set of objects, $A = \emptyset$, one may obtain $A' = \emptyset' = \{a, b, c, d\} = B$, and then by applying $(\cdot)'$ second time $B' = \emptyset$. Thus, the resulting concept is $(A, B) = (\emptyset, M)$.

Proposition 2. *Every formal concept of a context (G, M, I) has the form (X'', X') for some subset $X \subseteq G$ and the form (Y', Y'') for some subset $Y \subseteq M$. Vice versa all such pairs of sets are formal concepts.*

One may follow the naïve algorithm below:

1. $\underline{\mathfrak{B}}(G, M, I) := \emptyset$
2. For every subset X of G, add (X'', X') to $\underline{\mathfrak{B}}(G, M, I)$.

Exercise 5. 1. Prove proposition 2. 2. For the context of geometric figures from Example 3 find all formal concepts.   ☐

Since the total number of formal concept is equal to $2^{\min(|G|,|M|)}$ in the worst case, this naïve approach is quite inefficient even for small contexts. However, let us assume that now we know how find concepts and we are going to build the line diagram of a concept lattice.

1. Draw a rather small circle for each formal concept such that a circle for a concept is always depicted higher than the all circles for its subconcepts.
2. Connect each circle with the circles of its lower neighbors.

To label concepts by attribute and object names in a concise form, we need the notions of object and attributes concepts.

Definition 14. *Let (G, M, I) be a formal context, then for each object $g \in G$ there is the **object concept** $(\{g\}'', \{g\}')$ and for each attribute $m \in M$ the **attribute concept** is given by $(\{m\}', \{m\}'')$.*

So, if one has finished a line diagram drawing for some concept lattice, it is possible to label the diagram with attribute names: one needs to attach the attribute m to the circle representing the concept $(\{m\}', \{m\}'')$. Similarly for labeling by object names: one needs to attach each object g to the circle representing the concept $(\{g\}'', \{g\}')$. An example of such reduced labeling is given in Fig. 5.

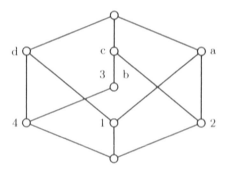

Fig. 5. An example of reduced labeling for the lattice of geometric figures

The naïve concept generation algorithm is not efficient since it enumerates all subsets of G (or M). For homogeneity, in what follows we reproduce the pseudocodes of the algorithms from [20]. There are different algorithms that compute closures for only some subsets of G and use an efficient test to check whether the current concept is generated first time (canonicity test). Thus, Ganter's Next Closure algorithm does not refer the list of generated concepts and uses little storage space.

Since the extent of a concept defines its intent in a unique way, to obtain the set of all formal concepts, it is enough to find closures either of subsets of objects or subsets of attributes.

We assume that there is a linear order ($<$) on G. The algorithm starts by examining the set consisting of the object maximal with respect to $<$ ($max(G)$), and finishes when the canonically generated closure is equal to G. Let A be a currently examined subset of G. The generation of A'' is considered canonical if $A'' \setminus A$ does not contain $g < max(A)$. If the generation of A'' is canonical (and A'' is not equal to G), the next set to be examined is obtained from A'' as follows:

$$A'' \cup \{g\} \setminus \{h | h \in A'' \wedge g < h\}, \text{ where } g = max(\{h | h \in G \setminus A''\}).$$

Otherwise, the set examined at the next step is obtained from A in a similar way, but the added object must be less (w.r.t. $<$) than the maximal object in A:

$$A'' \cup \{g\} \setminus \{h | h \in A \wedge g < h\}, \text{ where } g = max(\{h | h \in G \setminus A \wedge h < max(A)\}).$$

The pseudocode code is given in Algorithm 1 and the generation protocol of NEXTCLOSURE for the context of geometric figures is given in Table 1.

Algorithm 1. NextClosure

Input: $\mathbb{K} = (G, M, I)$ is a context
Output: L is the concept set
 1: $L := \emptyset, A := \emptyset, g := max(G)$
 2: **while** $A \neq G$ **do**
 3:     $A := A'' \cup \{g\} \setminus \{h | h \in A \wedge g < h\}$
 4:     **if** $\{h | h \in A \wedge g \leq h\} = \emptyset$ **then**
 5:         $L := L \cup \{(A'', A')\}$
 6:         $g := g = max(\{h | h \in G \setminus A''\})$
 7:         $A := A''$
 8:     **else**
 9:         $g = max(\{h | h \in G \setminus A \wedge h < g\})$
10:     **end if**
11: **end while**
12: **return** L

The NextClosure algorithm produces the set of all concepts in time $O(|G|^2 |M||L|)$ and has polynomial delay $O(|G|^2|M|)$.

We provide a simple recursive version of CBO. The algorithm generates concepts according to the lectic (lexicographic) order on the subsets of G (concepts whose extents are lectically less are generated first). By definition A is lectically less than B if $A \subseteq B$, or $B \not\subseteq A$ and $min((A \cup B) \setminus (B \cap A)) \in A$. Note that the NEXTCLOSURE algorithm computes concepts in a different lectic

Table 1. Generation protocol of NextClosure for the context of geometric figures

g	A	A''	formal concept (A, B)
4	$\{4\}$	$\{4\}$	$(\{4\}, \{2,4\})$
3	$\{3\}$	$\{3\}$	$(\{3\}, \{2,3\})$
4	$\{3,4\}$	$\{3,4\}$	$(\{3,4\}, \{2\})$
2	$\{2\}$	$\{1,2\}$	non-canonic generation
1	$\{1\}$	$\{1\}$	$(\{1\}, \{1,3,4\})$
4	$\{1,4\}$	$\{1,4\}$	$(\{1,4\}, \{4\})$
3	$\{1,3\}$	$\{1,2,3\}$	non-canonic generation
2	$\{1,2\}$	$\{1,2\}$	$(\{1,2\}, \{1,3\})$
4	$\{1,2,4\}$	$\{1,2,3,4\}$	non-canonic generation
3	$\{1,2,3\}$	$\{1,2,3\}$	$(\{1,2,3\}, \{3\})$
4	$\{1,2,3,4\}$	$\{1,2,3,4\}$	$(\{1,2,3,4\}, \{\})$

order: A is lectically less than B if $min((A \cup B) \setminus (B \cap A)) \in B$. The order in which concepts are generated by CBO is beneficial when the line diagram is constructed: the first generation of the concept is always canonical, which makes it possible to find a concept in the tree and to draw appropriate diagram edges. NEXTCLOSURE-like lectic order allows binary search, which is helpful when the diagram graph has to be generated after the generation of all concepts.

Algorithm 2. Close by One

Input: $\mathbb{K} = (G, M, I)$ is a context
Output: L is the concept set
1: $L := \emptyset$
2: **for all** $g \in G$ **do**
3:     Process($\{g\}, g, (\{g\}'', g)$)
4: **end for**
5: **return** L

The time complexity of CLOSE BY ONE (CBO) is $O(|G|^2|M||L|)$, and its polynomial delay is $O(|G|^3|M|)$.

The generation protocol of CBO in a tree-like form is given in Fig. 6. Each closed set of objects (extent) can be read from the tree by following the path from the root to the corresponding node. Square bracket] means that first prime operator has been applied after addition of the lectically next object g to the set A of the parent node and bracket) shows which object have been added after application of second prime operator, i.e. between] and) one can find $(A \cup \{g\})'' \setminus (A \cup \{g\})$. A non-canonic generation can be identified by simply checking whether there is an object between] and) that less than g w.r.t. $<$. One can note that the traverse of the generation tree is done in a depth-first search manner.

Algorithm 3. Process$(A, g, (C, D))$ with $C = A''$ and $D = A'$ and ¡ the lexical order on object names

Input: $\mathbb{K} = (G, M, I)$ is a context
Output: L is the concept set $C = A'', D = A'$
1: **if** $\{h|h \in C \setminus A \land g < h\} = \emptyset$ **then**
2:    $L := L \cup \{(C, D)\}$
3: **end if**
4: **for all** $f \in \{h|h \in G \setminus A \land g < h\}$ **do**
5:    $Z := C \cup \{f\}$
6:    $Z := D \cap \{f\}'$
7:    $X := Y'$
8:    Process$(Z, f, (X, Y))$
9: **end for**

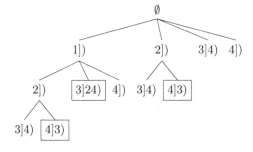

Fig. 6. The tree of CbO protocol for the context of geometric figures. Non-canonic generations are drawn in boxes.

After the inception of the first batch algorithms, the broadened FCA inventory includes efficient incremental algorithms [21] and the distributed versions of NextClosure and CbO for MapReduce [22, 23].

3.2 Many-Valued Contexts and Concept Scaling

Definition 15. *A **many-valued context** (G, M, W, I) consists of sets G, M and W and a ternary relation I between those three sets, i.e. $I \subseteq G \times M \times W$, for which it holds that $(g, m, w) \in I$ and $(g, m, v) \in I$ always imply $w = v$ The fact $(g, m, w) \in I$ means "the attribute m takes value w for object g", simply written as $m(g) = w$.*

Definition 16. *A (conceptual) **scale** for the attribute m of a many-valued context is a (one-valued) context $S_m = (G_m, M_m, I_m)$ with $m(G) = \{m(g)|\forall g \in G\} \subseteq G_m$. The objects of a scale are called **scale values**, the attributes are called **scale attributes**.*

Nominal scale is defined by the context $(W_m, W_m, =)$.

Table 2. Many-valued context of university subjects

G / M	Gender	Age	Subject	Mark
1	M	19	Math	8
2	F	20	CS	9
3	F	19	Math	7
4	M	20	CS	10
5	F	21	Data Mining	9

This type of scaling is suitable for binary representation of nominal (categorical) attributes like color. For the context of university subjects, the subjects can be scaled by nominal scaling as below.

=	Math	CS	DM
Math	×		
CS		×	
DM			×

A particular case of nominal scaling is the so called **dichotomic scaling**, which is suitable for attributes with two mutually exclusive values like "yes" and "no". In our example, the attribute Gender can be scaled in this way.

	M	F
M	×	
F		×

Ordinal scale is given by the context $(W_m, W_m, \leq)$ where denotes classical real number order. For our example, the attributes age and mark can be scaled by this type of scale.

	≤ 21	≤ 20	≤ 19
19	×	×	×
20		×	×
21			×

	≤ 7	≤ 8	≤ 9	≤ 10
7	×	×	×	×
8		×	×	×
9			×	×
10				×

Interordinal scale is given by $(W_m, W_m, \leq)|(W_m, W_m, \geq)$ where | denotes the apposition of two contexts.

This type of scale can be used as an alternative for ordinal scaling like in example below.

	≤ 7	≤ 8	≤ 9	≤ 10	≥ 7	≥ 8	≥ 9	≥ 10
7	×	×	×	×	×			
8		×	×	×	×	×		
9			×	×	×	×	×	
10				×	×	×	×	×

In some domains, e.g., in psychology or sociology there is similar biordinal (bipolar) scaling, which is a good representation of attributes with so called polarvalues "agree", "rather agree", "disagree", and "rather disagree".

There is a special type of scale, **contranominal scale**, which is rare case in real data, but has important theoretical meaning. Its context is given by inequality relation, i.e. $(\{1,\dots,n\}, \{1,\dots,n\}, \neq)$, and the example for $n = 4$ is given below.

$\neq$	1	2	3	4
1		×	×	×
2	×		×	×
3	×	×		×
4	×	×	×	

In fact, this type of contexts gives rise to 2^n formal concepts and can be used for testing purposes.

The resulting scaled (or plain) context for our university subjects example is below. Note that the Mark attribute is scaled by interordinal scale.

	M	F	≤ 19	≤ 20	≤ 21	Math	CS	DM	≤ 7	≤ 8	≤ 9	≤ 10	≥ 7	≥ 8	≥ 9	≥ 10
1	×		×	×	×	×				×	×	×	×	×		
2		×		×	×		×			×	×	×	×	×	×	
3		×	×	×	×	×		×	×	×	×	×				
4	×			×	×		×			×	×	×	×	×	×	×
5		×			×			×		×	×	×	×	×	×	

3.3 Attribute Dependencies

Definition 17. *Implication* $A \to B$, *where* $A, B \subseteq M$ *holds in context* (G, M, I) *if* $A' \subseteq B'$, *i.e., each object having all attributes from* A *also has all attributes from* B.

Example 4. For the context of geometric figures one may check that the following implication holds: $abc \to d$, $b \to c$, $cd \to b$. Note that for brevity we have omitted curly brackets around and commas between elements of a set attributes. □

Exercise 6. Find three more implications for the context of geometric figures. □

Implications satisfy **Armstrong rules** or inference axioms [24,25]:

$$\frac{}{X \to X} \text{ (reflexivity)}, \qquad \frac{X \to Y}{X \cup Z \to Y} \text{ (augmentation)},$$

$$\frac{X \to Y, Y \cup Z \to W}{X \cup Z \to W} \text{ (pseudotransitivity).}$$

An inference axiom is a rule that states if certain implications are valid in the context, then certain other implications are valid.

Example 5. Let us check that the first and second Armstrong axioms fulfill for implication over attributes.

Since $X' \subseteq X'$ it is always true that $X \to X$.

For the second rule we have $X' \subseteq Y'$. Applying property 4 from Proposition 1 we have: $(X \cup Z)' = X' \cap Z'$. Since $X' \cap Z' \subseteq X'$, we prove that $X' \cap Z' \subseteq Y'$. This implies $X \cup Z \to Y$. □

Exercise 7. 1. Prove by applying Armstrong rules that $A_1 \to B_1$ and $A_2 \to B_2$ imply $A_1 \cup A_2 \to B_1 \cup B_2$. 2. Check the third axiom by using implication definition. □

Definition 18. An **implication cover** *is a subset of implications from which all other implications can be derived by means of Armstrong rules.*

An **implication base** *is a minimal (by inclusion) implication cover.*

Definition 19. *A subset of attributes $D \subseteq M$ is a **generator** of a closed subset of attributes $B \subseteq M$, $B'' = B$ if $D \subseteq B$, $D'' = B = B''$.*

*A subset $D \subseteq M$ is a **minimal generator** if for any $E \subset D$ one has $E'' \neq D'' = B''$.*

*Generator $D \subseteq M$ is called **nontrivial** if $D \neq D'' = B''$.*

Denote the set of all nontrivial minimal generators of B by nmingen(B).

Generator implication cover *looks as follows:*

$$\{F \to (F'' \setminus F) \mid F \subseteq M, F \in \text{ nmingen } (F'')\}.$$

Example 6. For the context of geometric figures one may check that b is a minimal nontrivial generator for bc, The set ab is a minimal nontrivial generator for $abcd$, but abc, abd, and acd are its nontrivial generators. □

Exercise 8. For the context of geometric figures find all minimal generators and obtain its generator implication cover. □

Definition 20. *The **Duquenne-Guigues base** is an implication base where each implication is a pseudo-intent [26].*

*A subset of attributes $P \subseteq M$ is called a **pseudo-intent** if $P \neq P''$ and for any pseudo-intent Q such that $Q \subset P$ one has $Q'' \subset P$.*

The Duquenne-Guigues base looks as follows:

$$\{P \to (P'' \setminus P) \mid P \text{ is a pseudo-intent }\}.$$

The Duquenne-Guigues base is a minimum (cardinality minimal) implication base.

Example 7. Let us find all pseudo-intents for the context of geometric figures. We build a Table 3 with B and B''; it is clear that all closed sets are not pseudo-intents by the definition. Since we have to check the containment of a pseudo-intent in the generated pseudo-intents recursively, we should start with the smallest possible set, i.e. $\emptyset$.

Table 3. Finding pseudo-itents for the context of geometric figures

B	B'	B''	B is pseudo-intent?
$\emptyset$	1234	$\emptyset$	No, it's not
a	12	a	No, it's not
b	34	bc	Yes, it is
c	234	c	No, it's not
d	14	d	No, it's not
ab	$\emptyset$	$abcd$	No, it's not
ac	2	ac	No, it's not
ad	1	ad	No, it's not
bc	34	bc	No, it's not
bd	4	bcd	No, it's not
cd	4	bcd	Yes, it is
abc	$\emptyset$	$abcd$	Yes, it is
abd	$\emptyset$	$abcd$	No, it's not
acd	$\emptyset$	$abcd$	No, it's not
bcd	4	bcd	No, it's not
$abcd$	$\emptyset$	$abcd$	No, it's not

Thus, $\{b\}$ is the first non-closed set in our table and the second part of pseudo-intent definition fulfills trivially – there is no another pseudo-intent contained in $\{b\}$. So, the whole set of pseudo-intents is $\{b, cd, abc\}$. $\qquad\square$

Exercise 9. Write down the Duquenne-Guigues base for the context of geometric figures. Using Armstrong rules and the obtained Duquenne-Guigues base, deduce the rest implications of the original context. $\qquad\square$

For recent efficient algorithm of finding the Duquenne-Guigues base see [27].

Implications and Functional Dependencies. Data dependencies are one way to reach two primary purposes of databases: to attenuate data redundancy and enhance data reliability [25]. These dependencies are mainly used for data normalisation, i.e. their proper decomposition into interrelated tables (relations). The definition of functional dependency [25] in terms of FCA is as follows:

Definition 21. $X \rightarrow Y$ *is a **functional dependency** in a complete many-valued context* (G, M, W, I) *if the following holds for every pair of objects* g, $h \in G$:

$$(\forall m \in X \ m(g) = m(h)) \Rightarrow (\forall n \in Y \ n(g) = n(h)).$$

Example 8. For the example given in Table 2 the following functional dependencies hold: $Age \rightarrow Subject$, $Subject \rightarrow Age$, $Mark \rightarrow Gender$.   □

The first two functional dependencies may have sense since students of the same year may study the same subjects. However, the last one says Gender is functionally dependent by Mark and looks as a pure coincidence because of the small dataset.

The reduction of functional dependencies to implications:

Proposition 3. *For a many-valued context (G, M, W, I), one defines the context $\mathbb{K}_N := (\mathcal{P}_2(G), M, I_N)$, where $\mathcal{P}_2(G)$ is the set of all pairs of different objects from G and I_N is defined by*

$$\{g, h\} I_N m :\Leftrightarrow m(g) = m(h).$$

Then a set $Y \subseteq M$ is functionally dependent on the set $X \subseteq M$ if and only if the implication $X \rightarrow Y$ holds in the context $\mathbb{K}_N$.

Example 9. Let us construct the context $\mathbb{K}_N$ for the many-valued context of geometric figures.

	Gender	Age	Subject	Mark
{1,2}				
{1,3}		×	×	
{1,4}	×			
{1,5}				
{2,3}	×			
{2,4}		×	×	
{2,5}	×			×
{3,4}				
{3,5}	×			
{4,5}				

One may check that the following implications hold: $Age \rightarrow Subject$, $Subject \rightarrow Age$, $Mark \rightarrow Gender$, which are the functional dependencies that we so in Example 8.   □

An inverse reduction is possible as well.

Proposition 4. *For a context $\mathbb{K} = (G, M, I)$ one can construct a many-valued context $\mathbb{K}_W$ such that an implication $X \rightarrow Y$ holds if and only if Y is functionally dependent on X in $\mathbb{K}_W$.*

Example 10. To fulfill the reduction one may build the corresponding many-valued context in the following way:

1. Replace all "×" by 0s. 2. In each row, replace empty cells by the row number starting from 1. 3. Add a new row filled by 0s.

	a	b	c	d
1	0	1	1	0
2	0	2	0	2
3	3	0	0	3
4	4	0	0	0
5	0	0	0	0

□

Exercise 10. Check the functional dependencies from the previous example coincide with the implications of the context of geometric figures.   □

More detailed tutorial on FCA and fuctional dependencies is given in [28].

4   FCA Tools and Practice

In this section, we provide a short summary of ready-to-use software that supports basic functionality of Formal Concept Analysis.

– Software for FCA: Concept Explorer, Lattice Miner, ToscanaJ, Galicia, FCART etc.
– Exercises.

Concept Explorer. ConExp[4] is probably one of the most user-friendly FCA-based tools with basic functionality; it was developed in Java by S. Yevtushenko under Prof. T. Taran supervision in the beginning of 2000s [29]. Later on it has been improved several times, especially from lattice drawing viewpoint [30].

Now the features the following functionality:

– Context editing (tab separated and csv formats of input files are supported as well);
– Line diagrams drawing (allowing their import as image snapshots and even text files with nodes position, edges and attributes names, but vector-based formats are not supported);
– Finding the Duquenne-Guigues base of implications;
– Finding the base of association rules that are valid in a formal context;
– Performing attribute exploration.

It is important to note that the resulting diagram is not static and one may perform exploratory analysis in an interactive manner selecting interesting nodes, moving them etc. In Fig. 7, the line diagram of the concept lattice of interordinal scale for attribute Mark drawn by ConExp is shown. See more details in Fig. [31].

There is an attempt to reincarnate ConExp[5] by modern open software tools.

---

[4] http://conexp.sourceforge.net/.
[5] https://github.com/fcatools/conexp-ng/wiki.

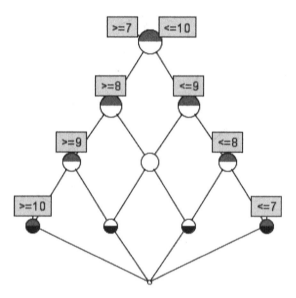

Fig. 7. The line diagram of the concept lattice for the interordinal scale of student marks drawn by ConExp.

ToscanaJ. The ToscanaJ[6] project is a result of collaboration between two groups from the Technical University of Darmstadt and the University of Queensland, which aim was declared as "to give the FCA community a platform to work with" [32] and "the creation of a professional tool, coming out of a research environment and still supporting research" [33].

This open project has a long history with several prototypes [34] and now it is a part of an umbrella framework for conceptual knowledge processing, Tockit[7]. As a result, it is developed in Java, supports different types of database connection via JDBC-ODBC bridge and contains an embedded database engine [33]. Apart from ConExp, it features work with multi-valued contexts, conceptual scaling, and nested line diagrams.

In Fig. 8 one can see the nested line diagram for two scales from the university subjects multi-valued context, namely for two attributes, Gender and Subject. Via PDF printing facilities it is possible to print out line diagrams in a vector graphic form.

Galicia. Galicia[8] was "intended as an integrated software platform including components for the key operations on lattices that might be required in practical applications or in more theoretically-oriented studies". Thus in addition to basic functionality of ConExp, it features work with multi-valued contexts

---

[6] http://toscanaj.sourceforge.net/.
[7] http://www.tockit.org/.
[8] http://www.iro.umontreal.ca/~galicia/.

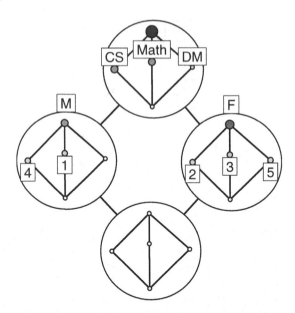

Fig. 8. The nested line diagram for the two one-attribute subcontexts of the context of university subjects. The outer diagram is for Gender attribute, and the inner one is for Subject.

and conceptual scaling, iceberg lattices (well-known in Data Mining community), Galois hierarchies and relational context families, which are popular in software engineering [35]. The software is open and its implementation in Java is cross-platform aimed at "adaptability, extensibility and reusability".

It is possible to navigate through lattice diagrams in an interactive manner; the resulting diagrams contain numbered nodes and this is different from the traditional way of line diagrams drawing. Another Galicia's unique feature is 3D lattice drawing. The diagram of the university subjects context after nominal scaling of all its attributes obtained in Galicia is depicted in Fig. 9. Galicia supports vector-based graphic formats, SVG and PDF. The authors of the program paid substantial attention to algorithmic aspects and incorporated batch and incremental algorithms into it. Various bases of implications and association rules can be generated by the tool. Nested line diagrams are in the to do list.

Lattice Miner. This is another attempt to establish basic FCA functionality and several specific features to the FCA community[9] [36].

The initial objective of the tool was "to focus on visualization mechanisms for the representation of concept lattices, including nested line diagrams"[10]. Thus, its interesting feature is multi-level nested line diagrams, which can help to explore comparatively large lattices.

---

[9] http://sourceforge.net/projects/lattice-miner/.
[10] https://en.wikipedia.org/wiki/Lattice_Miner.

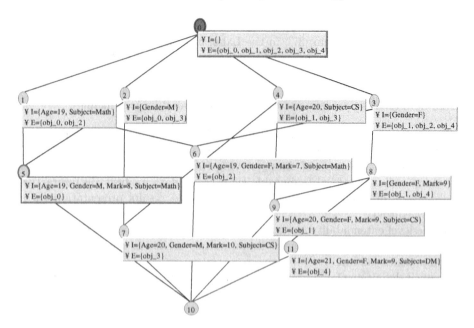

Fig. 9. The line diagram of concept lattice for the context of university subjects drawn by Galicia.

After more than a decade of development, FCA-based software having different features produced a lot of different formats thus requiring interoperability. To this end, in analogy to Rosetta Stone, FcaStone[11] was proposed. It supports convertation between commonly used FCA file formats (cxt, cex, csc, slf, bin.xml, and csx) and comma separated value (csv) files as well as convertation concept lattices into graph formats (dot, gxl, gml, etc. for use by graph editors such as yEd, jgraph, etc.) or into vector graphics formats (fig, svg, etc. for use by vector graphics editors such as Xfig, Dia, Inkscape, etc.). It can also be incorporated into a webpage script for generating lattices and line diagrams online. Another example of a web-based ported system with basic functionality including attribute exploration is OpenFCA[12].

FCART. Many different tools have been created and some of the projects are not developing anymore but the software is still available; an interested reader can refer Uta Priss's webpage to find dozens of tools[13]. However, new challenges such as handling large heterogeneous datasets (large text collections, social networks and media etc.) are coming and the community, which put a lot of efforts in the development of truly cross-platform and open software, needs a new wave of tools that adopts modern technologies and formats.

---

[11] http://fcastone.sourceforge.net/.
[12] https://code.google.com/p/openfca/.
[13] http://www.fcahome.org.uk/fcasoftware.html.

Inspired by the successful application of FCA-based technologies in text mining for criminology domain [37], in the Laboratory for Intelligent Systems and Structural Analysis, a tool named Formal Concept Analysis Research Toolbox (FCART) is developing.

FCART follows a methodology from [38] to formalise iterative ontology-driven data analysis process and to implement several basic principles:

1. Iterative process of data analysis using ontology-driven queries and interactive artifacts such as concept lattice, clusters, etc.
2. Separation of processes of data querying (from various data sources), data preprocessing (via local immutable snapshots), data analysis (in interactive visualizers of immutable analytic artifacts), and results presentation (in a report editor).
3. Three-level extendability: settings customisation for data access components, query builders, solvers and visualizers; writing scripts or macros; developing components (add-ins).
4. Explicit definition of analytic artifacts and their types, which enables integrity of session data and links artifacts for the end-users.
5. Availability of integrated performance estimation tools.
6. Integrated documentation for software tools and methods of data analysis.

Originally, it was yet another FCA-based "integrated environment for knowledge and data engineers with a set of research tools based on Formal Concept Analysis" [39,40] featuring in addition work with unstructured data (including texts with various metadata) and Pattern Structures [41]. In its current distributed version, FCART consists of the following parts:

1. AuthServer for authentication and authorisation.
2. Intermediate Data Storage (IDS) for storage and preprocessing of big datasets.
3. Thick Client for interactive data processing and visualisation in integrated graphical multi-document user interface.
4. Web-based solvers for implementing independent resource-intensive computations.

The workflow is shown in Fig. 10.

The main questions are the following: Whether the product has only technological advantages or it really has fruitful methodology? Can it become open in addition to its extendability? Can it finally handle big volumes of heterogeneous data in a suitable way for an FCART analyst? The answers to these posed questions seem to be forthcoming challenging steps.

CryptoLatt. This tool[14] was developed to help students and researchers from neighbouring domains (e.g., Data Mining) to recognise cryptomorphisms in lattice-based problems, i.e. to realise that a particular problem in one domain is "isomorphic" to some other in terms of lattice theory [42]. Thus, one of the

---

[14] http://www.cs.unic.ac.cy/florent/software.htm.

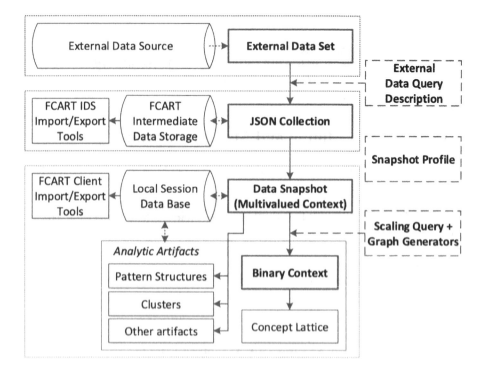

Fig. 10. FCART workflow

well-known cryptomorphisms in the FCA community is established between a lattice and a binary relation, also known as the basic theorem of FCA. Note that even a particular formal context, its concept lattice and set of implications represent the same information about the underlying dataset but in a different way.

Exercise 11. Practice with Concept Explorer:

1. Input the context of geometric figures, build its concept lattice diagram and find the Duquenne-Guigues base. Check whether the obtained base coincide with the base found before. Play with different layouts and other drawing options like labeling or node size.
2. Find real datasets where objects are described by nominal attributes and select about 10 objects and 10 attributes from it. Prepare the corresponding context, build the lattice diagram and find its implication base. Try to interpret found concepts and dependencies. □

Exercise 12. Practice with ToscanaJ:

1. Use Elba tool from the latest version of ToscanaJ for creating two scaled contexts for any two attributes of the context of university subjects. Save the contexts. Then upload them into ToscanaJ and draw their nested line diagram. The result should be similar to Fig. 8. □

Exercise 13. Practice with Galicia:

1. Perform tasks from Exercise 11.
2. Compose the context of university subjects. Scale it via *Algorithms→Multi-FCA→Interactive Multi-FCA* and build the lattice diargam. The result should be identical to Fig. 9.   □

5   FCA in Data Mining and Machine Learning

– Frequent Itemset Mining and Association Rules: FCA did it even earlier [43,44]
– Multimodal clustering (biclustering and triclustering) [45–47]
– FCA in Classification: JSM-method, version spaces*[15], and decision trees* [48]
– Pattern Structures for data with complex descriptions [49,50]
– FCA-based Boolean Matrix Factorisation [51]
– Educational Data Mining case study [52]
– Exercises with JSM-method in QuDA (Qualitative Data Analysis): solving classification task [53]

5.1   Frequent Itemset Mining and Association Rules

Knowledge discovery in databases (KDD) is introduced as the non-trivial extraction of valid, implicit, potentially useful and ultimately understandable information in large databases [54]. Data mining is a main step in KDD, and in its turn association rules and frequent itemset mining are among the key techniques in Data Mining. The original problem for association rules mining is market basket analysis. In early 90s, since the current level of technologies made it possible to store large amount of transactions of purchased items, companies started their attempts to use these data to facilitate their typical business decisions concerning "what to put on sale, how to design coupons, how to place merchandise on shelves in order to maximize the profit"[55]. So, firstly this market basket analysis problem was formalised in [55] as a task of finding frequently bought items together in a form of rules "if a customer buys items A, (s)he also buys items B". One of the first and rather efficient algorithms of that period was proposed in [43], namely Apriori. From the very beginning these rules are tolerant to some number of exceptions, they were not strict as implications in FCA. However, several years before, in [44], Michael Luxenburger introduced partial implications motivated by more general problem statement, "a generalisation of the theory of implications between attributes to partial implications" since "in data analysis the user is not only interested in (global) implications, but also in "implications with a few exceptions". The author proposed theoretical treatment of the problem in terms of Formal Concept Analysis and was guided by the idea of characterisation of "sets of partial implications which arise from real

---
[15] not covered here.

data" and "a possibility of an "exploration" of partial implications by a computer". In addition, he proposed a minimal base of partial implications known as Luxenburger's base of association rules as well.

Definition 22. *Let* $\mathbb{K} := (G, M, I)$ *be a context, where G is a set of objects, M is a set of attributes (items), $I \subseteq G \times M$* **An association rule** *of the context $\mathbb{K}$ is an expression $A \rightarrow B$, where $A, B \subseteq M$ and (usually) $A \cap B = \emptyset$.*

Definition 23. (Relative) support *of an association rule $A \rightarrow B$ defined as*

$$supp(A \rightarrow B) = \frac{|(A \cup B)'|}{|G|}.$$

The value of $supp(A \rightarrow B)$ shows which part of G contains $A \cup B$. Often support can be given in %.

Definition 24. (Relative) confidence *of an association rule $A \rightarrow B$ defined as*

$$conf(A \rightarrow B) = \frac{|(A \cup B)'|}{|A'|}.$$

This value $conf(A \rightarrow B)$ shows which part of objects that possess A also contains $A \cup B$. Often confidence can be given in %.

Example 11. An object-attribute table of transactions.

	Beer	Cakes	Milk	Müsli	Chips
c_1	×				×
c_2		×	×	×	
c_3	×		×	×	×
c_4	×	×	×		×
c_5		×	×	×	×

- $supp(\{\text{Beer}, \text{Chips}\}) = 3/5$
- $supp(\{\text{Cakes}, \text{Müsli}\} \rightarrow \{\text{ Milk }\}) = \frac{|(\{\text{Cakes,Müsli}\} \cup \{\text{Milk}\})'|}{|G|} = \frac{|\{C2,C5\}|}{5} =$ $2/5$
- $conf(\{\text{Cakes}, \text{Müsli}\} \rightarrow \{\text{ Milk }\}) = \frac{|(\{\text{Cakes,Müsli}\} \cup \{\text{Milk}\})'|}{|\{\text{Cakes,Müsli}\}'|} = \frac{|\{c_2,c_5\}|}{|\{c_2,c_5\}|} = 1$

□

The main task of association rules mining is formulated as follows: Find all association rules of a context, where support and confidence of the rules are greater than predefined thresholds, min-confidence and min-support, denoted as min_conf and min_supp, respectively [55].

Proposition 5. *(Association rules and implications)*
   Let $\mathbb{K}$ *be a context, then its associations rules under condition min_supp =* 0% *and min_conf =* 100% *are implications of the same context.*

Sometimes an association rule can be written as $A \xrightarrow[s]{c} B$, where c and s are confidence and support of the given rule.
   Two main steps of association rules mining are given below:

1. Find frequent sets of attributes (frequent itemsets), i.e. sets of attributes (items) that have support greater than *min_supp.*
2. Building association rules based on found frequent itemsets.

   The first step is the most expensive, the second one is rather trivial.
   The well-known algorithm for frequent itemset mining is Apriori [43] uses the antimonotony property to ease frequent itemsets enumeration.

Property 1. (Antimonotony property) For $\forall A, B \subseteq M$ and $A \subseteq B \Rightarrow supp(B) \leq supp(A)$.

   This property implies the following facts:

- The larger set, the smaller support it has or its support remains the same;
- Support of any itemset is not greater than a minimal support of any its subset;
- Aa itemset of size n is frequent if and only if all its $(n-1)$-subsets are frequent.

   The Apriori algorithm finds all frequent itemsets.
   It is check iteratively the set of all itemsets in a levelwise manner. At each iteration one level is considered, i.e. a subset of candidate itemsets C_i is composed by collecting the frequent itemsets discovered during the previous iteration (AprioriGen procedure). Then supports of all candidate itemsets are counted, and the infrequent ones are discarded.

Algorithm 4. Apriori(*Context, min_supp*)

Input: *Context, min_supp* is a minimal support
Output: all frequent itemsets I_F
 1: $C_1 \leftarrow$ 1-itemsets
 2: $i \leftarrow 1$
 3: **while** $C_i \neq \emptyset$ **do**
 4:    $SupportCount(C_i)$
 5:    $F_i \leftarrow \{f \in C_i \,|\, f.support \geq min\_supp\}$
 6:    $\{F_i$ is a set of frequent i-itemsets$\}$
 7:    $C_{i+1} \leftarrow AprioriGen(F_i)$
 8:    $\{C_i$ is a set of $(i+1)$-candidates$\}$
 9:    $i{+}{+}$
10: **end while**
11: $I_F \leftarrow \bigcup F_i$
12: **return** I_F

   For frequent itemsets of size i, procedure AprioriGen finds $(i+1)$-supersets and returns only the set of potentially frequent candidates.

Algorithm 5. AprioriGen(F_i)

Input: F_i is a set of frequent i-itemsets
Output: C_{i+1} is a set of $(i+1)$-itemsets candidates
 1: insert into C_{i+1} {union}
 2: select $p[1], p[2], \ldots, p[i], q[i]$
 3: from $F_i.p$, $F_i.q$
 4: where $p[1] = q[1], \ldots, p[i-1] = q[i-1], p[i] < q[i]$
 5: **for all** $c \in C_{i+1}$ **do**
 6:     {elimination}
 7:     $S \leftarrow (i-1)$-itemset c
 8:     **for all** $s \in S$ **do**
 9:         **if** $s \notin F_i$ **then**
10:             $C_{i+1} \leftarrow C_{i+1} \setminus c$
11:         **end if**
12:     **end for**
13: **end for**
14: **return** C_{i+1}

Example 12. Union and elimination steps of AprioriGen for a certain context.

- The set of frequent 3-itemsets: $F_3 = \{\{a, b, c\}, \{a, b, d\}, \{a, c, d\}, \{a, c, e\}, \{b, c, d\}\}$.
- The set of candidate 4-itemsets (union step): $C_4 = \{\{a, b, c, d\}, \{a, c, d, e\}\}$.
- The remaining candidate is $C_4 = \{\{a, b, c, d\}\}$, since is eliminated $\{a, c, d, e\}$ because $\{c, d, e\} \notin F_3$ (elimination step).   □

The worst-case computational complexity of the Apriori algorithm is $O(|G| |M|^2 2^{|M|})$ since all the itemsets may be frequent. However, it takes only $O(|M|)$ datatable scans compared to $O(2^|M|)$ for brute-force method.

Rules extraction is based on frequent itemsets.

Let F be a frequent 2-itemset. We compose a rule $f \rightarrow F \setminus f$ if

$$conf(f \rightarrow F \setminus f) = \frac{supp(F)}{supp(f)} \geq min\_conf, \text{where} f \subset F.$$

Property 2. Confidence $conf(f \rightarrow F \setminus f) = \frac{supp(F)}{supp(f)}$ is minimal, when $supp(f)$ is maximal.

- Confidence is maximal when rule consequent $F \setminus f$ consists of one attribute (1-itemset). The subsets of such an consequent have greater support and it turn smaller confidence.
- Recursive procedure of rules extraction starts with $(|F|-1)$-itemset f fulfilling min_conf and min_sup; then, it forms the rule $f \rightarrow F \setminus f$ and checks all its subsets $(|F|-2)$-itemset (if any) and so on.

Exercise 14. Find all frequent itemsets for the customers context with Apriori algorithm and $min\_sup = 1/3$.   □

Condensed Representation of Frequent Itemsets. According to basic results from Formal Concept Analysis, it is not necessary count the support of all frequent itemsets. Thus, it is possible to derive from some known supports the supports of all other itemsets: it is enough to know the support of all frequent concept intents. And it is also not necessary to compute all frequent itemsets for solving the association rule problem: it is sufficient to consider the frequent concept intents that also called closed itemsets in Data Mining. In fact, closed itemsets was independently discovered by three groups of researches in the late 90s [56–58].

Let $\mathbb{K} = (G, M, I)$ be a formal context.

Definition 25. *A set of attributes $FC \subseteq M$ is called **frequent closed itemset**, if $supp(FC) \geq min\_supp$ and there is no any F such that $F \supset FC$ and $supp(F) = supp(FC)$.*

Definition 26. *A set of attributes $MFC \subseteq M$ is called **maximal frequent itemset** if it is frequent and there is no any F such that $F \supset FMC$ and $supp(F) \geq min\_supp$.*

Proposition 6. *In a formal context $\mathbb{K}$, $\mathcal{MFC} \subseteq \mathcal{FC} \subseteq \mathcal{F}$, where $\mathcal{MFC}$ is the set of maximal frequent itemset, $\mathcal{FC}$ is the set of frequent closed itemsets, and $\mathcal{F}$ is the set of frequent itemsets of $\mathbb{K}$ with a minimal support min_supp.*

Proposition 7. *The concept lattice of a formal context $\mathbb{K}$ is (isomorphic to) its lattice of frequent closed itemsets with $min\_supp = 0$.*

One may check that the lattices, whose diagrams are depicted in Fig. 11, are isomorphic.

The set of all frequent concepts of the context $\mathbb{K}$ for the threshold min_sup is also known as the "iceberg concept lattice" [59], mathematically it corresponds to the order filter of the concept lattice. However, the idea of usage the size of concept's extent, intent (or even their different combinations) as a concept quality measure is not new in FCA [60].

Of course, the application domain is not restricted to market basket analysis; thus, the line diagram built in ConExp shows 25 largest concepts of visitors of HSE website in terms of news websites in 2006.

For real datasets, association rules mining usually results in the large number of rules. However, not all rules are necessary to present the information. Similar compact representation can be used here; thus, one can represent all valid association rules by their subsets that called bases. For example, the Luxenburger base is a set of association rules in the form

$$\{B_1 \to B_2 | (B_1', B_1) \text{ is an upper neighbour of concept } (B_2', B_2)\}.$$

The rest rules and their support and confidence can be derived by some calculus, which is not usually clear from the base definition.

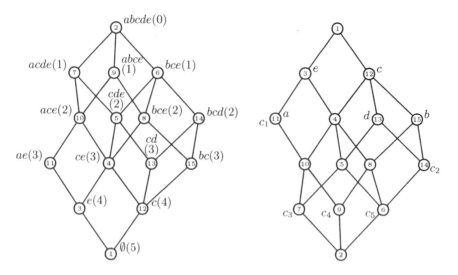

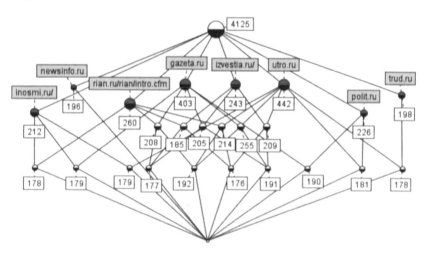

Fig. 11. The line diagrams of the lattice of closed itemsets (left, with support given in parentheses) and the concept lattice for the customers context (right, with reduced labeling)

Fig. 12. The line diagram of 25 largest concepts for the context of HSE web users

Exercise 15. 1. Find the Luxenburger base for the customers context with $min\_sup = 1/3$ and $min\_conf = 1/2$. 2. Check whether Concept Explorer generates the association rule base (for the same context) that consists of the Duquenne-Guigues base and the Luxenburger base.   □

One of the first algorithms that were explicitly designed to compute frequent closed itemsets is Close [56]. Inspired by Apriori it traverses the database in a level-wise manner and generates requent closed itemsets by computing the closures of all minimal generators. See more detailed survey in [61].

It is interesting that after the years of co-existence, one of the Apriori's authors has started to apply FCA in text mining [62]. FCA is also included into textbooks on data mining, see Chap. 8 & 9 in [63].

Another interesting subdomain of frequent pattern mining, where lattice-based methods are successfully used, is so called sequential pattern mining [41,64].

Multimodal Clustering (Biclustering and Triclustering). Clustering is an activity for finding homogeneous groups of instances in data. In machine learning, clustering is a part of so called unsupervised learning. The widely adopted idea of cluster relates to instances in a feature space. A cluster in this space is a subset of data instances (points) that are relatively close to each other but relatively far from other data points. Such feature space clustering algorithms are a popular tool in marketing research, bioinformatics, finance, image analysis, web mining, etc. With the growing popularity of recent data sources such as biomolecular techniques and Internet, other than instance-to-feature data appear for analysis.

One example is gene expression matrices, entries of which show expression levels of gene material captured in a polymerase reaction. Another example would be n-ary relations among several sets of entities such as:

- Folksonomy data [65] capturing a ternary relation among three sets: users, tags, and resources;
- Movies database IMDb(^{16}) describing a binary relation of "relevance" between a set of movies and a set keywords or a ternary relation between sets of movies, keywords and genres;
- product review websites featuring at least three itemsets (product, product features, product-competitor);
- job banks comprising at least four sets (jobs, job descriptions, job seekers, seeker skills).

For two-mode case other cluster approaches demonstrates growing popularity. Thus the notion of bicluster in a data matrix (coined by B. Mirkin in [66], p. 296) represents a relation between two itemsets. Rather than a single subset of entities, a bicluster features two subsets of different entities.

In general, the larger the values in the submatrix, the higher interconnection between the subsets, the more relevant is the corresponding bicluster. In the relational data, presence-absence facts represented by binary 1/0 values and this condition expresses the proportion of unities in the submatrix, its "density": the larger, the better. It is interesting that a bicluster of the density 1 is a formal concept if its constituent subsets cannot be increased without a drop in the density value, i.e. a maximal rectangle of 1 s in the input matrix w.r.t. permutations of its rows and columns [3]. Usually one of the related sets of

---

^{16} www.imdb.com.

entities is a set of objects, the other one is a set of attributes. So, in contrary to ordinary clustering, bicluster (A, B) captures similarity (homogeneity) of objects from A expressed in terms of their common (or having close values) attributes B, which usually embrace only a subset of the whole attribute space.

Obviously, biclusters form a set of homogeneous chunks in the data so that further learning can be organized within them. The biclustering techniques and FCA machinery are being developed independently in independent communities using different mathematical frameworks. Specifically, the mainstream in Formal Concept Analysis is based on ordered structures, whereas biclustering relies on conventional optimisation approaches, probabilistic and matrix algebra frameworks [67,68]. However, in fact these different frameworks considerably overlap in applications, for example: finding co-regulated genes over gene expression data [67–73], prediction of biological activity of chemical compounds [74–77], text summarisation and classification [78–82], structuring websearch results and browsing navigation in Information Retrieval [4,15,83,84], finding communities in two-mode networks in Social Network Analysis [85–89] and Recommender Systems [90–94].

For example, consider a bicluster definition from paper [70]. Bi-Max algorithm described in [70] constructs **inclusion-maximal biclusters** defined as follows:

Definition 27. *Given m genes, n situations and a binary table e such that $e_{ij} = 1$ (gene i is active in situation j) or $e_{ij} = 0$ (gene i is not active in situation j) for all $i \in [1, m]$ and $j \in [1, n]$, the pair $(G, C) \in 2^{\{1,\ldots,n\}} \times 2^{\{1,\ldots,m\}}$ is called an* **inclusion-maximal bicluster** *if and only if (1) $\forall i \in G, j \in C : e_{ij} = 1$ and (2) $\nexists (G_1, C_1) \in 2^{\{1,\ldots,n\}} \times 2^{\{1,\ldots,m\}}$ with (a) $\forall i_1 \in G_1, \forall j_1 \in C_1 : e_{i_1 j_1} = 1$ and (b) $G \subseteq G_1 \wedge C \subseteq C_1 \wedge (G_1, C_1) \neq (G, C)$.*

Let us denote by H the set of genes (objects in general), by S the set of situations (attributes in general), and by $E \subseteq H \times S$ the binary relation given by the binary table e, $|H| = m$, $|S| = n$. Then one has the following proposition:

Proposition 8. *For every pair (G, C), $G \subseteq H$, $C \subseteq S$ the following two statements are equivalent.*

1. *(G, C) is an inclusion-maximal bicluster of the table e;*
2. *(G, C) is a formal concept of the context (H, S, E).*

Exercise 16. Prove Proposition 8.   □

Object-Attribute-biclustering. Another example is OA-biclustering proposed in [95,96] as a reliable relaxation of formal concept.

Definition 28. *If $(g, m) \in I$, then (m', g') is called an object-attribute or OA-bicluster with density $\rho(m', g') = \frac{|I \cap (m' \times g')|}{|m'| \cdot |g'|}$.*

Here are some basic properties of OA-biclusters.

Proposition 9. *1.* $0 \leq \rho \leq 1$.
2. OA-bicluster (m', g') is a formal concept iff $\rho = 1$.
3. if (m', g') is a OA-bicluster, then $(g'', g') \leq (m', m'')$.

Exercise 17. a. Check that properties 1. and 2. from Proposition 9 follow directly
by definitions. b. Use antimonotonicity of $(\cdot)'$ to prove 3.   □

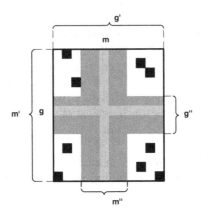

Fig. 13. Bicluster based on object and attribute closures

In Fig. 13 one can see the structure of the OA-bicluster for a particular pair
$(g, m) \in I$ of a certain context (G, M, I). In general, only the regions (g'', g')
and (m', m'') are full of non-empty pairs, i.e. have maximal density $\rho = 1$, since
they are object and attribute formal concepts respectively. Several black cells
indicate non-empty pairs which one may found in such a bicluster. It is quite
clear, the density parameter ρ would be a bicluster quality measure which shows
how many non-empty pairs the bicluster contains.

Definition 29. *Let $(A, B) \in 2^G \times 2^M$ be an OA-bicluster and ρ_{min} be a non-
negative real number, such that $0 \leq \rho_{min} \leq 1$, then (A, B) is called* dense *if it
satisfies the constraint $\rho(A, B) \geq \rho_{min}$.*

Order relation $\sqsubseteq$ on OA-biclusters is defined component-wise: $(A, B) \sqsubseteq (C, D)$
iff $A \subseteq C$ and $B \subseteq D$.
    Monotonicity (antimonotonicity) of constraints is often used in mining asso-
ciation rules for effective algorithmic solutions.

Proposition 10. *The constraint $\rho(A, B) \geq \rho_{min}$ is neither monotonic nor anti-
monotonic w.r.t. $\sqsubseteq$ relation.*

Exercise 18. 1. To prove Proposition 10 for context $\mathbb{K}$ consider OA-biclusters
$b_1 = (\{g_1, g_3, g_4, g_5\}, \{m_1, m_4, m_5\})$, $b_3 = (G, \{m_1, m_2, m_3\})$ and $b_2 = (G, M)$.
2. Find generating pairs (g, m) for all these three biclusters.   □

	m_1	m_2	m_3	m_4	m_5
g_1	×	×	×	×	×
g_2	×	×	×	×	
g_3	×			×	×
g_4	×			×	×
g_5	×			×	×

However, the constraint on ρ_{min} has other useful properties.

If $\rho = 0$, this means that we consider the set of all OA-biclusters of the context $\mathbb{K}$. For $\rho_{min} = 0$ every formal concept is "contained" in a OA-bicluster of the context $\mathbb{K}$, i.e., the following proposition holds.

Proposition 11. *For each* $(A_c, B_c) \in \mathfrak{B}(G, M, I)$ *there exists a OA-bicluster* $(A_b, B_b) \in \mathbf{B}$ *such that* $(A_c, B_c) \sqsubseteq (A_b, B_b)$.

Proof. Let $g \in A_c$, then by antimonotonicity of $(\cdot)'$ we obtain $g' \supseteq B_c$. Similarly, for $m \in B_c$ we have $m' \supseteq A_c$. Hence, $(A_b, B_b) \sqsubseteq (m', g')$. □

The number of OA-biclusters of a context can be much less than the number of formal concepts (which may be exponential in $|G| + |M|$), as stated by the following proposition.

Proposition 12. *For a given formal context* $\mathbb{K} = (G, M, I)$ *and* $\rho_{min} = 0$ *the largest number of OA-biclusters is equal to* $|I|$, *all OA-biclusters can be generated in time* $O(|I| \cdot (|G| + |M|))$.

Proposition 13. *For a given formal context* $\mathbb{K} = (G, M, I)$ *and* $\rho_{min} > 0$ *the largest number of OA-biclusters is equal to* $|I|$, *all OA-biclusters can be generated in time* $O(|I| \cdot |G| \cdot |M|)$.

Algorithm 6 is a rather straightforward implementation by definition, which takes initial formal context and minimal density threshold as parameters and computes biclusters for each (object, attribute) pair in relation I. However, in its latest implementations we effectively use hashing for duplicates elimination. In our experiments on web advertising data, the algorithm produces 100 times less patterns than the number of formal concepts. In general, for the worst case these values are $2^{\min(|G|,|M|)}$ vs $|I|$. The time complexity of our algorithm is polinomial $(O(|I||G||M|))$ vs exponential in the worst case for Bi-Max $(O(|I||G||L| \log |L|))$ or CbO $(O(|G|^2|M||L|))$, where $|L|$ is a number of generated concepts which is exponential in the worst case $(|L| = 2^{\min(|G|,|M|)})$.

Triadic FCA and Triclustering. As we have mentioned, there are such data sources as folksonomies, for example, a bookmarking website for scientific literature Bibsonomy[17] [97]; the underlying structure includes triples (user, tag, bookmark) like one in Fig. 14.

---

[17] http://www.bibsonomy.org.

Algorithm 6. OA-bicluster computation

Input: $\mathbb{K} = (G, M, I)$ is a formal context, ρ_{min} is a threshold density value of bicluster density

Output: $B = \{(A_k, B_k) | (A_k, B_k) \text{ is a bicluster}\}$

```
1: B ← ∅
2: if ρmin = 0 then
3:     for all (m, g) ∈ I do
4:         B.Add(m', g')
5:     end for
6: else
7:     for all (m, g) ∈ I do
8:         if ρ(m', g') ≥ ρmin then
9:             B.Add(m', g')
10:        end if
11:    end for
12: end if
13: B.RemoveDuplicates()
14: return B
```

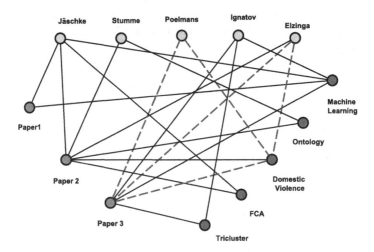

Fig. 14. An example of Bibsonomy relation for three paper, five authors and five tags.

Therefore, it can be useful to extend the biclustering and Formal Concept Analysis to process relations among more than two datasets. A few attempts in this direction have been published in the literature. For example, Zaki et al. [98] proposed Tricluster algorithm for mining biclusters extended by time dimension to real-valued gene expression data. A triclustering method was designed in [99] to mine gene expression data using black-box functions and parameters coming from the domain. In the Formal Concept Analysis framework, theoretic papers [100, 101] introduced the so-called Triadic Formal Concept Analysis. In [102], triadic formal concepts apply to analyse small datasets in a psychological domain.

Paper [45] proposed rather scalable method TRIAS for mining frequent triconcepts in Folksonomies. Simultaneously, a less efficient method on mining closed cubes in ternary relations was proposed by Ji et al. [103]. There are several recent efficient algorithms for mining closed ternary sets (triconcepts) and even more general algorithms than TRIAS. Thus, Data-Peeler [104] is able to mine n-ary formal concepts and its descendant mines fault-tolerant n-sets [105]; the latter was compared with DCE algorithm for fault-tolerant n-sets mining from [106]. The paper [107] generalises n-ary relation mining to multi-relational setting in databases using the notion of algebraic closure.

In triadic setting, in addition to set of objects, G, and set of attributes, M, we have B, a set of conditions. Let $\mathbb{K} = (G, M, B, I)$ be a **triadic context**, where G, M, and B are sets, and I is a ternary relation: $I \subseteq G \times M \times B$. The **triadic concepts** of an triadic context $(G, M, B, Y \subseteq G \times M \times B)$ are exactly the maximal 3-tuples (A_1, A_2, A_3) in $2^G \times 2^M \times 2^B$ with $A_1 \times A_2 \times A_3 \subseteq Y$ with respect to component-wise set inclusion [100, 101]. The notion of n-adic concepts can be introduced in the similar way to the triadic case [108].

Example 13. For the bibsonomy example, one of the triadic concepts is

$$(\{Poelmans, Elzinga\}, \{Domestic\ Violence\}, \{paper3\})$$

(see dotted edges on the graph in Fig. 14). It means that both users Poelmans and Elzinga marked paper 3 by the tag "Domestic Violence".   □

Guided by the idea of finding scalable and noise-tolerant triconcepts, we had a look at triclustering paradigm in general for a triadic binary data, i.e. for tricontexts as input datasets.

Suppose X, Y, and Z are some subsets of G, M, and B respectively.

Definition 30. *Suppose $\mathbb{K} = (G, M, B, I)$ is a triadic context and $Z \subseteq G$, $Y \subseteq M$, $Z \subseteq B$. A triple $T = (X, Y, Z)$ is called an OAC-tricluster. Traditionally, its components are called (tricluster) extent, (tricluster) intent, and (tricluster) modus, respectively.*

The *density* of a tricluster $T = (X, Y, Z)$ is defined as the fraction of all triples of I in $X \times Y \times Z$:

$$\rho(T) := \frac{|I \cap (X \times Y \times Z)|}{|X||Y||Z|}.$$

Definition 31. *The tricluster T is called dense iff its density is not less than some predefined threshold, i.e. $\rho(T) \geq \rho_{min}$.*

The collection of all triclusters for a given tricontext $\mathbb{K}$ is denoted by $\mathcal{T}$.

Since we deal with all possible cuboids in Cartesian product $G \times M \times B$, it is evident that the number of all OAC-triclusters, $|\mathcal{T}|$, is equal to $2^{|G| \cdot |M| \cdot |B|}$. However not all of them are supposed to be dense, especially for real data which are frequently quite sparse. Thus we have proposed two possible OAC-tricluster definitions, which give us an efficient way to find within polynomial time a number of (dense) triclusters not greater than the number $|I|$ of triples in the initial data.

In [109], we have compared a set of triclustering techniques proposed within Formal Concept Analysis and/or bicluster analysis perspectives: OAC-BOX [46], TRIBOX [110], SPECTRIC [47] and a recent OAC-PRIME algorithm. This novel algorithm, OAC-PRIME, overcomes computational and substantive drawbacks of the earlier formal-concept-like algorithms. In our spectral approach (SpecTric algorithm) we rely on an extension of the well-known reformulation of a bipartite graph partitioning problem to the spectral partitioning of a graph (see, e.g. [78]). For comparison purposes, we have proposed new developments in the following components of the experiment setting:

1. Evaluation criteria: The average density, the coverage, the diversity and the number of triclusters, and the computation time and noise tolerance for the algorithms.
2. Benchmark datasets: We use triadic datasets from publicly available internet data as well as synthetic datasets with various noise models.

A preceding work was done in [111].

As a result we have not defined an absolute winning methods, but the multicriteria choice allows an expert to decide which of the criteria are most important in a specific case and make a choice. Thus our experiments show that our Tribox and OAC-prime algorithms can be reasonable alternatives to triadic formal concepts and lead to Pareto-effective solutions. In fact TRIBOX is better with respect to noise-tolerance and the number of clusters, OAC-prime is the best on scalability to large real-world datasets. In paper [112], an efficient version of online OAC-prime has been proposed.

In our experiments we have used a context of top 250 popular movies from www.imdb.com, objects are movie titles, attributes are tags, whereas conditions are genres. Prime OAC-triclustering showed rather good results being one the fastest algorithm under comparison.

Example 14. Examples of Prime OAC triclusters with their density indication for the IMDB context are given below:

1. 36 %, {The Shawshank Redemption (1994), Cool Hand Luke (1967), American History X (1998), A Clockwork Orange (1971), The Green Mile (1999)}, {Prison, Murder, Friend, Shawshank, Banker}, {Crime, Drama}
2. 56, 67 %, {The Godfather: Part II (1974), The Usual Suspects (1995)}, {Cuba, New York, Business, 1920s, 1950s}, {Crime, Drama, Thriller}
3. 60 %, {Toy Story (1995), Toy Story 2 (1999)}, {Jealousy, Toy, Spaceman, Little Boy, Fight}, {Fantasy, Comedy, Animation, Family, Adventure}   □

5.2   FCA in Classification

It is a matter of fact that Formal Concept Analysis helped to algebraically rethink several models and methods in Machine Learning such as version spaces

[113], learning from positive and negative examples [48, 74], and decision trees [48]. It was also shown that concept lattice is a perfect search space for learning globally optimal decision trees [114]. Already in early 90 s both supervised and unsupervised machine learning techniques and applications based on Formal Concept Analysis were introduced in the machine learning community. E.g., in ML-related venues there were reported results on the concept lattice based clustering in GALOIS system that suited for information retrieval via browsing [115, 116]. [117] performed a comparison of seven FCA-based classification algorithms. [118] and [119] propose independently to use FCA to design a neural network architecture. In [120, 121] FCA was used as a data preprocessing technique to transform the attribute space to improve the results of decision tree induction. Note that FCA helps to perform feature selection via conceptual scaling and has quite evident relations with Rough Sets theory, a popular tool for feature selection in classification [122]. [123] proposed Navigala, a navigation-based approach for supervised classification, and applied it to noisy symbol recognition. Lattice-based approaches were also successfully used for classification of data with complex descriptions such as graphs or trees [75, 124]. Moreover, in [125] (Chap. 4, "Concept Learning") FCA is suggested as an alternative learning framework.

JSM-method of Hypothesis Generation. The JSM-method proposed by Viktor K. Finn in late 1970 s was proposed as attempt to describe induction in purely deductive form and thus to give at least partial justification of induction [126]. The method is named to pay respect to the English philosopher John Stuart Mill, who proposed several schemes of inductive reasoning in the 19th century. For example, his Method of Agreement, is formulated as follows: "If two or more instances of the phenomenon under investigation have only one circumstance in common, ... [it] is the cause (or effect) of the given phenomenon."

The method proved its ability to enable learning from positive and negative examples in various domains [127], e.g., in life sciences [74].

For RuSSIR audience, the example of the JSM-method application in paleography might be especially interesting [128]: JSM was used for dating birch-bark documents of 10–16 centuries of the Novgorod republic. There were five types of attributes: individual letter features, features common to several letters, handwriting, language features (morphology, syntax, and typical errors), style (letter format, addressing formulas and their key words).

Even though, the JSM-method was formulated in a mathematical logic setting, later on the equivalence between JSM-hypotheses and formal concepts was recognized [60].

The following definition of a hypothesis ("no counterexample-hypothesis") in FCA terms was given in [129].

Let $\mathbb{K} = (G, M, I)$ be a context. There are a **target attribute** $w \notin M$,

- **positive examples**, i.e. set $G_+ \subseteq G$ of objects known to have w,
- **negative examples**, i.e. set $G_- \subseteq G$ of objects known not to have w,

– **undetermined examples**, i.e. set $G_\tau \subseteq G$ of objects for which it is unknown whether they have the target attribute or do not have it.

There are three subcontexts of $\mathbb{K} = (G, M, I)$, the first two are used for the training sample: $\mathbb{K}_\varepsilon := (G_\varepsilon, M, I_\varepsilon)$, $\varepsilon \in \{-, +, \tau\}$ with respective derivation operators $(\cdot)^+$, $(\cdot)^-$, and $(\cdot)^\tau$.

Definition 32. *A **positive hypothesis** $H \subseteq M$ is an intent of $\mathbb{K}_+$ not contained in the intent g^- of any negative example $g \in G_-$: $\forall g \in G_- \quad H \not\subseteq g^-$. Equivalently,*

$$H^{++} = H, \quad H' \subseteq G_+ \cup G_\tau.$$

Negative hypotheses are defined similarly. An intent of $\mathbb{K}_+$ that is contained in the intent of a negative example is called a **falsified (+)-generalisation**.

Example 15. In Table 4, there is a many-valued context representing credit scoring data.

$G_+ = \{1, 2, 3, 4\}$, $G_- = \{5, 6, 7\}$, and $G_\tau = \{8, 9, 10\}$. The target attribute takes values $+$ and $-$ meaning "low risk" and "high risk" client, respectively.

Table 4. Many-valued classification context for credit scoring

G / M	Gender	Age	Education	Salary	Target
1	M	young	higher	high	+
2	F	middle	special	high	+
3	F	middle	higher	average	+
4	M	old	higher	high	+
5	M	young	higher	low	−
6	F	middle	secondary	average	−
7	F	old	special	average	−
8	F	young	special	high	τ
9	F	old	higher	average	τ
10	M	middle	special	average	τ

To apply JSM-method in FCA terms we need to scale the given data. One may use nominal scaling as below.

	M	F	Y	Mi	O	HE	Sp	Se	HS	A	L	w	$\bar{w}$
g_1	×		×			×			×			×	
g_2		×		×			×		×			×	
g_3		×		×		×				×		×	
g_4	×				×	×			×			×	
g_5	×		×			×					×		×
g_6		×		×				×		×			×
g_7		×			×		×			×			×

Then we need to find positive and negative non-falsified hypotheses. If Fig. 15 there are two lattices of positive and negative examples for the input context, respectively.

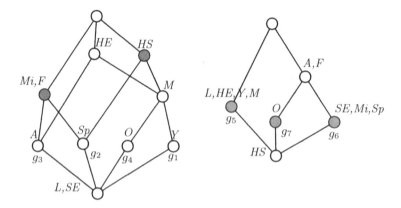

Fig. 15. The line diagrams of the lattice of positive hypotheses (left) and the lattice of negative hypotheses (right).

Shaded nodes correspond to maximal non-falsified hypotheses, i.e. they have no upper neighbors being non-falsified hypotheses.

For $\mathbb{K}_+$ hypothesis $\{HE\}$ is falsified since object g_5 provides a counterexample, i.e. $\{HE\} \subseteq g_5^- = \{M, Y, HE, L\}$.

For $\mathbb{K}_-$ hypothesis $\{A, F\}$ is falsified since there is a positive counterexample, namely $\{A, F\} \subseteq g_3^+ = \{F, M, HE, A\}$.   □

Undetermined examples g_τ from G_τ are classified as follows:

- If g_τ^T contains a positive, but no negative hypothesis, then g_τ is **classified positively** (presence of target attribute w predicted).
- If g_τ^T contains a negative, but no positive hypothesis, then g_τ **classified negatively** (absence of target attribute w predicted).
- If g_τ^T contains both negative and positive hypotheses, or if g_τ^T does not contain any hypothesis, then object classification is **contradictory** or **undetermined**, respectively.

It is clear, for performing classification it is enough to have only minimal hypotheses (w.r.t. $\subseteq$), negative and positive ones.

Exercise 19. For the credit scoring context, classify all undetermined examples.   □

There is a strong connection between hypotheses and implications.

Proposition 14. *A positive hypothesis h corresponds to an implication h →
{w} in the context* $K_+ = (G_+, M \cup \{w\}, I_+ \cup G_+ \times \{w\})$.
    A negative hypothesis h corresponds to an implication $h \to \{\bar{w}\}$ *in the context*
$K_- = (G_-, M \cup \{\bar{w}\}, I_- \cup G_- \times \{\bar{w}\})$.
    Hypotheses are implications which premises are closed (in K_+ *or in* K_- *).*

A detailed yet retrospective survey on JSM-method (in FCA-based and original
formulation) and its applications can be found in [14]. A further extension of
JSM-method to triadic data with target attribute in FCA-based formulation can
be found in [130, 131]; there, the triadic extension of JSM-method used CbO-like
algorithm for classification in Bibsonomy data.
    However, we saw that original data often need scaling, but, for example,
it is not evident what to do in case of learning with labeled graphs. To name
a few problems of this kind we would mention structure-activity relationship
problems for chemicals given by molecular graphs and learning semantics from
graph-based (XML, syntactic tree) text representations. Motivated by search of
possible extensions of original FCA machinery to analyse data with complex
structure, Ganter and Kuznetsov proposed so called Pattern Structures [132].

5.3 Pattern Structures for Data with Complex Descriptions

The basic definitions of Pattern Structures were proposed in [132].
    Let G be a set of objects and D be a set of all possible object descriptions.
Let $\sqcap$ be a similarity operator. It helps to work with objects that have non-
binary attributes like in traditional FCA setting, but those that have complex
descriptions like intervals [73], sequences [133] or (molecular) graphs [75]. Then
$(D, \sqcap)$ is a meet-semi-lattice of object descriptions. Mapping $\delta : G \to D$ assigns
an object g the description $d \in (D, \sqcap)$.
    A triple $(G, (D, \sqcap), \delta)$ is a pattern structure. Two operators $(\cdot)^\square$ define Galois
connection between $(2^G, \subseteq)$ and $(D, \sqcap)$:

$$A^\square = \bigsqcap_{g \in A} \delta(g) \text{ for } A \subseteq G \tag{1}$$

$$d^\square = \{g \in G | d \sqsubseteq \delta(g)\} \text{ for } d \in (D, \sqcap), \text{ where} \tag{2}$$
$$d \sqsubseteq \delta(g) \iff d \sqcap \delta(g) = d.$$

For a set of objects A operator 1 returns the common description (pattern)
of all objects from A. For a description d operator 2 returns the set of all objects
that contain d.
    A pair (A, d) such that $A \subseteq G$ and $d \in (D, \sqcap)$ is called a pattern concept
of the pattern structure $(G, (D, \sqcap), \delta)$ iff $A^\square = d$ and $d^\square = A$. In this case A
is called a pattern extent and d is called a pattern intent of a pattern concept
(A, d). Pattern concepts are partially ordered by $(A_1, d_1) \le (A_2, d_2) \iff A_1 \subseteq
A_2 (\iff d_2 \sqsubseteq d_1)$. The set of all pattern concepts forms a complete lattice
called a pattern concept lattice.

Intervals as Patterns. It is obvious that similarity operator on intervals should fulfill the following condition: two intervals should belong to an interval that contains them. Let this new interval be minimal one that contains two original intervals. Let $[a_1, b_1]$ and $[a_2, b_2]$ be two intervals such that $a_1, b_1, a_2, b_2 \in \mathbb{R}$, $a_1 \leq b_1$ and $a_2 \leq b_2$, then their similarity is defined as follows:

$$[a_1, b_1] \sqcap [a_2, b_2] = [\min(a_1, a_2), \max(b_1, b_2)].$$

Therefore

$$[a_1, b_1] \sqsubseteq [a_2, b_2] \iff [a_1, b_1] \sqcap [a_2, b_2] = [a_1, b_1]$$
$$\iff [\min(a_1, a_2), \max(b_1, b_2)] = [a_1, b_1]$$
$$\iff a_1 \leq a_2 \text{ and } b_1 \geq b_2 \iff [a_1, b_1] \supseteq [a_2, b_2]$$

Note that $a \in \mathbb{R}$ can be represented by $[a, a]$.

Interval Vectors as Patterns. Let us call p-adic vectors of intervals as interval vectors. In this case for two interval vectors of the same dimension $e = \langle [a_i, b_i] \rangle_{i \in [1,p]}$ and $f = \langle [c_i, d_i] \rangle_{i \in [1,p]}$ we define similarity operation via the intersection of the corresponding components of interval vectors, i.e.:

$$e \sqcap f = \langle [a_i, b_i] \rangle_{i \in [1,p]} \sqcap \langle [c_i, d_i] \rangle_{i \in [1,p]} \iff e \sqcap f = \langle [a_i, b_i] \sqcap [c_i, d_i] \rangle_{i \in [1,p]}$$

Note that interval vectors are also partially ordered:

$$e \sqsubseteq f \iff \langle [a_i, b_i] \rangle_{i \in [1,p]} \sqsubseteq \langle [c_i, d_i] \rangle_{i \in [1,p]} \iff [a_i, b_i] \sqsubseteq [c_i, d_i]$$

for all $i \in [1, p]$.

Example 16. Consider as an example Table 5 of movie ratings

Table 5. Movie rates

	The Artist	Ghost	Casablanca	Mamma Mia!	Dogma	Die Hard	Leon
User 1	4	4	5	0	0	0	0
User 2	5	5	3	4	3	0	0
User 3	0	0	0	4	4	0	0
User 4	0	0	0	5	4	5	3
User 5	0	0	0	0	0	5	5
User 6	0	0	0	0	0	4	4

Each user of this table can be described by vector of ratings' intervals. For example, $\delta(u_1) = \langle [4,4], [4,4], [5,5], [0,0], [0,0], [0,0], [0,0] \rangle$. If some new user

u likes movie Leon, a movie recommender system would reply who else like this movie by applying operator 2: $[4,5]^{\square}_{Leon} = \{u_5, u_6\}$. Moreover, the system would retrieve the movies that users 5 and 6 liked, hypothesizing that they have similar tastes with u. Thus, operator 1 results in $d = \{u_5, u_6\}^{\square} = \langle [0,0], [0,0], [0,0], [0,0], [0,0], [4,5], [4,5] \rangle$, suggesting that Die Hard is worth watching for the target user u.

Obviously, the pattern concept $(\{u_5, u_6\}, d)$ describes a small group of like-minded users and their shared preferences are stored in the vector d (cf. bicluster). □

Taking into account constant pressing of industry requests for Big Data tools, several ways of their fitting to this context were proposed in [50, 134]; thus, for Pattern Structures in classification setting, combination of lazy evaluation with projection approximations of initial data, randomisation and parallelisation, results in reduction of algorithmic complexity to low degree polynomial. This observations make it possible to apply pattern structures in text mining and learning from large text collections [135]. Implementations of basic Pattern Structures algorithms are available in FCART. □

Exercise 20. 1. Compose a small program, e.g. in Python, that enumerates all pattern concepts from the movie recommender example directly by definition or adapt CbO this end. 2. In case there is no possibility to perform 1., consider the subtable of the first four users and the first four movies from the movie recommender example. Find all pattern concepts by the definition. Build the line diagram of the pattern concept lattice. □

However, Pattern Structures is not the only attempt to fit FCA to data with more complex description than Boolean one. Thus, during the past years, the research on extending FCA theory to cope with imprecise and incomplete information made significant progress. The underlying model is a so called fuzzy concepts lattice; there are several definitions of such a lattice, but the basic assumption usually is that an object may posses attributes to some degree [136]. For example, in sociological studies age representation requires a special care: a person being a teenager cannot be treated as a truly adult one on the first day when his/her age exceeds a threshold of 18 years old (moreover, for formal reasons this age may differ in different countries). However, it is usually the case when we deal with nominal scaling; even ordinal scaling may lead to information loss because of the chosen granularity level. So, we need a flexible measure of being an adult and a teenage person at the same and it might be a degree lying in [0,1] interval for each such attribute. Another way to characterise this imprecision or roughness can be done in rough sets terms [137]. An interested reader is invited to follow a survey on Fuzzy and Rough FCA in [138]. The correspondence between Pattern Structures and Fuzzy FCA can be found in [139].

5.4   FCA-based Boolean Matrix Factorisation

Matrix Factorisation (MF) techniques are in the typical inventory of Machine Learning ([125], chapter Features), Data Mining ([63], chapter Dimensionality

Reduction) and Information Retrieval ([1], chapter Matrix decompositions and latent semantic indexing). Thus MF used for dimensionality reduction and feature extraction, and, for example, in Collaborative filtering recommender MF techniques are now considered industry standard [140].

Among the most popular types of MF we should definitely mention Singular Value Decomposition (SVD) [141] and its various modifications like Probabilistic Latent Semantic Analysis (PLSA) [142] and SVD++ [143]. However, several existing factorisation techniques, for example, non-negative matrix factorisation (NMF) [144] and Boolean matrix factorisation (BMF) [51], seem to be less studied in the context of modern Data Analysis and Information Retrieval.

Boolean matrix factorisation (BMF) is a decomposition of the original matrix $I \in \{0,1\}^{n \times m}$, where $I_{ij} \in \{0,1\}$, into a Boolean matrix product $P \circ Q$ of binary matrices $P \in \{0,1\}^{n \times k}$ and $Q \in \{0,1\}^{k \times m}$ for the smallest possible number of k. Let us define Boolean matrix product as follows:

$$(P \circ Q)_{ij} = \bigvee_{l=1}^{k} P_{il} \cdot Q_{lj}, \tag{3}$$

where $\bigvee$ denotes disjunction, and $\cdot$ conjunction.

Matrix I can be considered a matrix of binary relations between set X of objects (users), and a set Y of attributes (items that users have evaluated). We assume that xIy iff the user x evaluated object y. The triple (X, Y, I) clearly forms a formal context.

Consider a set $\mathcal{F} \subseteq \mathcal{B}(X, Y, I)$, a subset of all formal concepts of context (X, Y, I), and introduce matrices $P_{\mathcal{F}}$ and $Q_{\mathcal{F}}$:

$$(P_{\mathcal{F}})_{il} = \begin{cases} 1, i \in A_l, \\ 0, i \notin A_l, \end{cases} \quad (Q_{\mathcal{F}})_{lj} = \begin{cases} 1, j \in B_l, \\ 0, j \notin B_l. \end{cases}$$

where (A_l, B_l) is a formal concept from F.

We can consider decomposition of the matrix I into binary matrix product $P_{\mathcal{F}}$ and $Q_{\mathcal{F}}$ as described above. The following theorems are proved in [51]:

Theorem 2. (*Universality of formal concepts as factors*). For every I there is $\mathcal{F} \subseteq \mathcal{B}(X, Y, I)$, such that $I = P_{\mathcal{F}} \circ Q_{\mathcal{F}}$.

Theorem 3. (*Optimality of formal concepts as factors*). Let $I = P \circ Q$ for $n \times k$ and $k \times m$ binary matrices P and Q. Then there exists a set $\mathcal{F} \subseteq \mathcal{B}(X, Y, I)$ of formal concepts of I such that $|\mathcal{F}| \leq k$ and for the $n \times |\mathcal{F}|$ and $|\mathcal{F}| \times m$ binary matrices $P_{\mathcal{F}}$ and $Q_{\mathcal{F}}$ we have $I = P_{\mathcal{F}} \circ Q_{\mathcal{F}}$.

Example 17. Transform the matrix of ratings described above by thresholding ($geq3$), to a Boolean matrix, as follows:

$$\begin{pmatrix} 1 & 1 & 1 & 0 & 0 & 0 & 0 \\ 1 & 1 & 1 & 1 & 1 & 0 & 0 \\ 0 & 0 & 0 & 1 & 1 & 0 & 0 \\ 0 & 0 & 0 & 1 & 1 & 1 & 1 \\ 0 & 0 & 0 & 0 & 0 & 1 & 1 \\ 0 & 0 & 0 & 0 & 0 & 1 & 1 \end{pmatrix} = I.$$

The decomposition of the matrix I into the Boolean product of $I = A_{\mathcal{F}} \circ B_{\mathcal{F}}$ is the following:

$$
\begin{pmatrix}
1 & 1 & 1 & 0 & 0 & 0 & 0 \\
1 & 1 & 1 & 1 & 1 & 0 & 0 \\
0 & 0 & 0 & 1 & 1 & 0 & 0 \\
0 & 0 & 0 & 1 & 1 & 1 & 1 \\
0 & 0 & 0 & 0 & 0 & 1 & 1 \\
0 & 0 & 0 & 0 & 0 & 1 & 1
\end{pmatrix}
=
\begin{pmatrix}
1 & 0 & 0 \\
1 & 1 & 0 \\
0 & 1 & 0 \\
0 & 1 & 1 \\
0 & 0 & 1 \\
0 & 0 & 1
\end{pmatrix}
\circ
\begin{pmatrix}
1 & 1 & 1 & 0 & 0 & 0 & 0 \\
0 & 0 & 0 & 1 & 1 & 0 & 0 \\
0 & 0 & 0 & 0 & 0 & 1 & 1
\end{pmatrix}.
$$

Even this tiny example shows that the algorithm has identified three factors that significantly reduces the dimensionality of the data.   □

There are several algorithms for finding $P_{\mathcal{F}}$ and $Q_{\mathcal{F}}$ by calculating formal concepts based on these theorems [51]. Thus, the approximate algorithm (Algorithm 2 from [51]) avoids computation of all possible formal concepts and therefore works much faster than direct approach by all concepts generation. Its running time complexity in the worst case yields $O(k|G||M|^3)$, where k is the number of found factors, $|G|$ is the number of objects, $|M|$ is the number of attributes.

As for applications, in [120, 121], FCA-based BMF was used as a feature extraction technique for improving the results of classification. Another example closely relates to IR; thus, in [94, 145] BMF demonstrated comparable results to SVD-based collaborative filtering in terms of MAE and precision-recall metrics.

Further extensions of BMF to triadic and n-ary data were proposed in [146] and [147], respectively (the last one in not FCA-based). Triclustering, Triadic FCA and triadic tensor factorisation are useful techniques in Information Retrieval [109, 148, 149] for mining, ranking, and search in structured data like RDF or Folksonomies (see Sect. 6.8).

5.5   Case Study: Admission Process to HSE University

in this case study we reproduce results of our paper from [52]. Assuming probable confusion of the Russian educational system, we must say a few words about the National Research University Higher School of Economics[18] and its admission process.

Nowadays HSE is acknowledged as a leading university in the field of economics, management, sociology, business informatics, public policy and political sciences among Russian universities. Recently a number of bachelor programmes offered by HSE has been increased. In the year 2010 HSE offered 20 bachelor programmes. We consider only bachelor programmes in our investigation.

In order to graduate from school and enter a university or a college every Russian student must pass a Unified State Exam (Russian transcription: EGE), similar to US SAT–ACT or UK A-Level tests. During 2010 admission to U-HSE, entrants were able to send their applications to up to three programmes simultaneously. Some school leavers (major entrants of HSE bachelor programmes)

---

[18] http://www.hse.ru/en/.

chose only one programme, some chose two or three. Then entrants had to choose only one programme to study among successful applications.

We used data representing admission to HSE in 2010. It consists of information about 7516 entrants. We used mainly information about programmes (up to three) to which entrants apply[19]. Exactly 3308 entrants successfully applied at least to one programme, but just 1504 become students. Along with this data we also used the data of entrants' survey (76 % of entire assembly).

Further in the paper we mostly used data for the Applied Mathematics and Informatics programme to demonstrate some results. The total number of applications to the Applied Mathematics and Informatics programme was 843, of which 398 were successful but only 72 of them were actually accepted into the program. It might seem confusing only 72 out of 398 eligible prospective students decided to enroll, but since the admission process was set up in two stages, and at each stage only 72 entrants were eligible to attend the program, some of them decided to go for a different programme or university. As a result, the number of entrants whose applications were successful in any way came down to 398. Such situation is typical for all the bachelor programmes at HSE.

FCA requires object-attribute data. In our case objects are entrants and programmes they apply to are attributes. Together they are treated as a context. A series of contexts were constructed. Namely, we built a context for every programme where objects were entrants applying to that programme and attributes were other programmes they applied to. We built a separate context for every programme because it is meaningless to consider all programmes at once as programmes are very different in size and the resulting lattice would represent only the largest of them.

Likewise, we built a context for every programme where objects were entrants and attributes were programmes to which entrants successfully applied as well as the programmes that the entrants decided to enroll into, including those at other universities.

These contexts were then used to build concept lattices. Since the resulting lattices had too complicated a structure to interpret, we filtered concepts by their extent size (extent size is the number of objects, in our case it is the number of entrants), thus remaining concepts express only some of the more common patterns in entrants decisions.

To which programmes entrants often apply simultaneously? Trying to answer this question for every programme, we built diagrams[20] similar to Fig. 16. Such diagrams help us to reveal common patterns in entrants choices. Typical applications of FCA imply building formal concept lattices discussed earlier, but here we filter concepts by extent size to avoid complexity caused by noise in the data. Thus the order on remaining concepts is no longer a lattice, it is a partial order. Meaning of the labels on the diagram is obvious. A label above a node is a pro-

---

[19] HSE is a state university, thus most of student places are financed by government. In this paper we consider only such places.

[20] As any other data mining technique FCA implies an intensive use of software. All diagrams mentioned in this paper have been produced with meud (https://github.com/jupp/meud-wx).

gramme, a label below a node is a percent of entrants to Applied Mathematics and Informatics programme who also applied to programmes connected to a node from above. For example, the most left and bottom node on the diagram means that five percent of applied math's entrants also apply to Mathematics and Software Engineering. Then if we look at the nodes above the current node we may notice that ten percent Applied Mathematics and Informatics applicants also apply to Mathematics programme, and 70 percent also applied to Software Engineering.

Now let us try to interpret some knowledge unfolded by the diagram in Fig. 16. 70 percent of entrants who applied to Applied Mathematics and Informatics also apply to Software Engineering. The same diagram for Software Engineering states that 80 percent of Software Engineering applicants also apply to Applied Mathematics and Informatics. How this fact can be explained? Firstly it can easily be explained by the fact that these two programmes require to pass the same exams. Therefore there were not any additional obstacles to apply to both programmes simultaneously. Another possible explanation is that it is uneasy for entrants to distinguish these two programmes and successful application to any of them would be satisfactory result.

Analysing diagrams of other programmes' applications we found that equivalence of required exams is probably the most significant reason to apply to more than one programme.

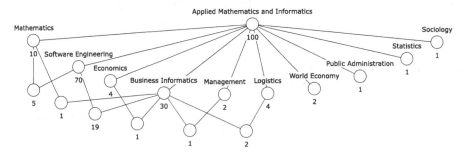

Fig. 16. Other programmes which entrants of Applied Mathematics and Informatics programme also applied.

Entrants' "Efficient" Choice. If an entrant successfully applied to more than one bachelor programme he or she must select a programme to study. Unlike the previous case, entrants have to select exactly one programme which gives us more precise information about entrants preferences. For that reason we define this situation as an efficient choice, efficient in the sense of more expressive about true entrants preferences.

Figure 17 presents the efficient choice of entrants to Applied Mathematics and Informatics programme. The meaning of diagram labels is almost the same

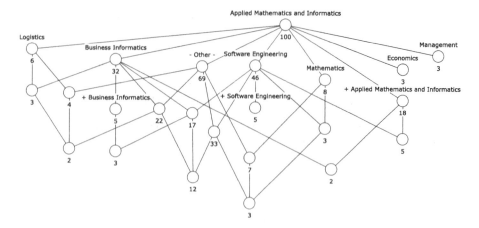

Fig. 17. "Efficient" choice of entrants to Applied Mathematics and Informatics programme.

as in Fig. 16. Programmes without plus sign (+) are successful applications, programmes with preceding plus sign are programmes chosen to study by entrants. Label "- Other -" means that the entrant canceled his application preferring another university or not to study this year altogether.

Together with diagram in Fig. 16 1 this diagram provides us with more precise knowledge about preferences of entrants to the Applied Mathematics and Informatics programme. More than two thirds of entrants who successfully apply to the Applied Math programme nevertheless prefer to study at another university. Whereas just 18 percent of successful applicants then become students on the Applied Mathematics and Informatics programme. Exactly 5 percent prefer to study Software Engineering and 5 percent of entrants who choose Applied Mathematics and Informatics also successfully applied to Software Engineering. It can be interpreted as equality of entrants preferences concerning these two programmes. Additionally, 5 percent prefer Business Informatics and only two percent of entrants who prefer Applied Mathematics and Informatics also successfully apply to Business Informatics, therefore in the pair Business Informatics and Applied Mathematics and Informatics the latter one is less preferable by entrants.

Here we should note that the sum of nodes percents with labels containing plus sign and node "- Other -" must equal to 100%, however here it does not because we excluded some nodes during filtering.

We built diagrams of "efficient" choice for every programme. Analysis of these diagrams helps us to recognise some relations between programmes in terms of entrants preferences. For example, some programmes in most cases is rather backup than actual entrants preference. Some programmes are close to each other by subject of study, these relations are also expressed by diagrams. With help of formalised survey data we found some possible factors of entrants' choice among some particular programmes. These knowledge can help our university to understand entrants' attitude to its undergraduate programmes and thus correct the structure and positioning of them.

Another Educational data mining case includes analysis of student achievements in two subsequent year for the same group by means of grading data [150].

5.6 Machine Learning Exercises with JSM-method in QuDA

QuDA was developed in early 2000s as "a software environment for those who want to learn Data Mining by doing" at the Intellectics group of the Darmstadt Technical University of Technology [53,151,152]. It includes various techniques, such as association rule mining, decision trees and rule-based learning, JSM-reasoning (including various reasoning scheme [153]), Bayesian learning, and interesting subgroup discovery. It also provides the experimenter with error estimation and model selection tools as well several preprocessing and post-processing utilities, including data cleansing tools, line diagrams, visualisation of attribute distributions, and a convenient rule navigator, etc. It was mostly aimed to support scientific and teaching activities in the field of Machine Learning and Data Mining. However, since QuDA has open architecture and support the most common data formats as well as the Predictive Model Markup Language (PMML)[21], it can be easily integrated into a working Data Mining circle. Originally, it was an acronym for "Qualitative Data Analysis". Now, since QuDA finally includes many quantitative methods integrated into it from WEKA[22], this name is a backronym[23] since it has lost its original meaning.

Exercise 21. Download QuDa[24]. Refer to QuDa's manual [151] for details and prepare the credit scoring context in csv format for opening in the QuDA environment. Perform nominal scaling of attributes and apply JSM classifier with basic setup. Compare the obtained rules with the hypotheses obtained manually.   □

Exercise 22. For zoo dataset available with QuDa (or some other dataset that suitable for classification from UCI ML repository[25]), perform nominal scaling and comparison of JSM-classification against all available methods (1) by splitting data into 80:20 training-to-test sample size ration (2) by 10-fold cross-validation. Compare learning curves and confusion matrices. Identify all non-covered examples by JSM-method. Change scaling type for attribute "number of legs". Reiterate comparison and check which methods have improved their classification quality.   □

6 FCA in Information Retrieval and Text Mining

Lattice-based models and FCA itself are not mainstream directions of modern IR; they attracted numerous researchers because of their interpretability and

---

[21] http://www.dmg.org/.

[22] http://www.cs.waikato.ac.nz/ml/weka/.

[23] http://en.wikipedia.org/wiki/Backronym.

[24] http://sourceforge.net/projects/quda/.

[25] http://archive.ics.uci.edu/ml/datasets.html.

human-centerdness, but their intrinsic complexity is a serious challenge to make them working on a Web scale.

Thus, from early works on Information Retrieval it is known that usage of a lattice as a search space requires treatment of the enormous number of subsets of documents: $10^{310,100}$ for a collection of one million documents [154]. At that time, in library classification domain, it was rather natural to consider documents and their categories, which may form requests as a combination of simple logical operations like AND, OR and NOT [155]. Thus, Mooers considered transformations $T : P \rightarrow L$, where P is the space of all possible document descriptors and L is the space of all possible document subsets [154]. Thus, T retrieves the largest set of documents from L according to a query (prescription) from P.

At that period in Soviet Russia, All-Soviet Institute for Scientific and Technical Information (VINITI) was organised to facilitate information interchange and fulfill growing scientific needs in cataloging and processing of scientific publications. Around the mid 1960s, Yulii A. Shreider, one of the leading researchers of VINITI, considered the problem of automatic classification of documents and their retrieval by means of a model featuring a triple (M, L, f), where M is a set of documents, L is a set of attributes and $f : M \rightarrow 2^L$ maps each document to a set attributes from L [156]. There, similarity of two documents was defined via non-emptiness of the intersection of their descriptions $f(d_1) \cap f(d_2)$. In that paper, Shreider mentioned the relevance of lattices to problems of document classification and retrieval, where he also cited the work of Soergel [157] on this issue.

Thus, these two introduced mappings, T and f highly resemble to conventional prime operators in FCA for the context of documents and their attributes (keywords, terms, descriptors) with "document-term containment" relation. In the middle of 80s, Godin et al. [158] proposed a lattice-based retrieval model for database browsing, where objects (documents, e.g. course syllabi) were described by associated keywords. The resulting (in fact, concept) lattice used for navigation by query modification using its generality/specificity relation.

In 90-s several FCA-based IR models and systems appeared, the reviews can be found [159, 160]. Thus, in [159], Carpineto and Romano classified main IR problems that can be solved by FCA means through review of their own studies by the year 2005. Uta Priss described a current state of FCA for IR domain [160] by the year 2004. Recently, a survey on FCA-based systems and methods for IR including prospective affordances was presented at FCA for IR workshop at ECIR 2013 [161], and our colleagues, Codocedo and Napoli, taking an inspiration, are summarising the latest work on the topic in a separate forthcoming survey.

Below, we shortly overview our own study on applying FCA-based IR methods for describing the state-of-the-art in FCA for IR field. The rest topics are spread among the most representative examples of FCA-based IR tasks and systems including the summary of the author's experience.

- Text Mining scientific papers: a survey on FCA-based IR applications [37]
- FCA-based meta-search engines (FOOCa, SearchSleuth, Credo etc.) [15, 83]
- FCA-based IR visualisation [15] and navigation (ImageSleuth, Camelis [162])

- FCA in criminology: text mining of police reports [37]
- FCA-based approach for advertising keywords in web search [96]
- FCA-based Recommender Systems [145]
- Triadic FCA for IR-tasks in Folksonomies [148]
- FCA-based approach for near-duplicate documents detection [81,163]
- Exploring taxonomies of web site users [164]
- Concept-based models in Crowdsourced platforms: a recommender system of like-minded persons, antagonists and ideas [165]

6.1 Text Mining Scientific Papers: A Survey on FCA-based IR Applications

In [4], we visually represented the literature on FCA and IR using concept lattices, in which the objects are the scientific papers and the attributes are the relevant terms available in the title, keywords and abstract of the papers. We developed an IR tool with a central FCA component that we use to index the papers with a thesaurus containing terms related to FCA research and to generate the lattices. It helped us to zoom in and give an extensive overview of 103 papers published between 2003 and 2009 on using FCA in information retrieval.

	browsing	mining	software	web services	FCA	Information Retrieval
$Paper1$	×	×	×		×	
$Paper2$			×		×	×
$Paper3$		×		×	×	
$Paper4$	×		×		×	
$Paper5$				×	×	×

We developed a knowledge browsing environment CORDIET to support our literature analysis process. One of the central components of our text analysis environment is the thesaurus containing the collection of terms describing the different research topics. The initial thesaurus was constructed based on expert prior knowledge and was incrementally improved by analyzing the concept gaps and anomalies in the resulting lattices. The layered thesaurus contains multiple abstraction levels. The first and finest level of granularity contains the search terms of which most are grouped together based on their semantic meaning to form the term clusters at the second level of granularity. The papers downloaded from the Web were converted to plain text and the abstract, title and keywords were extracted. The open source tool Lucene[26] was used to index the extracted parts of the papers using the thesaurus. The result was a cross table describing the relationships between the papers and the term clusters or research topics from the thesaurus. This cross table was used as a basis to generate the lattices.

The most relevant scientific sources that were used in the search for primary studies contain the work published in those journals, conferences and workshops which are of recognized quality within the research com-munity. These sources are: IEEE Computer Society, ACM Digital Library, Sciencedirect, Springerlink, EBSCOhost, Google Scholar, Conference repositories: ICFCA, ICCS and CLA conferences. Other important sources such as DBLP or CiteSeer were not explicitly included since they were indexed by some of the mentioned sources (e.g. Google Scholar). In the selected sources we used various search terms including "Formal Concept Analysis", "FCA", "concept lattices", "Information Retrieval". To identify the major categories for the literature survey we also took into account the number of citations of the FCA papers at CiteseerX.

The efficient retrieval of relevant information is promoted by the FCA representation that makes the inherent logical structure of the information transparent. FCA can be used for multiple purposes in IR [15,160]. First, FCA provides an elegant language for IR modeling and is an interesting instrument for browsing and automatic retrieval through document collections. Second, FCA can also support query refinement, ranking and enrichment by external resources. Because a document-term lattice structures the available information as clusters of related documents which are partially ordered, lattices can be used to make suggestions for query enlargement in cases where too few documents are retrieved and for query refinement in cases where too many documents are retrieved. Third, lattices can be used for querying and navigation supporting relevance feedback. An initial query corresponds to a start node in a document-term lattice. Users can then navigate to related nodes. Further, queries are used to "prune" a document-term lattice to help users focus their search (Carpineto et al. 1996b). For many purposes, some extra facilities are needed such as processing large document collections quickly, allowing more flexible matching operations, allowing ranked retrieval and give contextual answers to user queries. The past years many FCA researchers have also devoted attention to these issues.

86 % of the papers on FCA and information retrieval are covered by the research topics in Fig. 18. Further, in our study, we intuitively introduced the process of transforming data repositories into browsable FCA representations and performing query expansion and refinement operations. Then we considered 28 % of papers on using FCA for representation of and navigation in image, service, web, etc. document collections. Defining and processing complex queries covered 6 % of the papers and was described as well. The review of papers on contextual answers (6 % of papers) and ranking of query results (6 % of papers) concluded the case-study.

---

[26] https://lucene.apache.org/core/.

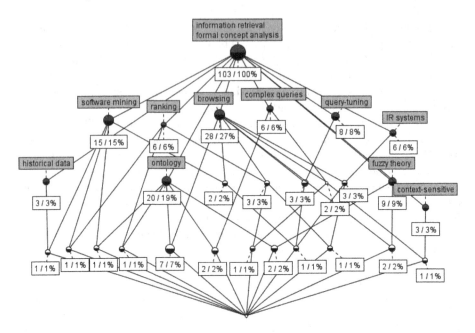

Fig. 18. Lattice containing 103 papers on using FCA in IR

Knowledge Representation and Browsing with FCA. In 28 % of the 103 selected papers, FCA is used for browsing and navigation through document collections. In more than half of these papers (18 % of total number of papers), a combination of navigation and querying based on the FCA lattices is pro-posed. Annotation of documents and finding optimal document descriptors play an important role in effective information retrieval (9 % of papers). All FCA-based approaches for information retrieval and browsing through large data repositories are based on the same underlying model. We first have the set D containing objects such as web pages, web services, images or other digitally available items. The set A of attributes can consist of terms, tags, descriptions, etc. These attributes can be related to certain objects through a relation $I \subseteq D \times A$ which indicates the terms, tags, etc. can be used to describe the data elements in D. This triple (D, A, I) is a formal context from which the concept lattice can be created.

Query Result Improvement with FCA. Search engines are increasingly being used by amongst others web users who have an information need. The intent of a concept corresponds to a query and the extent contains the search results. A query q features a set of terms T and the system returns the answer by evaluating T'. Upon evaluating a query q the system places itself on the concept (T', T'') which becomes the current concept c. For example, in Fig. 19, the intent of the current concept $B_c = \{t_1, t_2, t_3, t_4, t_5\}$ and the extent of the current concept $A_c = \{d_8, d_9\}$, where t stands for term and d stands for word. Since a query provided by a user only approximates a user's need, many techniques have been developed to expand and refine query terms and search results. Query tuning is the process of searching for the query that best approximates the information need of the user. Query refinements can help the user express his original need more clearly. Query refinement can be done by going to a lower neighbor of the current concept in the lattice by adding a new term to the query items. In **minimal conjunctive query refinement** the user can navigate for example to a subconcept $((B_c \cup \{t\})', (B_c \cup \{t\})'')$ by adding term t.

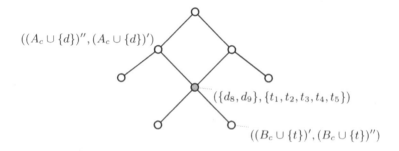

Fig. 19. Query modification in a concept lattice: a fish-eye view

Query enlargement, i.e. retrieving additional relevant web pages, can be performed by navigating to an upper neighbor of the current concept in the lattice by removing a term from the query items. The user can navigate for example to a superconcept $((B_c \cup \{d\})'', (B_c \cup \{d\})')$ by adding document d. The combination of subsequent refine and expand operations can be seen as navigation through the query space. Typically, navigation and querying are two completely separate processes, and the combination of both results in a more flexible and user-friendly method. These topics are investigated in 8 % of the IR papers. See the comprehensive survey on query refinement in [166].

Concept Lattice Based Ranking. 6 % of IR papers in our study are devoted concept lattice ranking.

Below we explain the concept lattice based ranking (CLR) proposed in [167] and compared with hierarchical clustering based (HCR) and best-first matching (BFR) rankings. The experiments with two public benchmark collections showed that CLR outperformed the two competitive methods when the ranked documents did no match the query and was comparable to BMF and better than HCR in the rest cases.

Let (D, T, I) be a document context, where D is the set of documents, T is the set of their terms, and $I \subseteq D \times T$. Consider the ordered set of all concepts $(\mathfrak{L}(D, T, I), \succ\!\prec)$ with the nearest neighbour relation $\succ\!\prec$, i.e., for $c_1, c_2 \in \mathfrak{L}(D, T, I), c_1 \succ\!\prec c_2$ iff $c_1 \succ c_2$ or $c_1 \prec c_2$. Then distance between concepts c_1 and c_2 is defined as the least natural number n as follows:

$$\exists c_{i_1}, \ldots, c_{i_n} \in \mathfrak{L}(D, T, I) \text{ such that } c_1 = c_{i_0} \succ\!\prec c_{i_1} \ldots \succ\!\prec c_{i_n} = c_2.$$

For each query q and $d_1, d_2 \in D$, d_1 is ranked higher than d_2 if the distance between (d_1'', d_1') and (q'', q') is shorter than than the distance between (d_2'', d_2') and (q'', q'). It means that we need less query transformations to obtain q from d_1 than to do that from d_2.

As the author of CLR admitted in [15], CLR is sensitive to addition new documents into the search collection. Further analysis of usage of concept neighbourhood based similarity for IR needs is given in [84].

Example 18. In Fig. 20 we provide an example of concept lattice based ranking for the previous context of papers and their terms. The underlying query is the conjunction of two terms: "browsing, FCA". The query concept is as follows:

$$(\{p_1, p_4\}, \{browsing, FCA, software\}).$$

The resulting ranking yields $p_4 < p_1 < p_2 < p_3 = p_5$. A curious reader may admit that concepts with the same ranks lie in concentric circles around the query concept at the same distance. Obviously, for concepts from the same such circle we need their subsequent ranking, e.g. by best-match ranking via dot product of the document and query profiles based on term frequency.

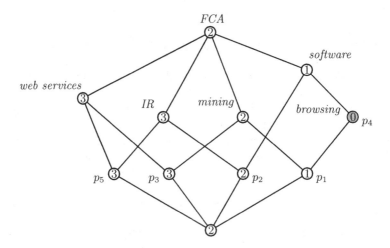

Fig. 20. Concept lattice based ranking for the query "browsing, FCA"; the distance values are given inside the corresponding circles.

An interested reader can find the rest sections in our survey:

- Web and email retrieval (partially covered in Sect. 6.2);
- Image, software and knowledge base retrieval (partially covered in Sect. 6.3);
- Defining and processing complex queries with FCA;
- Domain knowledge in search results: contextual answers & ranking.

6.2   FCA-based IR Visualisation and Meta-Search Engines

From the beginning 2000 s many independent IR developers proposed so called meta-search engines also known as search results clustering engines. To name a few, two project are still alive Carrots[227] and Nigma.ru[28]. See summarising survey on Web clustered search by Carpineto et al. in [168].

FCA has been used as the basis for many web-based knowledge browsing systems developed during the past years. Especially its comprehensible visualisation capabilities seem to be of interest to RuSSIR audience. The results returned by web search engines for a given query are typically formatted as a list of URLs accompanied by a document title and a snippet, i.e. a short summary of the document. Several FCA-based systems were developed for analyzing and exploring these search results. CREDO [169], FooCA [83] and SearchSleuth [170,171] build a context for each individual query which contains the result of the query as objects and the terms found in the title and summary of each result as attributes.

The CREDO system[29] then builds an iceberg lattice which is represented as a tree and can be interactively explored by the user (Fig. 21).

FooCA[30] shows the entire formal context to the user (Fig. 22) and offers a great degree of flexibility in exploring this table using the ranking of attributes, selecting the number of objects and attributes, applying stemming and stop word removal etc.

SearchSleuth does not display the entire lattice but focuses on the search concept, i.e. the concept derived from the query terms (Fig. 23). The user can easily navigate to its upper and lower neighbors and siblings. Nauer et al. [172] also propose to use FCA for iteratively and interactively analyzing web search results. The user can indicate which concepts are relevant and which ones are not for the retrieval task. Based on this information the concept lattice is dynamically modified. Their research resulted in the CreChainDo system[31]. Kim et al. [173] presented the FCA-based document navigation system KAnavigator for small web communities in specialized domains. Relevant documents can be annotated with keywords by the users. Kim et al. [174] extended the search functionality by combining lattice-based browsing with conceptual scales to reduce the complexity of the visualisation. Cigarran et al. [175] present the JBrainDead IR System which combines free-text search with FCA to organise the results of a query. Cole et al. [176] discuss a document discovery tool named Conceptual Email Manager (CEM) which is based on FCA. The program allows users to navigate through emails using a visual lattice. The paper also discusses how conceptual ontologies can support traditional document retrieval systems and aid knowledge discovery in document collections. The development of this software is based on

---

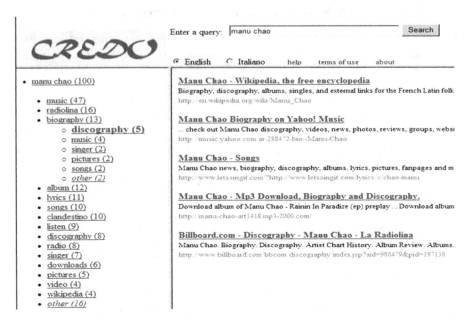

Fig. 21. An example of CREDO's web search interface

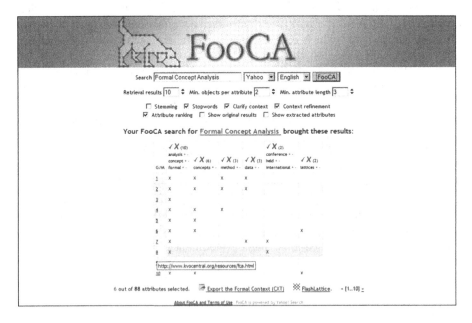

Fig. 22. An example of FooCA's web search interface. It processes the results of search queries to Yahoo or Google and organises them into the interactive cross-table.

earlier research on retrieval of information from semi-structured texts [177,178]. Building further on this work is the Mail-Sleuth software (Eklund et al. [179]) which can be used to mine large email archives. Eklund et al. [180] use FCA for displaying, searching and navigating through help content in a help system. Stojanovic [181] presents an FCA-based method for query refinement that provides a user with the queries that are "nearby" the given query. Their approach for query space navigation was validated in the context of searching medical abstracts. Stojanovic [181] presents the SMART system for navigation through an online product catalog. The products in the database are described by elements of an ontology and visualized with a lattice, in which users can navigate from a very general product-attribute cluster containing a lot of products to very specific clusters that seem to contain a few, but for the user highly relevant products. Spyratos et al. [182] describe an approach for query tuning that integrates navigation and querying into a single process. The FCA lattice serves for navigation and the attributes for query formulation. Le Grand et al. [183] present an IR method based on FCA in conjunction with semantics to provide

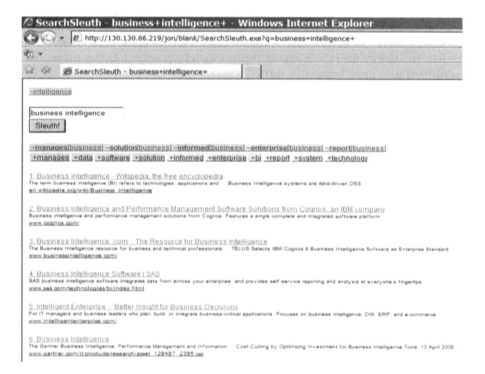

Fig. 23. An example of SearchSleuth's web interface. It processes the results of search queries to Yahoo. Passing to more general (more specific) categories is done by clicking -term(+term).

contextual answers to web queries. An overall lattice is built from tourism web pages. Then, users formulate their query and the best-matching concepts are returned, users may then navigate within the lattice by generalizing or refining their query. Eklund et al. [184] present AnnotationSleuth to extend a standard search and browsing interface to feature a conceptual neighborhood centered on a formal concept derived from curatorial tags in a museum management system. Cigarran et al. [185] focus on the automatic selection of noun phrases as documents descriptors to build an FCA based IR system. Automatic attribute selection is important when using FCA in a free text document retrieval framework. Optimal attributes as document descriptors should produce smaller, clearer and more browsable concept lattices with better clustering features. Recio-Garcia et al. [186] use FCA to perform semantic annotation of web pages with domain ontologies. Similarity matching techniques from Case Based Reasoning can be applied to retrieve these annotated pages as cases. Liu et al. [187] use FCA to optimise a personal news search engine to help users obtain the news content they need rapidly. The proposed technique combines the construction of user background using FCA, the optimisation of query keywords based on the user's background and a new layout strategy of search results based on a "Concept Tree". Lungley et al. [188] use implicit user feedback for adapting the underlying domain model of an intranet search system. FCA is used as an interactive interface to identify query refinement terms which help achieve better document descriptions and more browsable lattices (Figs. 21, 22 and 23).

6.3   FCA-based Image Retrieval and Navigation

FCA-based IR visualisation [15] and navigation (ImageSleuth, Camelis [162])

Ahmad et al. [189] build concept lattices from descriptions associated to images for searching and retrieving relevant images from a database. In the ImageSleuth project [190], FCA was also used for clustering of and navigation through annotated collections of images. The lattice diagram is not directly shown to the user. Only the extent of the present concept containing thumbnails, the intent containing image descriptions and a list of upper and lower neighbors is shown. In Ducrou [191], the author built an information space from the Amazon.com online store and used FCA to discover conceptually similar DVDs and explore their conceptual neighborhood. The system was called DVD-Sleuth. Amato et al. [192] start from an initial image given by the user and use a concept lattice for retrieving similar images. The attributes in this lattice are facets, i.e. an image similarity criterion based on e.g. texture, color or shape. The values in the context indicate for each facet how similar an image in the database is with respect to the user provided initial image. By querying, the user can jump to any cluster of the lattice by specifying the criteria that the sought cluster must satisfy. By navigation from any cluster, the user can move to a neighbor cluster, thus exploiting the ordering amongst clusters.

In [162] Ferre et al. proposed to use so called Logical Information Systems (LIS) for navigation through photo collections[32]. In fact LIS, similarly to Pattern Structures, exploit partially ordered object descriptions but expressed as logical formulas.

– location: Nizhniy Novgorod $\sqsubseteq$ Russia
– date: date = 18 Aug 2014 $\sqsubseteq$ date in Aug 2014 .. Jul 2015
– event: event is "summer school RuSSIR" $\sqsubseteq$ event contains "summer school"

Further it was extended for work with document collections [193]. Since Camelis uses lattice-based navigation and queriying by formulas, it overcomes current drawbacks of tree-like navigation imposed by current restrictions of file-systems.

Recently, the previous studies of Eklund et al. [84] in organising navigation through annotated collections of images in virtual museums resulted in an iPad application that allows users to explore an art collection via semantically linked pathways that are generated using Formal Concept Analysis[33]. In fact navigation in this application is organised by showing context and relationships among objects in a museum collection.

6.4 FCA in Criminology: Text Mining of Police Reports

In [37], we proposed an iterative and human-centred knowledge discovery methodology based on FCA. The proposed approach recognises the important role of the domain expert in mining real-world enterprise applications and makes use of specific domain knowledge, including human intelligence and domain-specific constraints. Our approach was empirically validated at the Amsterdam-Amstelland police to identify suspects and victims of human trafficking in 266,157 suspicious activity reports. Based on guidelines of the Attorney Generals of the Netherlands, we first defined multiple early warning indicators that were used to index the police reports.

Example 19. This is an example of a police report where some indicator-words are highlighted that are used for its contextual representation.   □

---

[32] Camelis, http://www.irisa.fr/LIS/ferre/camelis/.
[33] "A place for art", https://itunes.apple.com/au/app/a-place-for-art/id638054832?mt=8.

Report 1:

On the night of 23 of March
2008 we stopped a car with a
Bulgarian license plate for
routine motor vehicle inspection.
It was a *Mercedes GLK* with
license plate BL XXX. The car
was driving around in circles
in a *prostitution area.* On the
backseat of the car we noticed
two well dressed young girls. We
asked for their identification
papers but they did not speak
English nor Dutch. The driver
of the car was *in possession of
their papers* and told us that
they were *on vacation* in the
Netherlands for two weeks etc.

	Expensive cars	Prostitutes	Id-papers	Vacation	Former Eastern Europe	Information Retrieval
*Report*1	×	×	×	×	×	×
*Report*2	×	×	×	×		×
*Report*3	×	×	×			×
*Report*4		×			×	
*Report*5	×					×

Our method based on FCA consists of four main types of analysis which are carried out as follows:

1. Concept exploration of the forced prostitution problem of Amsterdam: In Poelmans et al. [194], this FCA-based approach for automatically detecting domestic violence in unstructured text police reports is described in detail.
2. Identifying potential suspects: concept lattices allow for the detection of potentially interesting links between independent observations made by different police officers.
3. Visual suspect profiling: some FCA-based methods such as temporal concept analysis (TCA) were developed to visually represent and analyse data with a temporal dimension [195]. Temporal concept lattices were used in Elzinga et al. [196] to create visual profiles of potentially interesting terrorism subjects. Elzinga et al. [197] used TCA in combination with nested line diagrams to analyse pedophile chat conversations.
4. Social structure exploration: concept lattices may help expose interesting persons related to each other, criminal networks, the role of certain suspects in these networks, etc. With police officers we discussed and compared various FCA-based visualisation methods of criminal networks.

In our investigations we also used the model that was developed by Bullens and Van Horn [198] for the identification of loverboys who typically force girls of Dutch nationality into prostitution. Loverboys use their love affair with a woman to force her to work in prostitution. Forcing girls and women into prostitution through a loverboy approach is seen as a special kind of human trafficking in the Netherlands (article 250a of the code of criminal law). This model is a resource used by the Amsterdam-Amstelland police during the trainings of police officers about this topic. A typical loverboy approach consists of three main phases which give rise to corresponding indicators:

1. Preparatory activities to recruit girls.
2. Forcing her into prostitution.
3. Keeping the girl in prostitution by emotional dependence or social isolation.

The pimp will also try to protect his organisation.

In our data-set, there were three reports available about girl H. The reports about this girl led us to the discovery of loverboy suspect B. The first report (26 November 2008) contains the notification of the police by a youth aid organisation in Alkmaar about girl H. They report a suspicious tattoo on her wrist containing the name B. This B refers to her boyfriend who carries the same first name, is 30 years old, and is of Surinamian origin. The second report was written by a police officer who works in the red light district and knows many women working in brothels or behind the windows. During a patrol he saw H working as a prostitute, had a conversation with her, and observed suspicious that his included in the report. The next report contains four suspicious facts recorded by the officer. First, an unbelievable story why she works as a prostitute: a bet between girlfriends if someone would dare to work as a prostitute. Second, the tattoos of which one tattoo is mentioned in the first report (B) and a new one on her belly. Third, the injuries, she has scratches on her arm (possibly from a fight) and burns on her leg.

According to the victim, she has dropped a hot iron on her leg and had an accident with a gourmet set. Fourth is the observation of making long working days. The third document (21 December 2008) showed an observation of the victim walking with the possible suspect. In this document the police officer reports that he saw the victim and a man walking close to each other. The police officer knows the man and knows that he is active in the world of prostitution. When the man saw the officer, he immediately took some distance of the victim. As soon as they have passed the officer, they walk close together and into a well-known street where prostitutes work behind the windows. The first name of the person is B, the same name which is tattooed on the victims wrist, and the description of the person is about the same as described by the youth aid organisation. This information signals that the man is the possible loverboy of the victim. The three reports together give serious presumptions of B being a loverboy with H being the victim. The next step is investigating B. We need serious indications that B is really involved in forced prostitution. Twelve observational reports were found for B and the resulting lattice is shown in Fig. 24.

Investigating these reports shows that he frequently visits the red light district and has strong relationships with other pimps. One of these pimps is the suspect of another loverboy case. From the six observations where B was seen in the red light district, four are violence related, including the observation of Hs suspicious burn wounds. The other violence-related observations are situations of fights with customers who are unwilling to leave or pay. Such violence-related

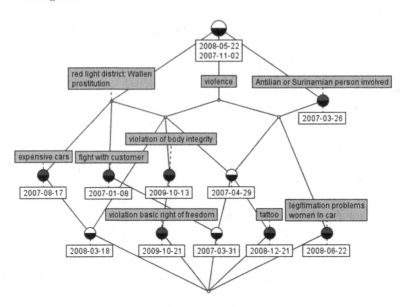

Fig. 24. Line diagram for the report context of loverboy suspect B

observations are related to pimps who want to protect their prostitutes from customers and competing gangs. Since in the Netherlands, prostitution is legal, each prostitute has the right to ask the police to protect her. The violence observations of the suspect strengthened the suspicion of B being the pimp of H. Moreover, we found another girl R who was also a potential victim of him. These indications were enough to create a summary report and send a request for using further investigation techniques to the public prosecutor.

Using concept lattices, we revealed numerous unknown human trafficking and loverboy suspects. Indepth investigation by the police resulted in a confirmation of their involvement in illegal activities resulting in actual arrestments been made. This approach was embedded into operational policing practice and is now successfully used on a daily basis to cope with the vastly growing amount of unstructured information.

There are other FCA-based studies in criminology, for example, terrorist activity modeling and analysis [199] and developments of lattice-based access policies for information systems [200, 201].

6.5   FCA-based Approach for Advertising Keywords in Web Search

Online advertising by keywords matching is bread and butter of modern web search companies like Google and Yandex. For our experimentation we used data of US Overture [202] (now, a part of Yahoo), which were first transformed in the standard context form. We consider the following context: $\mathbb{K}_{FT} = (F, T, I_{FT})$, where F is the set of advertising firms (companies), T is the set of advertising terms, or phrases, $fI_{FT}t$ means that firm $f \in F$ bought advertising term $t \in T$. In the context $|F| = 2000$, $|T| = 3000$, $|I_{FT}| = 92345$.

The data are typically sparse, thus the number of attributes per object is bounded as follows: $13 \leq |g'| \leq 947$. For objects per attribute we have $18 \leq |m'| \leq 159$. From this context we computed formal concepts of the form (advertisers, bids) that represent market sectors. Formal concepts of this form can be further used for recommendation to the companies on the market, which did not buy bids contained in the intent of the concept.

This can also be represented as association rules of the form "If an advertiser bought bid a, then this advertiser may buy term b" See [203] for the use of association rules in recommendation systems.

To make recommendations we used the following steps:

1. D-miner algorithm for detecting large market sectors as concepts and our biclustering algorithm;
2. Coron system for constructing association rules;
3. Construction of association metarules using morphological analysis;
4. Construction of association metarules using ontologies (thematic catalogs).

Detecting Large Market Sectors with D-miner and OA-biclustering. The D-miner algorithm [204] constructs the set of concepts satisfying given constraints on sizes of extents and intents (i.e. intersection of icebergs and dual icebergs). D-miner takes as input a context and two parameters: minimal admissible extent and intent sizes and outputs a "band" of the concept lattice: all concepts satisfying constraints given by parameter values ($|intent| \geq m$ and $|extent| \geq n$, where $m, n \in \mathbb{N}$, see Table 6).

Table 6. D-miner results.

Minimal extent size	Minimal intent size	Number of concepts
0	0	8 950 740
10	10	3 030 335
15	10	759 963
15	15	150 983
15	20	14 226
20	15	661
20	16	53
20	20	0

Example 20. We provide examples of two intents of formal concepts for the case $|L| = 53$, where $|L|$ is a number of formal concepts obtained by D-miner.

Hotel Market. {angeles hotel los, atlanta hotel, baltimore hotel, dallas hotel, denver hotel, hotel chicago, diego hotel san, francisco hotel san, hotel houston, hotel miami, hotel new orleans, hotel new york, hotel orlando, hotel philadelphia, hotel seattle, hotel vancouver}

Weight Loss Drug Market. {adipex buy, adipex online, adipex order, adipex prescription, buy didrex, buy ionamin, ionamin purchase, buy phentermine, didrex online, ionamin online, ionamin order, online order phentermine, online phentermine, order phentermine, phentermine prescription, phentermine purchase}   □

Applying the biclustering algorithm to our data we obtained 87 OA-biclusters ($\rho = 0.85$), which is much less than the number of concepts found by D-miner. Expert interpretation of these biclusters implies that each market described by formal concepts found by D-miner (where each market can be represented by several formal concepts) corresponds to a bicluster among these 87. The number of formal concepts generated by D-miner becomes feasible for human interpretation if there are no more than 20 firms and about 15 terms. For these thresholds D-miner could find only large markets and ignored important average-size markets. For our data these ignored markets were, e.g. car and flower markets, which were found using biclustering approach.

Example 21. Flower market OA-bicluster.
    ({ 24, 130, 170, 260, 344, 415, 530, 614, 616, 867, 926, 1017, 1153, 1160, 1220, 1361, 1410, 1538, 1756, 1893 }, {'anniversary flower', 'arrangement flower', 'birthday flower', 'bouquet flower', 'buy flower', 'buy flower online', 'delivery flower', 'flower fresh', 'flower gift', 'flower line', 'flower online', 'flower online order', 'flower online send', 'flower online shop', 'flower rose', 'flower send', 'flower shop', 'flower sympathy', 'red rose'}), with $\rho \approx 0.84$ □

Recommendations Based on Association Rules. Using the Coron system (see [205]) we construct the informative basis of association rules [206].

Example 22. Here are some examples of association rules:

- {*evitamin*} → {*cvitamin*}, supp=31 [1.55 %] and conf=0.86;
- {*gift graduation*} → {*anniversary gift*}, supp=41 [2.05 %] and conf=0.82.

                                                                      □

The value *supp* = 31 of the first rule means that 31 companies bought phrases "e vitamin" and "c vitamin". The value *conf* = 0.861 means that 86,1 % companies that bought the phrase "e vitamin" also bought the phrase "c vitamin".

To make recommendations for each particular company one may use an approach proposed in [203]. For company f we find all association rules, the antecedent of which contain all the phrases bought by the company, then we construct the set T_u of unique advertising phrases not bought by the company f before. Then we order these phrases by decreasing of confidence of the rules where the phrases occur in the consequences. If buying a phrase is predicted by multiple rules, we take the largest confidence.

Morphology-based Metarules. Each attribute of our context is either a word or a phrase. Obviously, synonymous phrases are related to same market sectors. The advertisers companies have usually thematic catalogs composed by experts, however due to the huge number of advertising terms manual composition of catalogs is a difficult task. Here we propose a morphological approach for detecting similar terms.

Let t be an advertising phrase consisting of several words (here we disregard the word sequence): $t = \{w_1, w_2, \ldots, w_n\}$. A stem is the root or roots of a word, together with any derivational affixes, to which inflectional affixes are added [207]. The stem of word w_i is denoted by $s_i = stem(w_i)$ and the set of stems of words of the phrase t is denoted by $stem(t) = \bigcup_i stem(w_i)$, where $w_i \in t$.

Consider the formal context $\mathbb{K}_{TS} = (T, S, I_{TS})$, where T is the set of all phrases and S is the set of all stems of phrases from T, i.e. $S = \bigcup_i stem(t_i)$. Then tIs denotes that the set of stems of phrase t contains s.

In this context we construct rules of the form $t \to s_i^{I_{TS}}$ for all $t \in T$, where $(.)^{I_{ts}}$ denotes the prime operator in the context $\mathbb{K}_{TS}$. Then the a morphology-based metarules of the context $\mathbb{K}_{TS}$ (we call it a metarule, because it is not based on experimental data, but on implicit knowledge resided in natural language constructions) corresponds to $t \xrightarrow{FT} s_i^{I_{TS}}$, an association rule of the context $\mathbb{K}_{FT} = (F, T, I_{FT})$. If the values of support and confidence of this rule in context $\mathbb{K}_{FT}$ do not exceed certain thresholds, then the association rules constructed from the context $\mathbb{K}_{FT}$ are considered not very interesting.

Example 23. An example of input contexts for morphological association rules (Tables 7 and 8). ??

□

Metarules of the following forms seem also to be reasonable. First, one can look for rules of the form $t \xrightarrow{FT} \bigcup_i s_i^{I_{TS}}$, i.e., rules, the consequent of which contain all terms containing at least one word with the stem common to a word in the antecedent term. Obviously, constructing rules of this type may result in the fusion of phrases related to different market sectors, e.g. "black jack" and

Table 7. A toy example of context $\mathbb{K}_{FT}$ for "long distance calling" market.

firm \ phrase	call distance long	distance calling long	calling distance long plan	carrier distance long	cheap distance long plan
f_1	x		x		x
f_2		x	x	x	
f_3				x	x
f_4		x	x		x
f_5	x	x		x	x

Table 8. A toy example of context $\mathbb{K}_{TS}$ for "long distance calling" market.

phrase \ stem	call	carrier	cheap	distanc	long	plan
call distance long	x			x	x	
calling distance long	x			x	x	
calling distance long plan	x			x	x	x
carrier distance long		x		x	x	
cheap distance long			x	x	x	

"black coat". Second, we considered rules of the form $t \xrightarrow{FT} (\bigcup_i s_i)^{ITS}$, i.e., rules with the consequent with the set of stems being the same as the set of stems of the antecedent. Third, we also propose to consider metarules of the form $t_1 \xrightarrow{FT} t_2$, where $t_2^{ITS} \subseteq t_1^{ITS}$. These are rules with the consequent being sets of stems that contain the set of stems of the antecedent.

Example 24. An example of metarules.

– $t \xrightarrow{FT} s_i^{ITS}$
  {*last minute vacation*} → {*last minute travel*}
  supp= 19 conf= 0,90
– $t \xrightarrow{FT} \bigcup_i s_i^{ITS}$
  {*mail order phentermine*} → {*adipex online order*, . . . ,
  phentermine purchase, phentermine sale}
  supp= 19   conf= 0,95
– $t \xrightarrow{FT} (\bigcup_i s_i)^{ITS}$
  {*distance long phone*} → {*call distance long phone*, . . . ,
  distance long phone rate, distance long phone service}
  supp= 37   conf= 0,88
– $t_1 \xrightarrow{FT} t_2$, $t_2^{ITS} \subseteq t_1^{ITS}$
  {*ink jet*} → {*ink*}, supp= 14   conf= 0,7   □

Experimental Validation. For validation of association rules and metarules we used an adapted version of cross-validation. The training set was randomly divided into 10 parts, 9 of which were taken as the training set and the remaining part was used as a test set. The confidence of rules averaged over the test set is almost the same as the min_conf for the training set, i.e., $(0.9 - 0.87)/0.9 \approx 0.03$.

Note that the use of morphology is completely automated and allows one to find highly plausible metarules without data on purchases. The rules with low support and confidence may be tested against recommendation systems such as Google AdWords, which uses the frequency of queries for synonyms. Thus 90 % of recommendations (words) for ontological rules (see [96]) were contained in the list of synonyms output by AdWords.

6.6   FCA-based Recommender Systems

Motivated by prospective applications of Boolean Matrix Factorisation (BMF) in the context of Recommender Systems (RS) we proposed an FCA-based approach which follows user-based k-nearest neighbours strategy [94]. Another approach similar to MF is biclustering, which has also been successfully applied in recommender system domain [96, 208]. As we have mentioned, FCA can be also used as a biclustering technique and there are several examples of its applications in the recommender systems domain [90, 92]. A parameter-free approach that exploits a neighbourhood of the object concept for a particular user also proved its effectiveness [209].

Belowe we discuss our recent studies in application of BMF for RS. In the recommender systems domain, the context is any auxiliary information concerning users (like gender, age, occupation, living place) and/or items (like genre of a movie, book or music), which shows not only a user's mark given to an item but explicitly or implicitly describes the circumstances of such evaluation (e.g., including time and place) [210].

From representational viewpoint an auxiliary information can be described by a binary relation, which shows that a user or an item posses a certain attribute-value pair.

As a result one may obtain a block matrix:

$$I = \begin{bmatrix} R & C_{user} \\ C_{item} & O \end{bmatrix},$$

where R is a utility matrix of users' ratings to items, C_{user} represents context information of users, C_{item} contains context iformation of items and O is a zero-filled matrix.

Example 25. An example of a rating matrix enriched by user-feature and item-feature auxiliary information.

$\square$

Table 9. Adding auxiliary information

| | Movies | | | | | | Gender | | Age | | |
	Brave Heart	Termi-nator	Gladi-ator	Slum-dog Million-aire	Hot Snow	God-father	M	F	0–20	21–45	46+
Anna	5		5	5		2		+	+		
Vladimir		5	5	3		5	+			+	
Katja	4		4	5		4		+		+	
Mikhail	3	5	5			5	+			+	
Nikolay			2		5	4	+				+
Olga	5	3	4	5				+	+		
Petr	5			4	5	4	+				+
Drama	+		+	+	+	+					
Action		+	+		+	+					
Comedy	+			+							

Example 26. In case of more complex rating's scale the ratings can be reduced to binary scale (e.g., *"like/dislike"*) by binary thresholding or by FCA-based scaling. □

Table 10. Derived Boolean utility matrix enriched by auxiliary information

	m_1	m_2	m_3	m_4	m_5	m_6	f_1	f_2	f_3	f_4	f_5
u_1	1	0	1	1	0	0	0	1	1	0	0
u_2	1	0	1	1	0	0	1	0	0	1	0
u_3	1	0	1	1	0	1	0	1	0	1	0
u_4	1	0	1	1	0	0	1	0	0	1	0
u_5	0	0	0	0	1	1	1	0	0	0	1
u_6	1	0	1	1	0	0	0	1	1	0	0
u_7	1	0	0	1	1	1	1	0	0	0	1
g_1	1	0	1	1	1	1	0	0	0	0	0
g_2	0	1	1	0	1	1	0	0	0	0	0
g_3	1	0	0	1	0	0	0	0	0	0	0

Once a matrix of ratings is factorized we need to learn how to compute recommendations for users and to evaluate whether a particular method handles this task well.

Given the factorized matrices already well-known algorithm based on the similarity of users can be applied, where for finding k nearest neighbors we use not the original matrix of ratings $R \in \mathbb{R}^{m \times n}$, but the matrix $I \in \mathbb{R}^{m \times f}$, where m is a number of users, and f is a number of factors. After the selection of k users, which are the most similar to a given user, based on the factors that are peculiar to them, it is possible, based on collaborative filtering formulas to calculate the prospective ratings for a given user.

After generation of recommendations the performance of the recommender system can be estimated by such measures as Mean Absolute Error (MAE), Precision and Recall.

Collaborative recommender systems try to predict the utility (in our case ratings) of items for a particular user based on the items previously rated by other users.

Memory-based algorithms make rating predictions based on the entire collection of previously rated items by the users. That is, the value of the unknown rating $r_{u,m}$ for a user u and item m is usually computed as an aggregate of the ratings of some other (usually, the k most similar) users for the same item m:

$$r_{u,m} = aggr_{\tilde{u} \in \tilde{U}} r_{\tilde{u},m},$$

where $\tilde{U}$ denotes a set of k users that are the most similar to user u, who have rated item m. For example, the function $aggr$ may have the following form [210]:

$$r_{u,m} = \frac{\sum\limits_{\tilde{u} \in \tilde{U}} sim(\tilde{u}, u) \cdot r_{\tilde{u},m}}{\sum\limits_{\tilde{u} \in \tilde{U}} sim(u, \tilde{u})}.$$

Similarity measure between users u and $\tilde{u}$, $sim(\tilde{u}, u)$, is essentially an inverse distance measure and is used as a weight, i.e., the more similar users c and $\tilde{u}$ are, the more weight rating $r_{\tilde{u},m}$ will carry in the prediction of $r_{\tilde{u},m}$.

Similarity between two users is based on their ratings of items that both users have rated. There are several popular approaches: Pearson correlation, cosine-based, and Hamming-based similarities.

We mainly use the cosine-based and normalised Hamming-based similarities.

To apply this approach in case of FCA-based BMF recommender algorithm we simply consider the user-factor matrices obtained after factorisation of the initial data as an input.

Example 27. For the input matrix in Table 10 one can find the following covering factors:

$$(\{u_1, u_3, u_6, u_7, g_1, g_2\}, \{m_1, m_4\}), (\{u_2, u_4\}, \{m_2, m_3, m_6, f_1, f_4\}),$$
$$(\{u_5, u_7\}, \{m_5, m_6, f_1, f_5\}), \quad (\{u_1, u_6\}, \{m_1, m_3, m_4, f_2, f_3\}),$$
$$(\{u_5, u_7, g_1, g_3\}, \{m_5, m_6\}), \quad (\{u_2, u_3, u_4\}, \{m_3, m_6, f_4\}),$$
$$(\{u_2, u_4, g_3\}, \{m_2, m_3, m_6\}), \quad (\{u_1, u_3, u_6, g_1\}, \{m_1, m_3, m_4\}),$$
$$(\{u_1, u_3, u_6\}, \{m_1, m_3, m_4, f_2\}).$$

The corresponding decomposition is below: □

$$
\begin{pmatrix}
1 & 0 & 0 & 1 & 0 & 0 & 0 & 1 & 1 \\
0 & 1 & 0 & 0 & 0 & 1 & 1 & 0 & 0 \\
1 & 0 & 0 & 0 & 0 & 1 & 0 & 1 & 1 \\
0 & 1 & 0 & 0 & 0 & 1 & 1 & 0 & 0 \\
0 & 0 & 1 & 0 & 1 & 0 & 0 & 0 & 0 \\
1 & 0 & 0 & 1 & 0 & 0 & 0 & 1 & 1 \\
1 & 0 & 1 & 0 & 1 & 0 & 0 & 0 & 0 \\
1 & 0 & 0 & 0 & 1 & 0 & 0 & 1 & 0 \\
0 & 0 & 0 & 0 & 1 & 0 & 1 & 0 & 0 \\
1 & 0 & 0 & 0 & 0 & 0 & 0 & 0 & 0
\end{pmatrix}
\circ
\begin{pmatrix}
1 & 0 & 0 & 1 & 0 & 0 & 0 & 0 & 0 & 0 \\
0 & 1 & 1 & 0 & 0 & 1 & 1 & 0 & 0 & 1 & 0 \\
0 & 0 & 0 & 0 & 1 & 1 & 1 & 0 & 0 & 0 & 1 \\
1 & 0 & 1 & 1 & 0 & 0 & 0 & 1 & 1 & 0 & 0 \\
0 & 0 & 0 & 0 & 1 & 1 & 0 & 0 & 0 & 0 & 0 \\
0 & 0 & 1 & 0 & 0 & 1 & 0 & 0 & 0 & 1 & 0 \\
0 & 1 & 1 & 0 & 0 & 1 & 0 & 0 & 0 & 0 & 0 \\
1 & 0 & 1 & 1 & 0 & 0 & 0 & 0 & 0 & 0 & 0 \\
1 & 0 & 1 & 1 & 0 & 0 & 0 & 1 & 0 & 0 & 0
\end{pmatrix}
$$

However, in this case in the obtained user profiles vectors most of the components are getting zeros, and thus we lose similarity information.

To smooth the loss effects we proposed the following weighted projection:

$$\tilde{P}_{uf} = \frac{I_{u\cdot} \cdot Q_{f\cdot}}{\|Q_{f\cdot}\|_1} = \frac{\sum\limits_{v \in V} I_{uv} \cdot Q_{fv}}{\sum\limits_{v \in V} Q_{fv}},$$

where $\tilde{P}_{uf}$ indicates whether factor f covers user u, $I_{u\cdot}$ is a binary vector describing profile of user u, $Q_{f\cdot}$ is a binary vector of items belonging to factor f (the corresponding row of Q in decomposition Eq. (3)). The coordinates of the obtained projection vector lie within $[0; 1]$.

Example 28. For Table 9 the weighted projection is as follows:

$$\tilde{P} = \begin{pmatrix} 1 & \frac{1}{5} & 0 & 1 & 0 & \frac{1}{3} & \frac{1}{3} & 1 & 1 \\ 0 & 1 & \frac{1}{2} & \frac{1}{5} & \frac{1}{2} & 1 & 1 & \frac{1}{3} & \frac{1}{4} \\ 1 & \frac{3}{5} & \frac{1}{4} & \frac{1}{4} & \frac{1}{2} & 1 & \frac{2}{3} & 1 & 1 \\ 0 & 1 & \frac{1}{2} & \frac{1}{5} & \frac{1}{2} & 1 & 1 & \frac{1}{3} & \frac{1}{4} \\ 0 & \frac{2}{5} & 1 & 0 & 1 & \frac{2}{3} & \frac{1}{3} & 0 & 0 \\ 1 & \frac{1}{5} & 0 & 1 & 0 & \frac{1}{3} & \frac{1}{3} & 1 & 1 \\ 1 & \frac{3}{5} & 1 & 1 & 1 & \frac{1}{3} & \frac{1}{3} & \frac{2}{3} & \frac{1}{3} \\ 1 & \frac{2}{5} & \frac{1}{2} & \frac{2}{5} & 1 & \frac{2}{3} & \frac{2}{3} & 1 & \frac{3}{4} \\ 0 & \frac{2}{5} & \frac{1}{2} & \frac{1}{5} & 1 & \frac{2}{3} & 1 & \frac{1}{3} & \frac{1}{4} \\ 1 & 0 & 0 & \frac{1}{5} & 0 & 0 & 0 & \frac{2}{3} & \frac{1}{2} \end{pmatrix}.$$

□

The proposed approach and compared ones have been implemented in C++[34] and evaluated on MovieLens-100k data set. This data set features 100000 ratings in five-star scale, 1682 Movies, Contextual information about movies (19 genres), 943 users (each user has rated at least 20 movies), and demographic info for the users (gender, age, occupation, zip (ignored)). The users have been divided into seven age groups: under 18, 18-25, 26-35, 36-45, 45-49, 50-55,56+.

Five star ratings are converted to binary scale by the following rule:

$$I_{ij} = \begin{cases} 1, & R_{ij} > 3, \\ 0, & \text{else} \end{cases}$$

The scaled dataset is split into two sets according to bimodal cross-validation scheme [211]: the training set and the test set with a ratio 80:20, and 20 % of ratings in the test set are hidden (Fig. 25).

We found out that MAE of our BMF-based approach is substantially lower than MAE of SVD-based approach for almost the same number of factors at fixed coverage level of BMF and of SVD (Fig. 25). The Precision of BMF-based approach is slightly lower when the number of neighbours is about a couple of dozens and comparable for the remaining part of the observed range. The Recall is lower that results in lower F-measure. It can be explained by different nature of factors in these factorisation models. The proposed weighted projection alleviates the information loss of original Boolean projection resulting in a substantial quality gain. We also revealed that the presence of auxiliary information results in a small quality increase (about 1-2 %) in terms of MAE, Recall and Precision. In our previous study, with the original BMF-based scheme (weighting is not used), we obtained comparable results in terms of MAE, and both Precision and Recall [94, 145].

6.7 FCA-based Approach for Near-Duplicate Documents Detection

From the dawn of web search engines, the problem of finding near-duplicate documents in the web search results is crucial for providing users with relevant documents [212–214].

---

[34] https://github.com/MaratAkhmatnurov/BMFCARS.

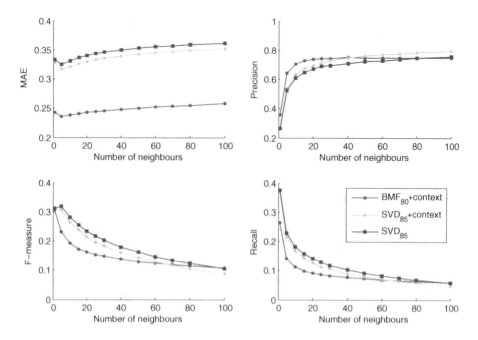

Fig. 25. Different approaches of matrix factorisation

Below we shortly describe our studies on near duplicate detection [81] within a competition "Internet mathematics" organised by Yandex and ROMIP (Russian Information Retrieval Evaluation Seminar) in 2004–2005: our project "Optimisation of search for near duplicates in the web: images and similarity" was selected, as the rest 33 projects, out of 252 applications[35].

As experimental data the ROMIP collection of web documents from narod.ru domain[36] was provided; it consists of 52 files of general size 4.04 GB. These files contained 530 000 web pages from narod.ru domain. Each document from the collection has size greater or equal to 10 words. For experiments the collection was partitioned into several parts consisting of three to 24 such files (from 5 % to 50 % percent of the whole collection). As an evaluation benchmark for recall and precision calculation we use the list of duplicate pairs provided by Yandex; the duplicate pairs were identified for all document pairs via Perl String::Similarity with 85 % similarity threshold.

For composing document images we followed a popular shingling approach [213]. For each text, the program **shingle** with two parameters (*length* and *offset*) generate contiguous subsequences of size *length* such that the distance between the beginnings of two subsequent substrings is *offset*. The set of sequences obtained in this way is hashed so that each sequence receives its own hash code. From the set of hash codes that corresponds to the document a fixed size (given by parameter)

---

[35] https://academy.yandex.ru/events/imat/grant2005/.
[36] http://romip.ru/en/collections/narod.html.

subset is chosen by means of random permutations described in [213, 215, 216]. The probability of the fact that minimal elements in permutations on hash code sets of shingles of documents A and B (these sets are denoted by F_A and F_B, respectively) coincide, equals to the similarity measure of these documents $sim(A, B)$:

$$sim(A, B) = P[min\{\pi(F_A)\} = min\{\pi(F_B)\}] = \frac{|F_A \cap F_B|}{|F_A \cup F_B|}.$$

Further we used FCA to define cluster of near duplicate documents.

Let $\mathbb{K}_{DF} = (D, F, I \subseteq D \times F)$ be a **context of documents**, where D is a set of documents, F is a set of hash codes (fingerprints), and I shows that a document d has an attribute f whenever dIf.

For a subset of documents $A \subseteq D$, A' describe their similarity in terms of common fingerprints, and the closed set A'' is a **cluster of similar documents**.

To find all near duplicate clusters, we need to enumerate all intents of the context $\mathbb{K}_{FD} = (F, D, I \subseteq F \times D)$ such that their common set of fingerprints exceeds a threshold set by user.

In fact, to this end we need to use nothing but frequent itemsets of documents. A set of documents $A \subseteq D$ is called k-**frequent** if $|A'| > k$, where k is a parameter.

Program Implementation. Software for experiments with syntactical representation comprise the units that perform the following operations:

1. XML Parser (provided by Yandex): it parses XML packed collections of web documents
2. Removing html-markup of the documents
3. Generating shingles with given parameters length-of-shingle, offset
4. Hashing shingles
5. Composition of document image by selecting subsets (of hash codes) of shingles by means of methods *n minimal elements in a permutation* and *minimal elements in n permutations*.
6. Composition of the inverted table the list of identifiers of documents shingle thus preparing data to the format of programs for computing closed itemsets.
7. Computation of clusters of *k-similar documents* with FPmax* algorithm: the output consists of strings, where the first elements are names (ids) of documents and the last element is the number of common shingles for these documents.
8. Comparing results with the existing list of duplicates (in our experiments with the ROMIP collection of web documents, we were supplied by a precomputed list of duplicate pairs).

This unit outputs five values: (1) the number of duplicate pairs in the ROMIP collection, (2) the number of duplicate pairs for our realisation, (3) the number of unique duplicate pairs in the ROMIP collection, (4) the number of unique duplicate pairs in our results, (5) the number of common pairs for the ROMIP collection and our results.

In step 7, we used a leader in time efficiency, the algorithm FPmax* [217], from the competition organised in series of workshops on Frequent Itemset Mining Implementations (FIMI)[37].

Experimental Results. In our experiments we used Cluto[38], a software package for clustering high-dimensional datasets including those from information retrieval domain, for comparison purposes. We chose the repeated-bisecting algorithm that uses the cosine similarity function with a 10-way partitioning (ClusterRB), which is mostly scalable according to its author [218]. The number of clusters was a parameter, documents were given by sets of attributes, fingerprints in our case. The algorithm outputs a set of disjoint clusters. Algorithms from FIMI repository can process very large datasets, however, to compare with Cluto (which is much more time consuming as we show below) we took collection narod.1.xml that contains 6941 documents.

Shingling parameters used in experiments were as follows: the number of words in shingles was 10 and 20, the offset was always taken to be 1 (which means that the initial set of shingles contained all possible contiguous word sequences of a given length). The sizes of resulting document images were taken in the interval 100 to 200 shingles. As frequency thresholds defining *frequent closed sets* (i.e., the numbers of common shingles in document images from one cluster) we experimentally studied different values in intervals, where the maximal value is equal to the number of shingles in the document image. For example, the interval [85, 100] for document images with 100 shingles, the interval [135, 150] for document images of size 150, etc. Obviously, choosing the maximal value of an interval, we obtain clusters where document images coincide completely.

For parameters taking values in these intervals we studied the relation between resulting clusters of duplicates and ROMIP collection of duplicates, which consists of pairs of web documents that are considered to be near duplicates. Similarity of each pair of documents in this list is based on Edit Distance measure, two documents were taken to be duplicates by authors of this testbed if the value of the Edit Distance measure exceeds threshold 0.85. As we show below, this definition of a duplicate is prone to errors, however making a testbed by manual marking duplication in a large web document collection is hardly feasible. Unfortunately, standard lists of near-duplicates were missing at that period even for standard corpora such as TREC or Reuters collection [219]. For validating their methods, researchers create ad-hoc lists of duplicates using slightly transformed documents from standard collections. Now the situation is drastically better, see, for example, workshop series on Plagiarism Analysis, Authorship Identification, and Near-Duplicate Detection (PAN)[39].

In our study for each such pair we found an intent that contains both elements of the pair, and vice versa, for each cluster of *very similar documents*

(i.e., for each corresponding closed set of documents with more than k common description units) we take each pair of documents in the cluster and looked for the corresponding pair in the ROMIP collection. As result we obtain the number of common number of near duplicate pairs found by our method and those in the ROMIP collection, and the number of unique pairs of HSE duplicates (document pairs occurring in a cluster of "very similar documents" and not occurring in the ROMIP collection). The results of our experiments showed that the ROMIP collection of duplicates, considered to be a benchmark, is far from being perfect. First, we detected that a large number of false duplicate pairs in this collection due to similar framing of documents. For example the pages with the following information in Table 11 about historical personalities 1 and 2 were declared to be near duplicates.

Table 11. Information about historical personalities

1. Garibald II, Duke of Bavaria	2. Giovanni, Duke of Milan
Short information:	Short information:
Full Name: Garibald	Full Name: Giovanni Visconti
Date of birth: unknown	Date of birth: unknown
Place of birth: unknown	Place of birth: unknown
Date of death: 610	Date of death: 1354
Place of death: unknown	Place of death: unknown
Father: Tassilo I Duke of Bavaria	Father: Visconti Matteo I, the Great Lord of Milan
Mother: uknown	Mother: uknown

However these pages, as well as many other analogous false duplicate pairs in ROMIP collection do not belong to concept-based (maximal frequent) clusters generated in our approach.

Second, in our study we also looked for *false duplicate clusters* in the ROMIP collection, caused by transitive closure of the binary relation "X is a duplicate of Y" (as in the typical definition of a document cluster in [216]). Since the similarity relation is generally not transitive, the clusters formed by transitive closure of the relation may contain absolutely dissimilar documents. Note that if clusters are defined via maximal frequent itemsets (subsets of attributes) there cannot be effects like this, because documents in these clusters share necessarily large itemsets (common subsets of attributes).

We analysed about 10000 duplicate document pairs and found four rather big *false duplicate clusters* about 50–60 documents each. Further studies on this collection see in [220].

We shortly summarise experimental results below and in Table 12:

- FPmax* (F-measure=0.61 and elapsed time 0.6 seconds), ClusterRB (F-measure=0.63 and elapsed time 4 hours);

- For FPMax* the number of single document cluster is 566, for ClusterRB 4227;
- The total number of clusters for FPmax* is 903 versus 5000 for Cluto 903;
- The number of NDD clusters for FPmax* is 337 versus 773 Cluto.

Table 12. Comparison of the obtained clusters in terms of pairs of near duplicate documents

The number of ROMIP duplicates:	2997
The number of NDD found by FPmax*:	2722
The number of NDD found by Cluto:	2897
The number of unique NND pairs of ROMIP:	1155
The number of unique NDD pairs found by FPmax*:	1001
The number of unique NDD pairs found by Cluto:	1055
The number of common NDD pairs for FPmax* and ROMIP:	1721
The number of common NDD pairs for Cluto and ROMIP:	1842

The obtained graphs and tables show that for 5000 clusters the output of ClusterRB has almost the same value of F-measure (0.63) as FPmax* for threshold 150 (F1=0,61). However, computations took 4 hours for ClusterRB and half a second for FPmax*.

We continued our developments of that time and developed GUI and duplicate document generator (for a provided text collection) for testing purposes [221].The archive of these projects is freely available at Bitbucket[40].

Later on, we proposed a prototype of near-duplicate detection system for web-shop owners. It's a typical situation for this online businesses to buy description of their goods from so-called copyrighters. A copyrighter may cheat from time to time and provide the owner with some almost identical descriptions for different items. In that study we demonstrated how we can use FCA for revealing and fast online clustering of such duplicates in a real online perfume shop. Our results were also applicable for near duplicate detection in collections of R&D project's documents [222].

6.8   Triadic FCA for IR-tasks in Folksonomies

Four our data mining studies on triclustering (see Sect. 5.1 and [109, 111, 112]) folksonomic data became a shootingrange since the first efficient FCA-based algorithm to mine tiradic data was proposed for mining communities in folksonomies [45].

---

[40] https://bitbucket.org/dimanomachine/nearduplicatesarch.

But it is a rich field with interactive resource-sharing systems like Bibsonomy[41], CiteULike[42], Flickr[43] and Delicious[44] that need fully-fledged IR functionality including retrieval, ranking and recommendations. For example, Bibsonomy is a social bookmarking and publication management system. Apart from DBLP[45] [223] that only collects, stores bibliographic data and provide publication and author search, Bibsonomy allows to create user's own lists of bibliographic bookmarks, use tags and social interactions (Fig. 26).

As we have mentioned in Sect. 5.1, the underlying Folksonomy structure is a formal tricontext $\mathbb{K} = (U, T, R, Y)$ with U being a set of users, T a set tags, and R a set of resources, where $Y \subseteq U \times T \times R$ relates entities from these three sets. Sometimes, a user-specific subtag/supertag-relation $\succ$ is also included into the definition, i.e. $\succ \subseteq U \times T \times T$.

We shortly discuss main IR tasks that folksonomic data give rise.

First of all, we have to say that traditional PageRank cannot be directly applied to folksonomies. The authors of paper [148] modified the PageRank algorithm for folksonomic data by considering the input triadic data as an undirected tripartite graph. The weights for each type of edge were assigned according to the occurrences of the third entity, e.g. an edge u, t being weighted with $|r \in R : (u, t, r) \in Y|$, the number occurrences of the related tags.

Formally, the weight spreading condition looks as follows:

$$w \leftarrow \alpha w + \beta A w + \gamma p, where$$

A is the row-stochastic version of the graph adjacency, p is a preference vector, $\alpha, \beta, \gamma \in [0, 1]$ are initial parameters with $\alpha + \beta + \gamma = 1$. Thus, α regulates the speed of convergence, while the proportion between β and γ controls the influence of the preference vector.

However, the first results on Delicious data were rather discouraging even with a term-frequency ranker combination, the resulting ranking was similar (though not identical) to the initial edge weights. It resulted in authors' own ranking algorithm FolkRank, which takes into account the difference in the resulting rankings with and without preference vector [148].

In that paper the authors formulated peculiar tasks:

– Documents that are of potential interest to a user can be suggested to him.
– Other related tags can be suggested to a user. Thus, FolkRank additionally considers the tagging behavior of other users and can be used for tag recommendations.
– Other users that work on related topics can be made explicit and this facilitates knowledge transfer and formation of user communities.

---

[41] http://www.bibsonomy.org/.
[42] http://www.citeulike.org/.
[43] https://www.flickr.com/.
[44] https://delicious.com/.
[45] http://dblp.uni-trier.de/.

Later on, they implemented (not only) all this features in the Bibsonomy systems [97].

Moreover, during those studies they admitted that search query logs naturally forms folksonomic data, $(users, queries, resources)$, where the resources are those that were clicked by a user after performing a query [224]. Predictably, they gave a name logsonomy to this new data structure. When Bibsonomy was at the early stages, it faced spam abuse problem and in 2008 ECML PKDD discovery challenge[46] addressed this problem. The year after the challenging problem[47] were recommendations for Bibsonomy and it resulted in new fruitful algorithms [225].

6.9   Exploring Taxonomies of Web Site Users

In 2006 we participated in analysis of web sites audience in collaboration with SpyLog company (now OpenStat[48])[226].

Owners of a web-site are often interested in analysing groups of users of their site. Information on these groups can help to optimise the structure and contents of the site. For example, interaction with members of each group may be organized in a special manner. In the performed study we used an approach based on formal concepts for constructing taxonomies of groups of web users (Fig. 26).

For our experiments we have chosen four target websites: the site of the State University Higher School of Economics, an e-shop of household equipment, the site of a large bank, and the site of a car e-shop (the names of the last three sites cannot be disclosed due to legal agreements).

Users of these sites are described by attributes that correspond to other sites, either external (from three groups of sites: finance, media, education) or internal (web-pages of the site). More precisely, initial "external" data consists of user records each containing the user id, the time when the user first entered this site, the time of his/her last visit, and the total number of sessions during the period under consideration. An "internal" user record, on the other hand, is simply a list of pages within the target website visited by a particular user.

By "external" and "internal" taxonomies we mean (parts of) concept lattices for contexts with either "external" or "internal" attributes. For example, the external context has the form $\mathbb{K}_e = (U, S_e, I_e)$, where U is the set of all users of the target site, S_e is the set of all sites from a sample (not including the target one), the incidence relation I_e is given by all pairs $(u, s)\colon u \in U, s \in S_e$, such that user u visited site s. Analogously, the internal context is of the form $\mathbb{K}_i = (U, S_i, I_i)$, where S_i is the set of all own pages of the target site.

A concept of this context is a pair (A, B) such that A is a group of users that visited together all other sites from B.

As we have mentioned, one of the target websites was the site of our university[49].

---

[46] http://www.kde.cs.uni-kassel.de/ws/rsdc08/.
[47] http://www.kde.cs.uni-kassel.de/ws/dc09/.
[48] https://www.openstat.com/.

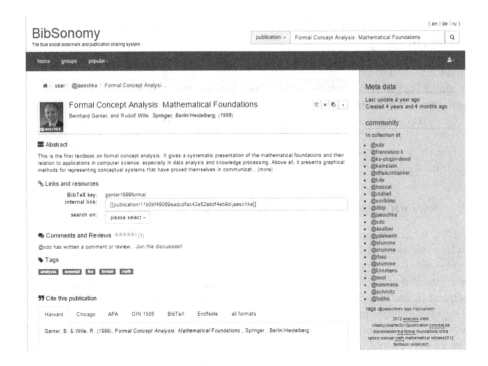

Fig. 26. An example of Bibsonomy interface

We received "external" data with the following fields for each user-site pair: (**user id, time of the first visit, time of the last visit, total number of sessions during the period**). "Internal" data have almost the same format with an additional field **url page**, which corresponds to a particular visited page of the target site.

The provided information was gathered from about 10000 sites of Russian segment of Internet (domain .ru). Describing users in terms of sites they visited, we had to tackle the problem of dimensionality, since the resulting concept lattices can be very large (exponential in the worst case in terms of objects or attributes). To reduce the size of input data we used the following techniques.

For each user we selected only those sites that were visited by more than a certain number of times during the observation period. This gave us information about permanent interests of particular users. Each target web site was considered in terms of sites of three groups: newspaper sites, financial sites, and educational sites.

However, even for large reduction of input size, concept lattices can be very large. For example, a context of size 4125×225 gave rise to a lattice with 57 329 concepts.

To choose interesting groups of users we employed stability index of a concept defined in [227, 228] and considered in [88] (in slightly different form) as a tool for constructing taxonomies. On one hand, stability index shows the independence of an intent on particular objects of extent (which may appear or not appear in the context depending on random factors). On the other hand, stability index of a concept shows how much extent of a concept is different from similar smaller extents (if this difference is very small, then its doubtful that extent refers to a "stable category"). For detailed motivation of stability indices see [88, 227, 228].

Definition 33. *Let* $\mathbb{K} = (G, M, I)$ *be a formal context and* (A, B) *be a formal concept of K. The stability index* σ *of* (A, B) *is defined as follows:*

$$\sigma(A, B) = \frac{|\{C \subseteq A | C' = B\}|}{2^{|A|}}.$$

Obviously, $0 \leq \sigma(A, B) \leq 1$.

The stability index of a concept indicates how much the concept intent depends on particular objects of the extent. A stable intent (with stability index close to 1) is probably "real" even if the description of some objects is "noisy". In application to our data, the stability index shows how likely we are to still observe a common group of interests if we ignore several users. Apart from being noise-resistance, a stable group does not collapse (e.g., merge with a different group, split into several independent subgroups) when a few members of the group stop attending the target sites.

In our experiments we used ConceptExplorer for computing and visualising lattices and their parts.

We compared results of taking most stable concepts (with stability index exceeding a threshold) with taking an "iceberg" lattice. The results look correlated, but nevertheless, substantially different. The set of stable extents contained very important, but not large groups of users.

In Figs. 12 and 27 we present parts of a concept lattice for the HSE web site described by "external" attributes which were taken to be Russian e-newspapers visited by users of www.hse.ru during one month (September 2006) more than 20 times. Figure 12 presents an iceberg with 25 concepts having largest extent. Many of the concepts correspond to newspapers that are in the middle of political spectrum, read "by everybody" and thus, not very interesting in characterising social groups.

Figure 27 presents an ordered set of 25 concepts having largest stability index. As compared to the iceberg, this part of the concept lattice contains several sociologically important groups such as readers of "ExpressGazeta" ("yellow press"), Cosmopolitan, Expert (high professional analytical surveys) etc.

6.10   FCA-based Models for Crowdsourcing

The success of modern collaborative technologies is marked by the appearance of many novel platforms for holding distributed brainstorming or carrying out so

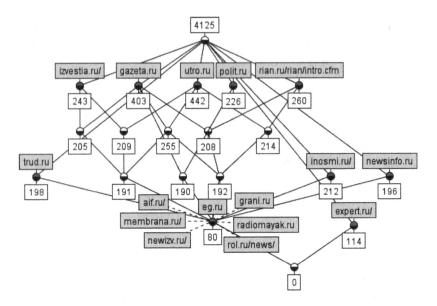

Fig. 27. Ordered set of 25 concepts with largest stability

called "public examination". There are a lot of such crowdsourcing companies in the USA (Spigit[50], BrightIdea[51], InnoCentive[52] etc.) and Europe (Imaginatik[53]). There is also the Kaggle platform[54] which is the most beneficial for data practitioners and companies that want to select the best solutions for their data mining problems. In 2011 Russian companies launched business in that area as well. The two most representative examples of such Russian companies are Witology[55] and Wikivote[56]. Several all-Russian projects have already been finished successfully (for example, Sberbank-21[57], National Entrepreneurial Initiative[58] etc.). The core of such crowdsourcing systems is a socio-semantic network [229,230], which data requires new approaches to analyze. Before we tried to accommodate FCA as a methodological base for the analysis of data generated by such collaborative systems [165].

As a rule, while participating in a project, users of such crowdsourcing platforms [231] discuss and solve one common problem, propose their ideas and evaluate ideas of each other as experts. Finally, as a result of the discussion and

[50] http://spigit.com/.

[51] http://www.brightidea.com/.

[52] http://www.innocentive.com/.

[53] http://www.imaginatik.com/.

[54] http://www.kaggle.com.

[55] http://witology.com/.

[56] http://www.wikivote.ru/.

[57] http://sberbank21.ru/.

[58] http://witology.com/en/clients_n_projects/3693/.

ranking of users and their ideas we get the best ideas and users (their generators). For deeper understanding of users's behavior, developing adequate ranking criteria and performing complex dynamic and statistic analyses, special means are needed. Traditional methods of clustering, community detection and text mining need to be adapted or even fully redesigned. Earlier we described models of data used in crowdsourcing projects in terms of FCA. Furthermore, we presented the collaborative platform data analysis system CrowDM (Crowd Data Mining), its architecture and methods underlying the key steps of data analysis [165].

The principles of these platforms' work are different from the work of online-shops or specialized music/films recommender websites. Crowdsourcing projects consist of several stages and results of each stage substantially depend on the previous stage results. That's why the existing models of the recommender systems should be adapted properly. In the accompanion paper [232] or in its shorter predecessors [233, 234], we present new methods for making recommendations based on FCA and OA-biclustering (see Sect. 5.1): The original methods of idea recommendation (for voting stage), like-minded persons recommendation (for collaboration) and antagonists recommendation (for counteridea generation stage). The last recommendation type is very important for stimulating user's activity on Witology platform during the stage of counteridea generation.

7   FCA in Ontology Modeling and Attribute Exploration

Applications of FCA in ontology modeling and its relations with Semantic Web deserve a special treatment. However, we shortly mention several interesting approaches and showcase an interactive technique which can be used for ontology and knowledge bases refinement and building.

- Attribute exploration as an expert knowledge acquisition method [235]
- FCA in ontology building and refining [79, 236]

7.1   Attribute Exploration

Attribute exploration is an interactive knowledge acquisition procedure based on implications and counter examples [235] that was initially applied for knowledge acquisition in mathematics itself and still a suitable tool up to date [237].

The basic algorithm is as follows:

- Start with any (possibly empty) set of objects.
- Generate an implication valid in the current subcontext.
- If the implication is not valid in the entire context, provide an object that violates it (a counterexample).
- Go to the next implication and so on.

A sophisticated algorithm implementation can follow the Duquenne-Guigues base to ask minimal number of questions.

Example 29. Attribute exploration for the context of transportation means.

The resulting concept means are enumerated with respect to their attributes (surface, air, water, underwater)?

To this end we start attribute exploration with composing the corresponding formal context (Fig. 28).

	surface	air	water	underwater
plane		×		
amphibian car	×		×	
catamaran			×	
car			×	×
submarine			×	×

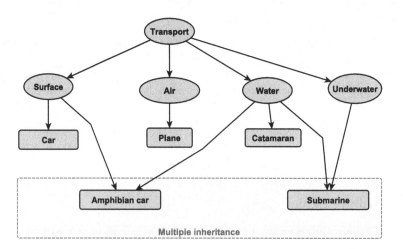

Fig. 28. The taxonomy of transportation means as an example of not a tree-like (multiple) inheritance

The main steps of attribute exploration, as a dialog between system A and expert E for transport context, is as follows:

- Step 1. A Question: Is it true that, when an object has attribute "Can move underwater", it also has attribute "Can move by water"?
- Step 1. E Answer: Yes, it is. The expert knows that it is true for submarines and there are no other types of underwater transport.
- Step 2. A Question: Is it true that, when an object has attributes "Can move by air" and "Can move by water" have attributes "Can move by surface" and "Can move underwater"?

- Step 2. *E* Answer: No, it is not. There is a counterexample, $\{hydroplane\}' = \{air, water\}$.
- Step 3. *A* Question: Is it true that, when an object has attributes "Can move by air", "Can move by water" "Can move underwater" have attributes "Can move by surface"?
- Step 3. *E* Answer: Yes, it is. $\{air, water, underwater\}' = \emptyset$.
- Steps 4, 5, 6 Trivial questions. □

The resulting concept lattice can be considered as a non-tree like taxonomy of transportation means since it allows multiple inheritance in the concept hierarchy. An example of non-tree taxonomy for our case is shown in Fig. 28. If the expert suppose that the not only objects but attributes are missed then object exploration can be done in similar manner, e.g. by the same proce- dure on the transposed context.

Exercise 23. 1. Compare the concept lattices from the previous example before starting and after completion of the attribute exploration. What is/are new concept(s) that we have obtained? How can it/they be interpreted? 2. Perform attribute exploration with ConceptExplorer for a slightly modified context from [238] □

	Asian	EU	European	G7	Mediterranean
France		×	×	×	×
Turkey	×		×		×
Germany		×	×	×	

7.2   FCA in Ontology Building and Refining

Often, the notion of Ontology in Computer Science is introduced as related sets of concepts and the typical relation can be "is-a", "has-a", "part-of", or super/subconcept relation. Concept lattices could be seen as ontology-like structures since they feature hierarchically related concepts by super/subconcept order (cf. subsumption order in Descriptive logic). However, because of their simplicity tree-like ontologies seem to be more popular, thus in the early paper of Cimiano et al. [79], the way to transform concept lattices built from text collections to tree-like ontologies was proposed.

Exercise 24. Build concept lattice from the context of terms extracted from texts (left). Find the transformation that resulted in the tree-like ontology of terms on the right side.

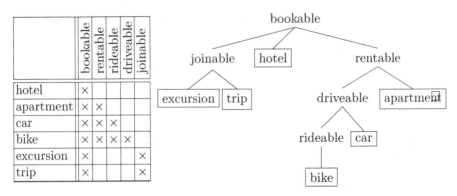

	bookable	rentable	rideable	driveable	joinable
hotel	×				
apartment	×	×			
car	×	×	×		
bike	×	×	×	×	
excursion	×				×
trip	×				×

Another example where FCA can help is ontology merging: The authors of [236] successfully tested their FCA-based merging approach on two text collection from touristic domain.

There is also strong connection between Description Logic, Ontologies and Formal Concept Analysis [238].

Thus OntoComP[59] [239] is a Protégé[60] 4 plugin for OWL ontologies completion. It enables the user to check whether an OWL ontology contains "all relevant information" about the application domain, and extend the ontology appropriately otherwise. It asks the users questions like "are instances of classes C_1 and C_2 also instances of the class C_3?". If the user replies positively, then a new axiom of the application domain (that does not follow from the ontology) has been discovered, and this axiom should be added to the ontology. If the user provides a counterexample to this question, i.e., an object that is an instance of C_1, C_2 and not C_3. When all such questions (about the initially given classes) have been answered, the ontology is supposed to be complete.

Obviously, this approach that was originally introduced in [240] for completing Description Logic knowledge bases uses attribute exploration.

It seems that attribute exploration is a fruitful technique for ontology building and refinement. Two more examples, Rudolph [241] proposed its extension for relation exploration in ontological modeling for knowledge specification and recently in combination with machine learning techniques attribute exploration was used for ontology refinement [242]. You probably have seen from Exercise 23, that attribute exploration may be uneasy because of laborious fact checking. However, to help potential users, in [243] the authors paired attribute exploration with web information retrieval, in particular by posing appropriate queries to search engines[61].

8   Conclusion

In the end of the invited talk at the "FCA meets IR" workshop 2013, Prof. Carpineto has summarised strengths and limitations of FCA for IR. It seems

---

[59] http://code.google.com/p/ontocomp/.
[60] http://www.co-ode.org/downloads/protege-x/.
[61] https://github.com/rjoberon/web-attribute-exploration.

to be evident that IR will be increasingly relying on contextual knowledge and structured data and FCA can improve both query pre-processing and query post-processing of modern IR systems. Among the mentioned technologies that could benefit from FCA are query expansion, web search diversification, ontology-based information retrieval, querying and navigating RDF (there is a progress to this date [244]), and many others. However, the community needs to endeavour (by theoretical advances and system engineering) to deploy a comprehensive FCA-based tool for information retrieval and integrate it with existing search and indexing taking into account both the intrinsic complexity issues and the problem of good features generation.

Even in an extensive tutorial it is not possible to cover all models and applications of Formal Concept Analysis. For example, concept lattices and its applications in social sciences including Social Network Analysis deserve a special treatment. The grounding steps have been done by Vincent Duquenne [85], Linton Freeman [86] and their collaborators (see also [89] for our SNA-related study). Another large and interesting domain is Software Engineering [245,246]. For these two and many other topics, we also refer the readers to the recent surveys [5,6].

Overall, we hope that this introductory material with many examples and exercises will help the reader not only to understand the theory basics, but having this rich variety of tools and showcases to use FCA in practice.

Acknowledgments. The author would like to thank all colleagues who have made this tutorial possible: Jaume Baixeries, Pavel Braslavsky, Peter Becker, Radim Belohlavek, Aliaksandr Birukou, Jean-Francois Boulicaut, Claudio Carpineto, Florent Domenach, Fritjhof Dau, Vincent Duquenne, Bernhard Ganter, Katja Hofmann, Robert Jaeshke, Evgenia Revne (Il'ina), Nikolay Karpov, Mehdy Kaytoue, Sergei Kuznetsov, Rokia Missaoui, Elena Nenova, Engelbert Mephu Nguifo, Alexei Neznanov, Lhouari Nourin, Bjoern Koester, Natalia Konstantinova, Amedeo Napoli, Sergei Obiedkov, Jonas Poelmans, Nikita Romashkin, Paolo Rosso, Sebastian Rudolph, Alexander Tuzhilin, Pavel Serdyukov, Baris Serkaya, Dominik Slezak, Marcin Szchuka, and, last but not least, the brave listeners. The author would also like to commemorate Ilya Segalovich who inspired the author's enthusiasm in Information Retrieval studies, by giving personal explanations of near duplicate detection techniques in 2005, in particular.

Special thank should go to my grandmother, Vera, who has been hosting me in a peaceful countryside place, Prechistoe, during the last two weeks of the final preparations.

The author was partially supported by the Russian Foundation for Basic Research grants no. 13-07-00504 and 14-01-93960 and prepared the tutorial within the project "Data mining based on applied ontologies and lattices of closed descriptions" supported by the Basic Research Program of the National Research University Higher School of Economics.

References

1. Manning, C.D., Raghavan, P., Schütze, H.: Introduction to Information Retrieval. Cambridge University Press, Cambridge (2008)

2. Wille, R.: Restructuring lattice theory: An approach based on hierarchies of concepts. In: Rival, I. (ed.) Ordered Sets. NATO Advanced Study Institutes Series, vol. 83, pp. 445–470. Springer, Heidelberg (1982)

3. Ganter, B., Wille, R.: Formal Concept Analysis: Mathematical Foundations, 1st edn. Springer-Verlag New York Inc, Secaucus, NJ, USA (1999)

4. Poelmans, J., Ignatov, D.I., Viaene, S., Dedene, G., Kuznetsov, S.O.: Text mining scientific papers: A survey on FCA-based information retrieval research. In: Perner, P. (ed.) ICDM 2012. LNCS, vol. 7377, pp. 273–287. Springer, Heidelberg (2012)

5. Poelmans, J., Kuznetsov, S.O., Ignatov, D.I., Dedene, G.: Formal concept analysis in knowledge processing: A survey on models and techniques. Expert Syst. Appl. **40**(16), 6601–6623 (2013)

6. Poelmans, J., Ignatov, D.I., Kuznetsov, S.O., Dedene, G.: Formal concept analysis in knowledge processing: A survey on applications. Expert Syst. Appl. **40**(16), 6538–6560 (2013)

7. Serdyukov, P., Braslavski, P., Kuznetsov, S.O., Kamps, J., Rüger, S.M., Agichtein, E., Segalovich, I., Yilmaz, E. (eds.): Advances in Information Retrieval. LNCS, vol. 7814. Springer, Heidelberg (2013)

8. Arnauld, A., Nicole, P.: Logic or the Art of Thinking, translated by Jill V. Cambridge University Press, Buroker (1996)

9. Birkhoff, G.: Lattice Theory, 3rd edn. American Mathematical Society, Providence (1967)

10. Ore, O.: Galois connexions. Trans. Amer. Math. Soc. **55**(3), 494–513 (1944)

11. Barbut, M., Monjardet, B.: Ordre et Classification. Hachette, Paris (1970)

12. Duquenne, V.: Latticial structures in data analysis. Theor. Comput. Sci. **217**(2), 407–436 (1999)

13. Wolski, M.: Galois connections and data analysis. Fundam. Inform. **60**(1–4), 401–415 (2004)

14. Kuznetsov, S.O.: Galois connections in data analysis: Contributions from the soviet era and modern russian research. In: Formal Concept Analysis, Foundations and Applications, pp. 196–225 (2005)

15. Carpineto, C., Romano, G.: Concept data analysis - theory and applications. Wiley, Chichester (2005)

16. Davey, B.A., Priestley, H.A.: Introduction to Lattices and Order. Cambridge University Press, Cambridge (2002)

17. Dominich, S.: The Modern Algebra of Information Retrieval, 1st edn. Springer Publishing Company, Heidelberg (2008). Incorporated

18. Wolff, K.E.: A first course in formal concept analysis how to understand line diagrams. In: Faulbaum, F. (ed.), vol. 4 of SoftStat 1993. Advances in Statistical Software, pp. 429–438 (1993)

19. Belohlávek, R.: Introduction to Formal Concept Analysis. Palacky University, Olomouc (2008)

20. Kuznetsov, S.O., Obiedkov, S.A.: Comparing performance of algorithms for generating concept lattices. J. Exp. Theor. Artif. Intell. **14**(2–3), 189–216 (2002)

21. Kourie, D.G., Obiedkov, S.A., Watson, B.W., van der Merwe, D.: An incremental algorithm to construct a lattice of set intersections. Sci. Comput. Program. **74**(3), 128–142 (2009)

22. Krajca, P., Vychodil, V.: Distributed algorithm for computing formal concepts using Map-reduce framework. In: Siebes, A., Boulicaut, J.-F., Robardet, C., Adams, N.M. (eds.) IDA 2009. LNCS, vol. 5772, pp. 333–344. Springer, Heidelberg (2009)

23. Xu, B., de Fréin, R., Robson, E., Ó Foghlú, M.: Distributed formal concept analysis algorithms based on an iterative MapReduce framework. In: Ignatov, D.I., Poelmans, J., Domenach, F. (eds.) ICFCA 2012. LNCS, vol. 7278, pp. 292–308. Springer, Heidelberg (2012)

24. Armstrong, W.: Dependency structures of data base relationships. Inf. Process. **74**, 580–583 (1974)

25. Maier, D.: The Theory of Relational Databases. Computer Science Press, Rockville (1983)

26. Guigues, J.L., Duquenne, V.: Familles minimales d'implications informatives rsultant d'un tableau de donnes binaires. Math. et Sci. Humaines **95**(1), 5–18 (1986). In French

27. Bazhanov, K., Obiedkov, S.A.: Optimizations in computing the duquenne-guigues basis of implications. Ann. Math. Artif. Intell. **70**(1–2), 5–24 (2014)

28. Baixeries, J., Kaytoue, M., Napoli, A.: Characterization of database dependencies with FCA and pattern structures. In: Ignatov, D.I., Khachay, M.Y., Panchenko, A., Konstantinova, N., Yavorsky, R.E. (eds.) AIST 2014. CCIS, vol. 436, pp. 3–14. Springer, Heidelberg (2014)

29. Yevtushenko, S.A.: System of data analysis "concept explorer". (in russian). In: Proceedings of the 7th National Conference on Artificial Intelligence KII-2000, pp. 127–134 (2000)

30. Yevtushenko, S.: Computing and Visualizing Concept Lattices. Ph.D. thesis, TU Darmstadt, Fachbereich Informatik (2004)

31. Yevtushenko, S.A.: Concept Explorer. The User Guide, September 12 2006

32. Becker, P.: Numerical analysis in conceptual systems with ToscanaJ. In: Eklund, P. (ed.) ICFCA 2004. LNCS (LNAI), vol. 2961, pp. 96–103. Springer, Heidelberg (2004)

33. Becker, P., Correia, J.H.: The toscanaj suite for implementing conceptual information systems. In: Formal Concept Analysis, Foundations and Applications, pp. 324–348 (2005)

34. Vogt, F., Wille, R.: TOSCANA – A graphical tool for analyzing and exploring data. In: Tamassia, R., Tollis, I.G. (eds.) Graph Drawing. LNCS, vol. 894, pp. 226–233. Springer, Heidelberg (1995)

35. Valtchev, P., Grosser, D., Roume, C., Hacene, M.R.: Galicia: an open platform for lattices. In: de Moor, A., Ganter, B., (ed.), Using Conceptual Structures: Contributions to 11th International Conference on Conceptual Structures, pp. 241–254 (2003)

36. Lahcen, B., Kwuida., L.: Lattice miner: A tool for concept lattice construction and exploration. In: Suplementary Proceeding of International Conference on Formal Concept Analysis (ICFCA 2010) (2010)

37. Poelmans, J., Elzinga, P., Ignatov, D.I., Kuznetsov, S.O.: Semi-automated knowledge discovery: identifying and profiling human trafficking. Int. J. Gen. Syst. **41**(8), 774–804 (2012)

38. Poelmans, J., Elzinga, P., Neznanov, A., Viaene, S., Kuznetsov, S., Ignatov, D., Dedene, G.: Concept relation discovery and innovation enabling technology (cordiet). In: Proceedings of 1st International Workshop on Concept Discovery in Unstructured Data. vol. 757 of CEUR Workshop proceedings (2011)

39. Neznanov, A., Ilvovsky, D., Kuznetsov, S.O.: Fcart: A new fca-based system for data analysis and knowledge discovery. In: Contributions to the 11th International Conference on Formal Concept Analysis, TU Dresden, pp. 31–44 (2013)

40. Neznanov, A.A., Parinov, A.A.: FCA analyst session and data access tools in FCART. In: Agre, G., Hitzler, P., Krisnadhi, A.A., Kuznetsov, S.O. (eds.) AIMSA 2014. LNCS, vol. 8722, pp. 214–221. Springer, Heidelberg (2014)

41. Buzmakov, A., Neznanov, A.: Practical computing with pattern structures in FCART environment. In: Proceedings of the International Workshop "What can FCA do for Artificial Intelligence?" (FCA4AI at IJCAI 2013), pp. 49–56. Beijing, China, August 5 2013

42. Domenach, F.: CryptoLat - a pedagogical software on lattice cryptomorphisms and lattice properties. In: Proceedings of the Tenth International Conference on Concept Lattices and Their Applications, La Rochelle, France, October 15–18, 2013, pp. 93–103 (2013)

43. Agrawal, R., Srikant, R.: Fast algorithms for mining association rules in large databases. In: Bocca, J.B., Jarke, M., Zaniolo, C. (eds.): VLDB, Morgan Kaufmann, pp. 487–499 (1994)

44. Luxenburger, M.: Implications partielles dans un contexte. Mathématiques, Informatique et Sci. Humaines **29**(113), 35–55 (1991)

45. Jäschke, R., Hotho, A., Schmitz, C., Ganter, B., Stumme, G.: Trias-an algorithm for mining iceberg tri-lattices. In: Proceedings of the Sixth International Conference on Data Mining. ICDM 2006, pp. 907–911. IEEE Computer Society, Washington, DC, USA (2006)

46. Ignatov, D.I., Kuznetsov, S.O., Magizov, R.A., Zhukov, L.E.: From triconcepts to triclusters, vol. 247 257–264

47. Ignatov, D.I., Kuznetsov, S.O., Poelmans, J., Zhukov, L.E.: Can triconcepts become triclusters? Int. J. Gen. Syst. **42**(6), 572–593 (2013)

48. Kuznetsov, S.O.: Machine learning and formal concept analysis, vol. 248, 287–312

49. Ganter, B., Grigoriev, P.A., Kuznetsov, S.O., Samokhin, M.V.: Concept-based data mining with scaled labeled graphs. In: Delugach, H.S., Wolff, K.E., Pfeiffer, H.D. (eds.) ICCS 2004. LNCS (LNAI), vol. 3127, pp. 94–108. Springer, Heidelberg (2004)

50. Kuznetsov, S.O.: Fitting pattern structures to knowledge discovery in big data. In: Cellier, P., Distel, F., Ganter, B. (eds.) ICFCA 2013. LNCS, vol. 7880, pp. 254–266. Springer, Heidelberg (2013)

51. Belohlávek, R., Vychodil, V.: Discovery of optimal factors in binary data via a novel method of matrix decomposition. J. Comput. Syst. Sci. **76**(1), 3–20 (2010)

52. Romashkin, N., Ignatov, D.I., Kolotova, E.: How university entrants are choosing their department? mining of university admission process with FCA taxonomies. In: Proceedings of the 4th International Conference on Educational Data Mining, Eindhoven, The Netherlands, July 6–8, 2011, pp. 229–234 (2011)

53. Grigoriev, P.A., Yevtushenko, S.A.: Quda: Applying formal concept analysis in a data mining environment. vol. 248, pp. 386–393

54. Han, J., Kamber, M.: Data Mining Concepts and Techniques. Morgan Kaufmann, San Francisco (2000)

55. Agrawal, R., Imieliński, T., Swami, A.: Mining association rules between sets of items in large databases. ACM SIGMOD Rec. **22**(2), 207–216 (1993). ACM

56. Pasquier, N., Bastide, Y., Taouil, R., Lakhal, L.: Efficient mining of association rules using closed itemset lattices. Inf. Syst. **24**(1), 25–46 (1999)

57. Zaki, M.J., Hsiao, C.J.: Charm: An efficient algorithm for closed association rule mining. Technical Report, Computer Science, Rensselaer Polytechnic Institute (1999)

58. Stumme, G.: Conceptual knowledge discovery with frequent concept lattices. Technical Report FB4- Preprint 2043, TU Darmstadt (1999)

59. Stumme, G., Taouil, R., Bastide, Y., Pasquier, N., Lakhal, L.: Computing iceberg concept lattices with T. Data Knowl. Eng. **42**(2), 189–222 (2002)
60. Kuznetsov, S.: Mathematical aspects of concept analysis. J. Math. Sci. **80**(2), 1654–1698 (1996)
61. Lakhal, L., Stumme, G.: Efficient mining of association rules based on formal concept analysis. In: Formal Concept Analysis, Foundations and Applications, pp. 180–195 (2005)
62. Agrawal, R., Christoforaki, M., Gollapudi, S., Kannan, A., Kenthapadi, K., Swaminathan, A.: Mining videos from the web for electronic textbooks. In: Formal Concept Analysis - 12th International Conference, ICFCA 2014, Cluj-Napoca, Romania, June 10–13, 2014. Proceedings, pp. 219–234 (2014)
63. Zaki, M.J., Wagner Meira, J.: Data Mining and Analysis: Fundamental Concepts and Algorithms. Cambridge University Press, Cambridge (2014)
64. Zaki, M.J.: Spade: An efficient algorithm for mining frequent sequences. Mach. Learn. **42**, 31–60 (2001)
65. Vander Wal, T.: Folksonomy coinage and definition. (2007). http://vanderwal.net/folksonomy.html. Accessed on 12.03.2012
66. Mirkin, B.: Math. Classif. Clustering. Kluwer, Dordrecht (1996)
67. Madeira, S.C., Oliveira, A.L.: Biclustering algorithms for biological data analysis: A survey. IEEE/ACM Trans. Comput. Biology Bioinform. **1**(1), 24–45 (2004)
68. Eren, K., Deveci, M., Kktun, O., atalyrek, M.V.: A comparative analysis of biclustering algorithms for gene expression data. Briefings in Bioinformatics (2012)
69. Besson, J., Robardet, C., Boulicaut, J.F., Rome, S.: Constraint-based concept mining and its application to microarray data analysis. Intell. Data Anal. **9**(1), 59–82 (2005)
70. Barkow, S., Bleuler, S., Prelic, A., Zimmermann, P., Zitzler, E.: Bicat: a biclustering analysis toolbox. Bioinformatics **22**(10), 1282–1283 (2006)
71. Tarca, A.L., Carey, V.J., wen Chen, X., Romero, R., Drăghici, S.: Machine learning and its applications to biology. PLoS Comput. Biol. **3**(6), e116 (2007)
72. Hanczar, B., Nadif, M.: Bagging for biclustering: application to microarray data. In: Sebag, M., Balcázar, J.L., Gionis, A., Bonchi, F. (eds.) ECML PKDD 2010, Part I. LNCS, vol. 6321, pp. 490–505. Springer, Heidelberg (2010)
73. Kaytoue, M., Kuznetsov, S.O., Napoli, A., Duplessis, S.: Mining gene expression data with pattern structures in formal concept analysis. Inf. Sci. **181**(10), 1989–2001 (2011)
74. Blinova, V.G., Dobrynin, D.A., Finn, V.K., Kuznetsov, S.O., Pankratova, E.S.: Toxicology analysis by means of the jsm-method. Bioinformatics **19**(10), 1201–1207 (2003)
75. Kuznetsov, S.O., Samokhin, M.V.: Learning closed sets of labeled graphs for chemical applications. In: Kramer, S., Pfahringer, B. (eds.) ILP 2005. LNCS (LNAI), vol. 3625, pp. 190–208. Springer, Heidelberg (2005)
76. DiMaggio, P.A., Subramani, A., Judson, R.S., Floudas, C.A.: A novel framework for predicting in vivo toxicities from in vitro data using optimal methods for dense and sparse matrix reordering and logistic regression. Toxicol. Sci. **118**(1), 251–265 (2010)
77. Asses, Y., Buzmakov, A., Bourquard, T., Kuznetsov, S.O., Napoli, A.: A hybrid classification approach based on FCA and emerging patterns - an application for the classification of biological inhibitors. In: Proceedings of The 9th International Conference on Concept Lattices and Their Applications, pp. 211–222 (2012)

78. Dhillon, I.S.: Co-clustering documents and words using bipartite spectral graph partitioning. In: Proceedings of the Seventh ACM SIGKDD International Conference on Knowledge Discovery and Data Mining. KDD 2001, New York, NY, USA, pp. 269–274. ACM (2001)

79. Cimiano, P., Hotho, A., Staab, S.: Learning concept hierarchies from text corpora using formal concept analysis. J. Artif. Intell. Res. (JAIR) **24**, 305–339 (2005)

80. Banerjee, A., Dhillon, I.S., Ghosh, J., Merugu, S., Modha, D.S.: A generalized maximum entropy approach to bregman co-clustering and matrix approximation. J. Mach. Learn. Res. **8**, 1919–1986 (2007)

81. Ignatov, D.I., Kuznetsov, S.O.: Frequent itemset mining for clustering near duplicate web documents. [249] 185–200

82. Carpineto, C., Michini, C., Nicolussi, R.: A concept lattice-based kernel for SVM text classification. In: Rudolph, S., Ferré, S. (eds.) ICFCA 2009. LNCS, vol. 5548, pp. 237–250. Springer, Heidelberg (2009)

83. Koester, B.: Conceptual knowledge retrieval with FooCA: improving web search engine results with contexts and concept hierarchies. In: Perner, P. (ed.) ICDM 2006. LNCS (LNAI), vol. 4065, pp. 176–190. Springer, Heidelberg (2006)

84. Eklund, P.W., Ducrou, J., Dau, F.: Concept similarity and related categories in information retrieval using formal concept analysis. Int. J. Gen. Syst. **41**(8), 826–846 (2012)

85. Duquenne, V.: Lattice analysis and the representation of handicap associations. Soc. Netw. **18**(3), 217–230 (1996)

86. Freeman, L.C.: Cliques, Galois lattices, and the structure of human social groups. Soc. Netw. **18**, 173–187 (1996)

87. Latapy, M., Magnien, C., Vecchio, N.D.: Basic notions for the analysis of large two-mode networks. Soc. Netw. **30**(1), 31–48 (2008)

88. Roth, C., Obiedkov, S.A., Kourie, D.G.: On succinct representation of knowledge community taxonomies with formal concept analysis. Int. J. Found. Comput. Sci. **19**(2), 383–404 (2008)

89. Gnatyshak, D., Ignatov, D.I., Semenov, A., Poelmans, J.: Gaining insight in social networks with biclustering and triclustering. In: Aseeva, N., Babkin, E., Kozyrev, O. (eds.) BIR 2012. LNBIP, vol. 128, pp. 162–171. Springer, Heidelberg (2012)

90. du Boucher-Ryan, P., Bridge, D.G.: Collaborative recommending using formal concept analysis. Knowl. Based Syst. **19**(5), 309–315 (2006)

91. Symeonidis, P., Nanopoulos, A., Papadopoulos, A.N., Manolopoulos, Y.: Nearest-biclusters collaborative filtering based on constant and coherent values. Inf. Retr. **11**(1), 51–75 (2008)

92. Ignatov, D.I., Kuznetsov, S.O.: Concept-based recommendations for internet advertisement. In: Belohlavek, R., Kuznetsov, S.O. (eds.): Proc. CLA 2008. Vol. 433 of CEUR WS., Palack University, Olomouc, 2008, pp. 157–166 (2008)

93. Nanopoulos, A., Rafailidis, D., Symeonidis, P., Manolopoulos, Y.: Musicbox: personalized music recommendation based on cubic analysis of social tags. IEEE Trans. Audio, Speech Lang. Process. **18**(2), 407–412 (2010)

94. Ignatov, D.I., Nenova, E., Konstantinova, N., Konstantinov, A.V.: Boolean matrix factorisation for collaborative filtering: An FCA-based approach. In: Agre, G., Hitzler, P., Krisnadhi, A.A., Kuznetsov, S.O. (eds.) AIMSA 2014. LNCS, vol. 8722, pp. 47–58. Springer, Heidelberg (2014)

95. Ignatov, D.I.: Mathematical Models, Algorithms and Software Tools of Biclustering Based on Closed Sets. Ph.D. thesis, National Research University Higher School of Economics (2010)

96. Ignatov, D.I., Kuznetsov, S.O., Poelmans, J.: Concept-based biclustering for internet advertisement. In: ICDM Workshops, IEEE Computer Society, 123–130 (2012)
97. Benz, D., Hotho, A., Jäschke, R., Krause, B., Mitzlaff, F., Schmitz, C., Stumme, G.: The social bookmark and publication management system bibsonomy - A platform for evaluating and demonstrating web 2.0 research. VLDB J. **19**(6), 849–875 (2010)
98. Zhao, L., Zaki, M.J.: Tricluster: An effective algorithm for mining coherent clusters in 3D microarray data. In: Özcan, F. (ed.): SIGMOD Conference, pp. 694–705. ACM (2005)
99. Li, A., Tuck, D.: An effective tri-clustering algorithm combining expression data with gene regulation information. Gene Regul. Syst. Biol. **3**, 49–64 (2009)
100. Wille, R.: The basic theorem of triadic concept analysis. Order **12**, 149–158 (1995)
101. Lehmann, F., Wille, R.: A triadic approach to formal concept analysis. In: Ellis, G., Levinson, R., Rich, W., Sowa, J.F. (eds.) Conceptual Structures: Applications, Implementation and Theory. LNCS, vol. 954, pp. 32–43. Springer, Heidelberg (1995)
102. Krolak-Schwerdt, S., Orlik, P., Ganter, B.: Tripat: a model for analyzing three-mode binary data. In Bock, H.H., Lenski, W., Richter, M. (eds.): Information Systems and Data Analysis. Studies in Classification, Data Analysis, and Knowledge Organization. Springer, Berlin Heidelberg, pp. 298–307 (1994)
103. Ji, L., Tan, K.L., Tung, A.K.H.: Mining frequent closed cubes in 3d datasets. In: Proceedings of the 32nd International Conference on Very Large Data Bases. VLDB 2006, VLDB Endowment, pp. 811–822 (2006)
104. Cerf, L., Besson, J., Robardet, C., Boulicaut, J.F.: Closed patterns meet n-ary relations. ACM Trans. Knowl. Discov. Data **3**, 3:1–3:36 (2009)
105. Cerf, L., Besson, J., Nguyen, K.N., Boulicaut, J.F.: Closed and noise-tolerant patterns in n-ary relations. Data Min. Knowl. Discov. **26**(3), 574–619 (2013)
106. Georgii, E., Tsuda, K., Schölkopf, B.: Multi-way set enumeration in weight tensors. Mach. Learn. **82**(2), 123–155 (2011)
107. Spyropoulou, E., De Bie, T., Boley, M.: Interesting pattern mining in multi-relational data. Data Min. Knowl. Disc. **28**(3), 808–849 (2014)
108. Voutsadakis, G.: Polyadic concept analysis. Order **19**(3), 295–304 (2002)
109. Ignatov, D., Gnatyshak, D., Kuznetsov, S., Mirkin, B.: Triadic formal concept analysis and triclustering: searching for optimal patterns. Mach. Learn. **42**, 1–32 (2015)
110. Mirkin, B., Kramarenko, A.V.: Approximate bicluster and tricluster boxes in the analysis of binary data. [247] 248–256
111. Gnatyshak, D., Ignatov, D.I., Kuznetsov, S.O.: From triadic fca to triclustering: Experimental comparison of some triclustering algorithms. [250] 249–260
112. Gnatyshak, D.V., Ignatov, D.I., Kuznetsov, S.O., Nourine, L.: A one-pass triclustering approach: Is there any room for big data? In: CLA 2014 (2014)
113. Ganter, B., Kuznetsov, S.O.: Hypotheses and version spaces. In: Ganter, B., de Moor, A., Lex, W. (eds.) Conceptual Structures for Knowledge Creation and Communication. LNCS, vol. 2746, pp. 83–95. Springer, Heidelberg (2003)
114. Belohlávek, R., Baets, B.D., Outrata, J., Vychodil, V.: Inducing decision trees via concept lattices. Int. J. Gen. Syst. **38**(4), 455–467 (2009)
115. Carpineto, C., Romano, G.: Galois: An order-theoretic approach to conceptual clustering. In: Proceeding of ICML93, Amherst, pp. 33–40 (1993)
116. Carpineto, C., Romano, G.: A lattice conceptual clustering system and its application to browsing retrieval. Mach. Learn. **24**, 95–122 (1996)

117. Fu, H., Fu, H., Njiwoua, P., Nguifo, E.M.: A comparative study of FCA-based supervised classification algorithms. In: Eklund, P. (ed.) ICFCA 2004. LNCS (LNAI), vol. 2961, pp. 313–320. Springer, Heidelberg (2004)

118. Rudolph, S.: Using FCA for encoding closure operators into neural networks. In: Proceedings of the 15th International Conference on Conceptual Structures, ICCS 2007, Sheffield, UK, July 22–27, 2007, pp. 321–332 (2007)

119. Tsopzé, N., Nguifo, E.M., Tindo, G.: CLANN: concept lattice-based artificial neural network for supervised classification. In: Proceedings of the 5th International Conference on Concept Lattices and Their Applications, CLA 2007 (2007)

120. Outrata, J.: Boolean factor analysis for data preprocessing in machine learning. In: The Ninth International Conference on Machine Learning and Applications, ICMLA 2010, Washington, DC, USA, 12–14 December 2010, pp. 899–902 (2010)

121. Belohlávek, R., Outrata, J., Trnecka, M.: Impact of boolean factorization as preprocessing methods for classification of boolean data. Ann. Math. Artif. Intell. **72**(1–2), 3–22 (2014)

122. Ganter, B., Kuznetsov, S.O.: Scale coarsening as feature selection. In: Medina, R., Obiedkov, S. (eds.) Formal Concept Analysis. LNCS, vol. 4933, pp. 217–228. Springer, Heidelberg (2008)

123. Visani, M., Bertet, K., Ogier, J.: Navigala: an original symbol classifier based on navigation through a Galois lattice. IJPRAI **25**(4), 449–473 (2011)

124. Zaki, M.J., Aggarwal, C.C.: Xrules: An effective algorithm for structural classification of XML data. Mach. Learn. **62**(1–2), 137–170 (2006)

125. Flach, P.: Machine Learning: The Art and Science of Algorithms That Make Sense of Data. Cambridge University Press, New York (2012)

126. Finn, V.: On machine-oriented formalization of plausible reasoning in f.bacon-j.s.mill style. Semiotika i Informatika **20**, 35–101 (1983). (in Russian)

127. Kuznetsov, S.: Jsm-method as a machine learning. Method. Itogi Nauki i Tekhniki, ser. Informatika **15**, 17–52 (1991). (in Russian)

128. Gusakova, S.: Paleography with jsm-method. Technical Report, VINITI (2001)

129. Ganter, B., Kuznetsov, S.: Formalizing hypotheses with concepts. In: Ganter, B., Mineau, G. (eds.) Conceptual Structures: Logical, Linguistic, and Computational Issues. Lecture Notes in Computer Science, vol. 1867, pp. 342–356. Springer, Heidelberg (2000)

130. Zhuk, R., Ignatov, D.I., Konstantinova, N.: Concept learning from triadic data. In: Proceedings of the Second International Conference on Information Technology and Quantitative Management, ITQM 2014, National Research University Higher School of Economics (HSE), Moscow, Russia, June 3–5, 2014, pp. 928–938 (2014)

131. Ignatov, D.I., Zhuk, R., Konstantinova, N.: Learning hypotheses from triadic labeled data. In: 2014 IEEE/WIC/ACM International Joint Conferences on Web Intelligence (WI) and Intelligent Agent Technologies (IAT), Warsaw, Poland, August 11–14, 2014 - vol. I, pp. 474–480 (2014)

132. Ganter, B., Kuznetsov, S.O.: Pattern structures and their projections. In: Delugach, H.S., Stumme, G. (eds.) ICCS 2001. LNCS (LNAI), vol. 2120, pp. 129–142. Springer, Heidelberg (2001)

133. Buzmakov, A., Egho, E., Jay, N., Kuznetsov, S.O., Napoli, A., Raïssi, C.: On projections of sequential pattern structures (with an application on care trajectories). [250] 199–208

134. Kuznetsov, S.O.: Scalable knowledge discovery in complex data with pattern structures. In: Maji, P., Ghosh, A., Murty, M.N., Ghosh, K., Pal, S.K. (eds.) PReMI 2013. LNCS, vol. 8251, pp. 30–39. Springer, Heidelberg (2013)

135. Strok, F., Galitsky, B., Ilvovsky, D., Kuznetsov, S.: Pattern structure projections for learning discourse structures. In: Agre, G., Hitzler, P., Krisnadhi, A.A., Kuznetsov, S.O. (eds.) AIMSA 2014. LNCS, vol. 8722, pp. 254–260. Springer, Heidelberg (2014)
136. Belohlávek, R.: What is a fuzzy concept lattice? II. [247] 19–26
137. Kent, R.E.: Rough concept analysis: A synthesis of rough sets and formal concept analysis. Fundam. Inform. **27**(2/3), 169–181 (1996)
138. Poelmans, J., Ignatov, D.I., Kuznetsov, S.O., Dedene, G.: Fuzzy and rough formal concept analysis: a survey. Int. J. Gen. Syst. **43**(2), 105–134 (2014)
139. Pankratieva, V.V., Kuznetsov, S.O.: Relations between proto-fuzzy concepts, crisply generated fuzzy concepts, and interval pattern structures. Fundam. Inform. **115**(4), 265–277 (2012)
140. Koren, Y., Bell, R., Volinsky, C.: Matrix factorization techniques for recommender systems. Computer **42**(8), 30–37 (2009)
141. Elden, L.: Matrix methods in data mining and pattern recognition. In: Society for Industrial and Applied Mathematics (2007). http://www.amazon.com/Methods-Pattern-Recognition-Fundamentals-Algorithms/dp/0898716268
142. Hofmann, T.: Unsupervised learning by probabilistic latent semantic analysis. Mach. Learn. **42**(1–2), 177–196 (2001)
143. Koren, Y.: Factorization meets the neighborhood: A multifaceted collaborative filtering model. In: Proceedings of the 14th ACM SIGKDD International Conference on Knowledge Discovery and Data Mining. KDD 2008, pp. 426–434. New York, NY, USA, ACM (2008)
144. Lin, C.J.: Projected gradient methods for nonnegative matrix factorization. Neural Comput. **19**(10), 2756–2779 (2007)
145. Nenova, E., Ignatov, D.I., Konstantinov, A.V.: An fca-based boolean matrix factorisation for collaborative filtering. In: International Workshop FCA meets IR at ECIR 2013. vol. 977, CEUR Workshop Proceeding, pp. 57–73 (2013)
146. Belohlávek, R., Glodeanu, C., Vychodil, V.: Optimal factorization of three-way binary data using triadic concepts. Order **30**(2), 437–454 (2013)
147. Miettinen, P.: Boolean tensor factorization. In: Cook, D., Pei, J., Wang, W., Zaïane, O., Wu, X. (eds.) ICDM 2011, 11th IEEE International Conference on Data Mining, pp. 447–456. Canada, IEEE Computer Society, CPS, Vancouver (2011)
148. Hotho, A., Jäschke, R., Schmitz, C., Stumme, G.: Information retrieval in folksonomies: search and ranking. In: Domingue, J., Sure, Y. (eds.) ESWC 2006. LNCS, vol. 4011, pp. 411–426. Springer, Heidelberg (2006)
149. Zhiltsov, N., Agichtein, E.: Improving entity search over linked data by modeling latentsemantics. In: 22nd ACM International Conference on Information and Knowledge Management, CIKM 2013, San Francisco, CA, USA, October 27 - November 1, 2013, pp. 1253–1256 (2013)
150. Ignatov, D.I., Mamedova, S., Romashkin, N., Shamshurin, I.: What can closed sets of students and their marks say? In: Proceedings of the 4th International Conference on Educational Data Mining, Eindhoven, The Netherlands, July 6–8, 2011, pp. 223–228 (2011)
151. Grigoriev, P., Yevtushenko, S., Grieser, G.: QuDA, a Data Miners Discovery Environment. Technical Report AIDA-03-06, Technische Universität Darmstadt (2003)
152. Grigoriev, P.A., Yevtushenko, S.A.: Elements of an agile discovery environment. In: Grieser, G., Tanaka, Y., Yamamoto, A. (eds.) Discovery Science. LNCS, vol. 2843, pp. 311–319. Springer, Heidelberg (2003)

153. Grigoriev, P., Kuznetsov, S., Obiedkov, S., Yevtushenko, S.: On a version of mill's method of difference. In: Proceedings of the ECAI 2002 Workshop on Concept Lattices in Data Mining, Lyon, pp. 26–31 (2002)

154. Mooers, C.N.: A mathematical theory of language symbols in retrieval. In: Proceedings of the International Conference Scientific Information, Washington D.C. (1958)

155. Fairthorne, R.A.: The patterns of retrieval. Am. Documentation **7**(2), 65–70 (1956)

156. Shreider, Y.: Mathematical model of classification theory, pp. 1–36. VINITI, Moscow (1968). (in Russian)

157. Soergel, D.: Mathematical analysis of documentation systems. Inf. Stor. Retr. **3**, 129–173 (1967)

158. Godin, R., Saunders, E., Gecsei, J.: Lattice model of browsable data spaces. Inf. Sci. **40**(2), 89–116 (1986)

159. Carpineto, C., Romano, G.: Using concept lattices for text retrieval and mining. In: Formal Concept Analysis, Foundations and Applications, pp. 161–179 (2005)

160. Priss, U.: Formal concept analysis in information science. ARIST **40**(1), 521–543 (2006)

161. Valverde-Albacete, F.J., Pelaez-Moreno, C.: Systems vs. methods: an analysis of the affordances of formal concept analysis for information retrieval? In: Proceedings of the of International Workshop on FCA for IR at ECIR 2013, HSE, Moscow (2013)

162. Ferr, S.: Camelis: Organizing and browsing a personal photo collection with a logical information system. In: Eklund, P.W., Diatta, J., Liquiere, M. (eds.): CLA. vol. 331 of CEUR Workshop Proceedings. CEUR-WS.org (2007)

163. Ignatov, D.I., Konstantinov, A.V., Chubis, Y.: Near-duplicate detection for online-shops owners: An fca-based approach. [7] 722–725

164. Kuznetsov, S.O., Ignatov, D.I.: Concept stability for constructing taxonomies of web-site users. In: Obiedkov, S., Roth, C. (eds.) Proceedings of the Social Network Analysis and Conceptual Structures: Exploring Opportunities, Clermont-Ferrand (France), February 16, 2007 (2007)

165. Ignatov, D.I., Kaminskaya, A.Y., Bezzubtseva, A.A., Konstantinov, A.V., Poelmans, J.: FCA-based models and a prototype data analysis system for crowd-sourcing platforms. In: Pfeiffer, H.D., Ignatov, D.I., Poelmans, J., Gadiraju, N. (eds.) ICCS 2013. LNCS, vol. 7735, pp. 173–192. Springer, Heidelberg (2013)

166. Carpineto, C., Romano, G.: A survey of automatic query expansion in information retrieval. ACM Comput. Surv. **44**(1), 1 (2012)

167. Carpineto, C., Romano, G.: Order-theoretical ranking. JASIS **51**(7), 587–601 (2000)

168. Carpineto, C., Osinski, S., Romano, G., Weiss, D.: A survey of web clustering engines. ACM Comput. Surv. **41**(3), 1–38 (2009)

169. Carpineto, C., Romano, G.: Exploiting the potential of concept lattices for information retrieval with CREDO. J. UCS **10**(8), 985–1013 (2004)

170. Ducrou, J., Eklund, P.W.: Searchsleuth: The conceptual neighbourhood of an web query. In: Proceedings of the Fifth International Conference on Concept Lattices and Their Applications, CLA 2007, Montpellier, France, October 24–26, 2007 (2007)

171. Dau, F., Ducrou, J., Eklund, P.: Concept similarity and related categories in searchsleuth. In: Eklund, P., Haemmerlé, O. (eds.) ICCS 2008. LNCS (LNAI), vol. 5113, pp. 255–268. Springer, Heidelberg (2008)

172. Nauer, E., Toussaint, Y.: Crechaindo: an iterative and interactive web information retrieval system based on lattices. Int. J. Gen. Syst. **38**(4), 363–378 (2009)

173. Kim, M., Compton, P.: Evolutionary document management and retrieval for specialized domains on the web. Int. J. Hum.-Comput. Stud. **60**(2), 201–241 (2004)

174. Kim, M.H., Compton, P.: A hybrid browsing mechanism using conceptual scales. In: Richards, D., Hoffmann, A., Tsumoto, S., Kang, B.-H. (eds.) PKAW 2006. LNCS (LNAI), vol. 4303, pp. 132–143. Springer, Heidelberg (2006)

175. Cigarrán, J.M., Gonzalo, J., Peñas, A., Verdejo, F.: Browsing search results via formal concept analysis: Automatic selection of attributes. [248] 74–87

176. Cole, R.J., Eklund, P.W., Stumme, G.: Document retrieval for e-mail search and discovery using formal concept analysis. Appl. Artif. Intell. **17**(3), 257–280 (2003)

177. Cole, R.J., Eklund, P.W.: Browsing semi-structured web texts using formal concept analysis. In: Conceptual Structures: Broadening the Base, 9th International Conference on Conceptual Structures, ICCS 2001, Stanford, CA, USA, July 30-August 3, 2001, Proceedings, pp. 319–332 (2001)

178. Eklund, P.W., Cole, R.J.: A knowledge representation for information filtering using formal concept analysis. Electron. Trans. Artif. Intell. **4**(C), 51–51 (2000)

179. Eklund, P.W., Ducrou, J., Brawn, P.: Concept lattices for information visualization: Can novices read line-diagrams? [248] 57–73

180. Eklund, P., Wormuth, B.: Restructuring help systems using formal concept analysis. In: Godin, R., Ganter, B. (eds.) ICFCA 2005. LNCS (LNAI), vol. 3403, pp. 129–144. Springer, Heidelberg (2005)

181. Stojanovic, N.: On the query refinement in the ontology-based searching for information. Inf. Syst. **30**(7), 543–563 (2005)

182. Spyratos, N., Meghini, C.: Preference-based query tuning through refinement/enlargement in a formal context. In: Hegner, S.J., Dix, J. (eds.) FoIKS 2006. LNCS, vol. 3861, pp. 278–293. Springer, Heidelberg (2006)

183. Le Grand, B., Aufaure, M.-A., Soto, M.: Semantic and conceptual context-aware information retrieval. In: Yetongnon, K., Chbeir, R., Damiani, E., Dipanda, A. (eds.) SITIS 2006. LNCS, vol. 4879, pp. 247–258. Springer, Heidelberg (2009)

184. Eklund, P., Ducrou, J.: Navigation and annotation with formal concept analysis. In: Kang, B.-H., Richards, D. (eds.) PKAW 2008. LNCS, vol. 5465, pp. 118–121. Springer, Heidelberg (2009)

185. Cigarrán, J.M., Peñas, A., Gonzalo, J., Verdejo, M.F.: Automatic selection of noun phrases as document descriptors in an FCA-based information retrieval system. In: Ganter, B., Godin, R. (eds.) ICFCA 2005. LNCS (LNAI), vol. 3403, pp. 49–63. Springer, Heidelberg (2005)

186. Recio-García, J.A., Gómez-Martín, M.A., Díaz-Agudo, B., González-Calero, P.A.: Improving annotation in the semantic web and case authoring in textual CBR. In: Göker, M.H., Roth-Berghofer, T.R., Güvenir, H.A. (eds.) ECCBR 2006. LNCS (LNAI), vol. 4106, pp. 226–240. Springer, Heidelberg (2006)

187. Liu, M., Shao, M., Zhang, W., Wu, C.: Reduction method for concept lattices based on rough set theory and its application. Comput. Math. Appl. **53**(9), 1390–1410 (2007)

188. Lungley, D., Kruschwitz, U.: Automatically maintained domain knowledge: initial findings. In: Berrut, C., Soule-Dupuy, C., Mothe, J., Boughanem, M. (eds.) ECIR 2009. LNCS, vol. 5478, pp. 739–743. Springer, Heidelberg (2009)

189. Ahmad, I., Jang, T.: Old fashion text-based image retrieval using FCA. ICIP **3**, 33–36 (2003)

190. Ducrou, J., Vormbrock, B., Eklund, P.: FCA-based browsing and searching of a collection of images. In: Øhrstrøm, P., Hitzler, P., Schärfe, H. (eds.) ICCS 2006. LNCS (LNAI), vol. 4068, pp. 203–214. Springer, Heidelberg (2006)

191. Ducrou, J.: DVDSleuth: a case study in applied formal concept analysis for navigating web catalogs. In: Priss, U., Hill, R., Polovina, S. (eds.) ICCS 2007. LNCS (LNAI), vol. 4604, pp. 496–500. Springer, Heidelberg (2007)

192. Amato, G., Meghini, C.: Faceted content-based image retrieval. In: 19th International Workshop on Database and Expert Systems Applications (DEXA 2008), 1–5 September 2008, Turin, Italy, pp. 402–406 (2008)

193. Ferré, S.: Camelis: a logical information system to organise and browse a collection of documents. Int. J. Gen. Syst. **38**(4), 379–403 (2009)

194. Poelmans, J., Elzinga, P., Viaene, S., Dedene, G.: Formally analysing the concepts of domestic violence. Expert Syst. Appl. **38**(4), 3116–3130 (2011)

195. Wolff, K.E.: States, transitions, and life tracks in temporal concept analysis. In: Formal Concept Analysis, Foundations and Applications, pp. 127–148 (2005)

196. Elzinga, P., Poelmans, J., Viaene, S., Dedene, G., Morsing, S.: Terrorist threat assessment with formal concept analysis. In: IEEE International Conference on Intelligence and Security Informatics, ISI 2010, Vancouver, BC, Canada, May 23–26, 2010, Proceedings, pp. 77–82 (2010)

197. Elzinga, P., Wolff, K.E., Poelmans, J.: Analyzing chat conversations of pedophiles with temporal relational semantic systems. In: 2012 European Intelligence and Security Informatics Conference, EISIC 2012, Odense, Denmark, August 22–24, 2012, pp. 242–249 (2012)

198. Bullens, R., Van Horn, J.: Daad uit liefde: Gedwongen prostitutie van jonge meisjes. Justitiele Verkenningen **26**(6), 25–41 (2000)

199. Koester, B., Schmidt, S.: Information superiority via formal concept analysis. In: Argamon, S., Howard, N. (eds.) Computational Methods for Counterterrorism, pp. 143–171. Springer, Heidelberg (2009)

200. Obiedkov, S.A., Kourie, D.G., Eloff, J.H.P.: Building access control models with attribute exploration. Comput. Secur. **28**(1–2), 2–7 (2009)

201. Dau, F., Knechtel, M.: Access policy design supported by FCA methods. [249] 141–154

202. Zhukov, L.E.: Spectral clustering of large advertiser datasets. Overture R&D, Technical Report (2004)

203. Sarwar, B.M., Karypis, G., Konstan, J.A., Riedl, J.: Analysis of recommendation algorithms for e-commerce. In: ACM Conference on Electronic Commerce, pp. 158–167 (2000)

204. Besson, J., Robardet, C., Boulicaut, J.F., Rome, S.: Constraint-based bi-set mining for biologically relevant pattern discovery in microarray data. Intell. Data Anal. J. **9**(1), 59–82 (2005)

205. Szathmary, L., Napoli, A.: CORON: A Framework for Levelwise Itemset Mining Algorithms. In: Supplements Proceedings of ICFCA 2005, Lens, France, pp. 110–113, February 2005

206. Szathmary, L., Napoli, A., Kuznetsov, S.O.: ZART: a multifunctional itemset mining algorithm. In: Proceedings of the 5th International Conference on Concept Lattices and Their Applications (CLA 2007), pp. 26–37. Montpellier, France, October 2007

207. Crystal, D.: A dictionary of linguistics and phonetics, 3rd edn. Blackwell Publishers, Oxford (1991)

208. Symeonidis, P., Ruxanda, M.M., Nanopoulos, A., Manolopoulos, Y.: Ternary semantic analysis of social tags for personalized music recommendation. In: Bello, J.P., Chew, E., Turnbull, D. (eds.): ISMIR, pp. 219–224 (2008)
209. Alqadah, F., Reddy, C., Hu, J., Alqadah, H.: Biclustering neighborhood-based collaborative filtering method for top-n recommender systems. Knowl. Inf. Syst. **44**, 1–17 (2014)
210. Adomavicius, G., Tuzhilin, A.: Toward the next generation of recommender systems: A survey of the state-of-the-art and possible extensions. IEEE Trans. Knowl. Data Eng. **17**(6), 734–749 (2005)
211. Ignatov, D.I., Poelmans, J., Dedene, G., Viaene, S.: A new cross-validation technique to evaluate quality of recommender systems. In: Kundu, M., Mitra, S., Mazumdar, D., Pal, S. (eds.) Perception and Machine Intelligence. LNCS, vol. 7143, pp. 195–202. Springer, Heidelberg (2012)
212. Brin, S., Davis, J., García-Molina, H.: Copy detection mechanisms for digital documents. SIGMOD Rec. **24**(2), 398–409 (1995)
213. Broder, A.Z., Glassman, S.C., Manasse, M.S., Zweig, G.: Syntactic clustering of the web. Comput. Netw. **29**(8–13), 1157–1166 (1997)
214. Ilyinsky, S., Kuzmin, M., Melkov, A., Segalovich, I.: An efficient method to detect duplicates of web documents with the use of inverted index. In: Proceedings of the 11th International World Wide Web Conference (WWW 2002), Honolulu, Hawaii, USA, 7–11 May 2002, ACM (2002)
215. Broder, A.Z., Charikar, M., Frieze, A.M., Mitzenmacher, M.: Min-wise independent permutations (extended abstract). In: Proceedings of the Thirtieth Annual ACM Symposium on the Theory of Computing, Dallas, Texas, USA, May 23–26, 1998, pp. 327–336 (1998)
216. Broder, A.: Identifying and filtering near-duplicate documents. In: Giancarlo, R., Sankoff, D. (eds.) CPM 2000. LNCS, vol. 1848, pp. 1–10. Springer, Heidelberg (2000)
217. Grahne, G., Zhu, J.: Efficiently using prefix-trees in mining frequent itemsets. In: FIMI 2003, Frequent Itemset Mining Implementations, Proceedings of the ICDM 2003 Workshop on Frequent Itemset Mining Implementations, 19 December 2003, Melbourne, Florida, USA (2003)
218. Karypis, G.: Cluto. a clustering toolkit. Technical Report: 2–017 MN 55455, University of Minnesota, Department of Computer Science Minneapolis, November 28 2003
219. Potthast, M., Stein, B.: New issues in near-duplicate detection. In: Data Analysis, Machine Learning and Applications - Proceedings of the 31st Annual Conference of the Gesellschaft für Klassifikation e.V., Albert-Ludwigs-Universität Freiburg, March 7–9, 2007, pp. 601–609 (2007)
220. Zelenkov, Y.G., Segalovich, I.V.: Comparative analysis of near-duplicate detection methods of web documents. In: Proceedings of the 9th All-Russian Scientific Conference Digital Libraries: Advanced Methods and Technologies, Digital Collections, Pereslavl-Zalessky, pp. 166–174 (2007) (in Russian)
221. Ignatov, D.I., Jánosi-Rancz, K.T., Kuznetzov, S.O.: Towards a framework for near-duplicate detection in document collections based on closed sets of attributes. Acta Univ. Sapientiae Inf. **1**(2), 215–233 (2009)
222. Ignatov, D., Kuznetsov, S., Lopatnikova, V., Selitskiy, I.: Development and aprobation of near duplicate detection system for collections of r&d documents. Bus. Inf. **4**, 21–28 (2008). (in Russian)
223. Ley, M.: DBLP - some lessons learned. PVLDB **2**(2), 1493–1500 (2009)

224. Benz, D., Hotho, A., Jäschke, R., Krause, B., Stumme, G.: Query logs as folksonomies. Datenbank-Spektrum **10**(1), 15–24 (2010)
225. Doerfel, S., Jäschke, R.: An analysis of tag-recommender evaluation procedures. In: Seventh ACM Conference on Recommender Systems, RecSys 2013, Hong Kong, China, October 12–16, 2013, pp. 343–346 (2013)
226. Kuznetsov, S.O., Ignatov, D.I.: Concept stability for constructing taxonomies of web-site users. In: Obiedkov, S., Roth, C. (eds.), Proceedings of the Social Network Analysis and Conceptual Structures: Exploring Opportunities, Clermont-Ferrand (France), February 16, 2007, pp. 19–24 (2007)
227. Kuznetsov, S.: Stability as an estimate of the degree of substantiation of hypotheses derived on the basis of operational similarity. Nauchn. Tekh. Inf. Ser. **2**(12), 21–29 (1990). (Automat. Document. Math. Linguist.)
228. Kuznetsov, S.O.: On stability of a formal concept. Ann. Math. Artif. Intell. **49** (1–4), 101–115 (2007)
229. Roth, C., Cointet, J.P.: Social and semantic coevolution in knowledge networks. Soc. Netw. **32**, 16–29 (2010)
230. Yavorsky, R.: Research challenges of dynamic socio-semantic networks. In: Ignatov, D., Poelmans, J., Kuznetsov, S. (eds.): CEUR Workshop proceedings vol. 757, CDUD 2011 - Concept Discovery in Unstructured Data, pp. 119–122 (2011)
231. Howe, J.: The rise of crowdsourcing. Wired, San Francisco (2006)
232. Ignatov, D.I., Mikhailova, M., Kaminskaya, A.Y.Z., Malioukov, A.: Recommendation of ideas and antagonists for crowdsourcing platform witology. In: Proceedings of 8th RuSSIR, Springer (2014) (this volume)
233. Ignatov, D.I., Kaminskaya, A.Y., Konstantinova, N., Malyukov, A., Poelmans, J.: FCA-based recommender models and data analysis for crowdsourcing platform witology. In: Hernandez, N., Jäschke, R., Croitoru, M. (eds.) ICCS 2014. LNCS, vol. 8577, pp. 287–292. Springer, Heidelberg (2014)
234. Ignatov, D.I., Kaminskaya, A.Y., Konstantinova, N., Konstantinov, A.V.: Recommender system for crowdsourcing platform witology. In: 2014 IEEE/WIC/ACM International Joint Conferences on Web Intelligence (WI) and Intelligent Agent Technologies (IAT), Warsaw, Poland, August 11–14, 2014, vol. II, pp. 327–335 (2014)
235. Ganter, B.: Attribute exploration with background knowledge. Theor. Comput. Sci. **217**(2), 215–233 (1999). ORDAL'96
236. Stumme, G., Maedche, A.: Fca-merge: Bottom-up merging of ontologies. In: Nebel, B. (ed.): IJCAI, Morgan Kaufmann, pp. 225–234 (2001)
237. Revenko, A., Kuznetsov, S.O.: Attribute exploration of properties of functions on sets. Fundam. Inform. **115**(4), 377–394 (2012)
238. Sertkaya, B.: A survey on how description logic ontologies benefit from FCA. In: Proceedings of the 7th International Conference on Concept Lattices and Their Applications, Sevilla, Spain, October 19–21, 2010, pp. 2–21 (2010)
239. Sertkaya, B.: Ontocomp: A protégé plugin for completing OWL ontologies. In: The Semantic Web: Research and Applications, 6th European Semantic Web Conference, ESWC 2009, Heraklion, Crete, Greece, May 31-June 4, 2009, Proceedings, pp. 898–902 (2009)
240. Baader, F., Ganter, B., Sertkaya, B., Sattler, U.: Completing description logic knowledge bases using formal concept analysis. In: IJCAI 2007, Proceedings of the 20th International Joint Conference on Artificial Intelligence, Hyderabad, India, January 6–12, 2007, pp. 230–235 (2007)

241. Rudolph, S.: Relational exploration: combining description logics and formal concept analysis for knowledge specification. Ph.D. thesis, Dresden University of Technology (2006)
242. Potoniec, J., Rudolph, S., Lawrynowicz, A.: Towards combining machine learning with attribute exploration for ontology refinement. In: Proceedings of the ISWC 2014 Posters & Demonstrations Track a track within the 13th International Semantic Web Conference, ISWC 2014, Riva del Garda, Italy, October 21, 2014, pp. 229–232 (2014)
243. Jäschke, R., Rudolph, S.: Attribute exploration on the web. In: Cellier, P., Distel, F., Ganter, B. (eds.): Contributions to the 11th International Conference on Formal Concept Analysis, Technische Universit Dresden, pp. 19–34, May 2013
244. Codocedo, V., Lykourentzou, I., Napoli, A.: A semantic approach to concept lattice-based information retrieval. Ann. Math. Artif. Intell. **72**(1–2), 169–195 (2014)
245. Tilley, T., Cole, R., Becker, P., Eklund, P.: A survey of formal concept analysis support for software engineering activities. In: Wille, R., Stumme, G., Ganter, B. (eds.) Formal Concept Analysis. LNCS (LNAI), vol. 3626, pp. 250–271. Springer, Heidelberg (2005)
246. Arévalo, G., Desnos, N., Huchard, M., Urtado, C., Vauttier, S.: Formal concept analysis-based service classification to dynamically build efficient software component directories. Int. J. Gen. Syst. **38**(4), 427–453 (2009)
247. Mirkin, B.G., Kuznetsov, S.O., Slkezak, D., Hepting, D.H. (eds.): RSFDGrC 2011. LNCS, vol. 6743. Springer, Heidelberg (2011)
248. Eklund, P., Ducrou, J., Brawn, P.: Concept lattices for information visualization: can novices read line-diagrams? In: Eklund, P. (ed.) ICFCA 2004. LNCS (LNAI), vol. 2961, pp. 57–73. Springer, Heidelberg (2004)
249. Dau, F., Rudolph, S., Kuznetsov, S.O. (eds.): ICCS 2009. LNCS, vol. 5662. Springer, Heidelberg (2009)
250. Ojeda-Aciego, M., Outrata, J. (eds.): Proceedings of the tenth international conference on concept lattices and their applications. La Rochelle, France, October 15–18, 2013. CLA. vol. 1062 of CEUR Workshop Proceedings, CEUR-WS.org (2013)

Visualization and Data Mining for High Dimensional Data
–With Connections to Information Retrieval

Alfred Inselberg[1]([⊠]) and Pei Ling Lai[2]

[1] School of Mathematical Sciences, Tel Aviv University, Tel Aviv, Israel
aiisreal@post.tau.ac.il
http://www.cs.tau.ac.il/~aiisreal
[2] Department of Electronic Engineering,
Southern Taiwan University of Science and Technology, Tainan, Taiwan
pllai@stust.edu.tw

1 Introduction

The first, and still more popular application, of parallel coordinates is in exploratory data analysis (EDA); discovering data subsets (relations) satisfying given objectives. A dataset with M items has 2^M subsets anyone of which may be the one we really want. With a good data display our fantastic pattern-recognition ability can cut great swaths searching through this combinatorial explosion and also extract insights from the visual patterns. These are the core reasons for data visualization. With parallel coordinates (abbr.‖-coords) the search for relations in multivariate datasets is transformed into a 2-D pattern recognition problem. Guidelines and strategies for knowledge discovery are illustrated on several real datasets one with hundreds of variables. A geometric classification algorithm is presented and applied to complex datasets. It has low computational complexity providing the classification rule explicitly and *visually*. The minimal set of variables required to state the rule is found and ordered by their predictive value. Multivariate relations can be modeled as hypersurfaces and used for decision support. A model of a country's economy reveals sensitivies, impact of constraints, trade-offs and sectors unknowingly competing for the same resources. Foundational background is provided where needed.

The penultimate section of the tutorial is devoted to an investigation into the well-known parallel between artificial neural networks and statistics and prepared by the second author.

1.1 Origins

For the visualization of multivariate problems numerous mappings encoding multidimensional information visually into 2-D or 3-D (see [6, 21–23]) have been

---

A large collection of methodologies tracing the development of the field can be found in [6].

© Springer International Publishing Switzerland 2015
P. Braslavski et al. (eds.): RuSSIR 2014, CCIS 505, pp. 142–184, 2015.
DOI: 10.1007/978-3-319-25485-2_4

invented to augment our perception, which is limited by our 3-dimensional habitation. Wonderful successes like Minard's "Napoleon's March to Moscow", Snow's "dot map" and others are *ad hoc* (i.e. one-of-a-kind) and exceptional. Succinct multivariate relations are rarely apparent from **static** displays; **interactivity** is essential. In turn, this raises the issues of effective *GUI* – Graphic User Interface, queries, exploration strategies and information preserving displays but we are getting ahead of ourselves.

1.2 The Case for Visualization

Searching a dataset with M items for interesting, depending on the objectives, properties is inherently hard. There are 2^M possible subsets anyone of which may satisfy the objectives. The *visual cues*, our eyes can pick from a good data display, help navigate through this combinatorial explosion. How this is done is part of the story. Clearly, if the transformation : *data* → *picture* clobbers information a great deal is lost right at the start. We postulate that a display of datasets with N variables suitable for *exploration* satisfy the following requirements:

1. **should preserve information** – the dataset can be completely reconstructed from the picture,
2. **has low representational complexity** – the computational cost of constructing the display is low,
3. **works for any N** – not limited by the dimension,
4. **treats every variable uniformly**,
5. **has invariance under projective transformations** – the dataset can be recognized after rotations, translations, scalings and perspective transformations,
6. **reveals multivariate relations in the dataset** – the most important and controversial single criterion,
7. **is based on a rigorous mathematical and algorithmic methodology** – to eliminate ambiguity in the results.

No completeness or uniqueness is claimed for this list which should invite criticism and changes. Commentary for each item, illustrated by examples and comparisons, is listed in the same order.

1. The numerical values for each N-tuple (i.e.each data point) are recoverable from the scatterplot matrix (abbr. *SM*) and the ∥-coords display. This may not be necessary or desirable for presentation graphics (e.g. pie-charts, histograms).
2. In the pairwise scatterplot matrix the N variables appear $N(N-1)/2$ times. By contrast there are only N axes in ∥-coords though there is an additional preprocessing cost pointed out in the ensuing. For $N \geq 10$, the practical barriers due to the required display *space* and visual difficulty limit the use of *SM*. These barriers are less stringent for ∥-coords.

3. Even with a nice perspective, orthogonal coordinates are inherently limited to $N = 3$ due to the dimensionality of our existence. With added aspects the illusion of a few more dimensions can be added. By contrast, for ∥-coords implementation capabilities rather than conceptual barriers determine the maximum useable N.

4. An example is the "Chernoff Faces" display where each variable corresponds to a specific facial feature and treated accordingly. The correspondence *facial feature → variable* is arbitrary. Choosing a different correspondence gives a different display and the fun part is there is no way to know that the **two different** displays portray the **same dataset**. Of course, this is true for general *glyphs* displays.

5. This is true for *SM* and for ∥-coords though it has not been implemented in general. Incidentally, this is a wonderful M.Sc. thesis topic.

6. The real value of visualization, in my opinion, is not seeing "zillions of objects" but rather recognizing **relations** among them. We know that projections lose information which can possibly be recovered with interactivity. Nevertheless, important *clues which can guide the exploration* are lost. So it is preferable to *start* with a display containing all the information though albeit it may be dicy to uncover. The visual cues available and how they guide the exploration are crucial deciding factors.

7. The value of rigor is self-evident.

These and additional issues comprising the discovery process are better appreciated via the exploration of real datasets. The basic queries are introduced at first with an example of satellite and then a much larger astronomical dataset. Subsequently, they are combined with boolean operators to form complex queries applied to financial data. An example with several hundred variables is discussed briefly before moving to automatic classification. Visualization and ∥-coords play key roles in the geometric algorithm's conception, internal function and visual presentation of the classification rule. The minimal set of the variables needed to state the rule is found and ordered according to their predictive value.

Mathematical background is interspersed to provide a deeper understanding and wiser use of ∥-coords and its applications. Specifically,

1. learning the *patterns* correspoding to the basic relations and seek them out for EDA,

2. understanding the design and use of the *querries*,

3. motivating further sophisticated applications to Statistics like *Response Surfaces* [8], and

4. applications to *Regression* – Section on Proximate Planes,

5. understanding that the relational information resides in the *crossings*,

6. **concentrating the relational information in the data in clear patterns eliminating the polygonal lines altogether** as with the "proximate planes" is feasible. encouraging research on efficient (parallel) algorithms for accomplishing this on general datasets. Hence eliminating the "clutter" by reducing the display into patterns corresponding, at least approximately, to the multivariate existing in the data.

Before entering the nitty gritties we pose a visualization challenge which we ask the reader to ponder. For a plane

$$\pi : c_1 x_1 + c_2 x_2 + c_3 x_3 = c_0 , \tag{1}$$

allow the coefficients to vary each within a small interval. This generates a family of "close" let's call them *proximate*) planes:

$$\Pi = \{\pi : c_1 x_1 + c_2 x_2 + c_3 x_3 = c_0, \quad c_i \in [c_i^-, c_i^+], \quad i = 0, 1, 2, 3\}. \tag{2}$$

These are the planes generated by small rotations and translations of π with respect to the 3 coordinates axes. Altogether they form a "twisted slab" which even in 3-D with *orthogonal axes* is difficult to visualize. Conversely given lots of points in 3-D how can it be discovered, using **any** general visual method you like, that they lie on a twisted slab and how such a creature can be visualized and described precisely; for $N = 3$ and then for *any N*? In the meantime, you can project, pursue, scale, reduce, scintillate, regress, corollate or tribulate to your heart's content but please do not peek at the answer given at the end.

2   Exploratory Data Analysis with ‖-coords

2.1   Multidimensional Detective

Parallel coordinates transform multivariate relations into 2-D patterns suitable for exploration and analysis. For this reason they are included in lots of software tools. The queries "'parallel coordinates + Software" on Google returned about 31,000 "hits" and "Scatterplot matrix + Software" about 15,000. Irrespective of the apparent *2:1* relative ratio, the comparable numbers for the two astounded me having heard the appellations: "esoteric", "unnatural", "difficult", "squiggly" and more for ‖-coords after their introduction.

The exploration[1] paradigm is that of a *detective*, starting from the data, searching for clues leading to conjectures, testing, backtracking until *voila* ... the "culprit" is discovered. The task is especially intricate when many variables (i.e. dimensions) are involved calling for the employment of a *multidimensional* detective (abbr. *MD*). As if there were any doubts, our display of choice is ‖-coords where the data appears in the by now familiar squiggly blotches and which, by means of *queries*, the *MD* skilfully dissects to find precious hidden secrets.

During the ensuing interaction think, dear reader, how similar queries can be done using other exploration methodologies including the ubiquitous spreadsheets. More important, what visual clues are available that would **prompt** use of such queries. This is a good place to point out a few basics. In ‖-coords due to the *point* ↔ *line* and other dualities, some but *not* all actions are best performed in the dual. The queries, which are the "cutting tools", operate on the display

---

[1] The venerable name "Exploratory Data Analysis" *EDA* is used interchangeably with the currently more fashionable "Visual Data Mining".

i.e. *dual*. Their design should exploit the methodology's strengths and avoid its weaknesses; rather than mimic the action of queries operating on standard "non-dual" displays. As a surgeon's many specialized cutting tools, one of our early software versions had lots of specialized queries. Not only was it hard to classify and remember them but they still could not handle all situations encountered. After experimentation, I opted for a few(3) intuitive queries called **atomic** which can be combined via *boolean* operations to form complex intricate cuts. Even for relatively small datasets the ‖-coords display can look uninformative and intimidating. Lack of understanding the basics of the underlying geometry and poor choice of queries limits the use of ‖-coords to unrealistically small datasets. Summarizing, the requirements for successful exploratory data analysis are:

– an informative display *without loss of information* of the data,
– good choice of queries, and
– skillful interaction with the display.

2.2   An Easy Case Study – Satellite Data

The first admonition is

– **do not let the picture intimidate you,**

as can easily happen by taking an uninformed look at Fig. 2 showing the dataset to be explored. It consists of over 9,000 measurements with 9 variables, the first two (X, Y) specify the location on the map in Fig. 1 (left), a portion of Slovenia, where 7 types of ground emissions are measured by satellite. The ground location, (X, Y), of one data item is shown in Fig. 1 (right), which corresponds to the map's

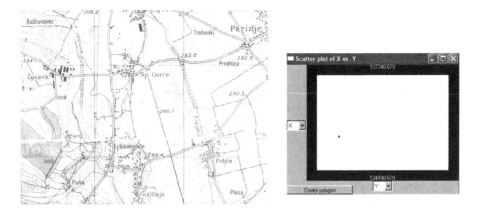

Fig. 1. Seven types of ground emissions were measured on this region of Slovenia. Measurements recorded by the LandSat Thematic Mapper are shown in subsequent figures. Thanks to Dr. Ana Tretjak and Dr. Niko Schlamberger, statistics office of Slovenia for providing the data. (Right) the display is the map's rectangular region, the dot marks the position where the 7-tuple shown in the next figure was measured.

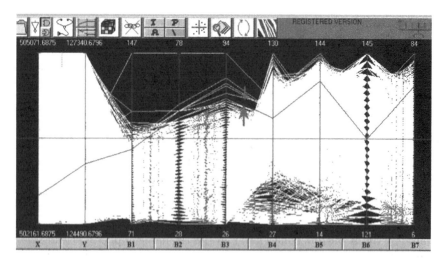

Fig. 2. Query on Parallax showing a single data item. The X, Y (position, also shown on the right of Fig. 1), and values of the 7-tuple $(B1, B2, B3, B4, B5, B6, B7)$ at that point.

region and remains open during the exploration. The query, shown in Fig. 2, used to select the data item is called *Pinch*. It is activated by the button **P** on the tool bar. By means of this query, a bunch of polygonal lines (i.e. data items) can be chosen by being "pinched" *in-between* the axes. The cursor's movement changes the position of the *selected* arrow-head which is the larger of the two shown. In due course various parts of the *GUI* are illustrated(*Parallax*[2]).

Aside from the starting the exploration without biases it is essential

– **to understand the objectives.**

Here the task is the detection and location of various ground features (i.e. built-up areas, vegetation, water etc.) on the map. There is a prominent lake, on the lower-left corner with an unusual shape the upward pointing "finger". This brings up the next admonition, that no matter how messy it looks

– **carefully scrutinize the data display for clues and patterns.**

Follow up on anything that catches the eyes, gaps, regularities, holes, twists, peaks & valleys, density contrasts like the one at the lower values of $B3$ through $B7$. Using the *Interval* query, activated by the **I** button, starting at the minimum we grab the low range of $B4$ (between the arrowheads) stopping at the dense part as shown in Fig. 3. The result, on the left of Fig. 4, is amazing. Voila we found the water, the lake is clearly visible together with two other regions which in the map turn up to be small streams. Our scrutiny having been rewarded we recall the adage

---

[2] MDG's Ltd proprietary software–All Rights Reserved, is used by permission.

– a good thing may be worth repeating.

Examining for density variations now *within the selected lower interval of B4* we notice another. The lowest part is much denser. Experimenting a bit, appreciating the importance of interactivity, we select the sparse portion, Fig. 5, which defines the water's edge (right) 4 and in fact more. By dropping the lower arrow we see the lake filling up starting from the edge i.e. shallow water first. So the lower values of $B4$ reveal the water and the lowest "measure" the water's depth; not bad for few minutes of playing around. But all this pertains to a single variable when we are supposed to be demonstrating *multivariate* exploration. This is a valid point but we did *pick B4* among several variables. Further, this is a nice "warm-up" for the subsequent more involved examples enabling us to show two of the queries. The astute observer must have already noticed the regularity, the vertical bands, between the $B1, B2$ and $B3$ axes. This is where the *angle* query, activated by the **A** button, comes into play. As the name implies it selects groups of lines within a user-specified angle range. A data subset is selected between the $B2, B3$ axes as shown, with enlarged inter-axes distance better showing the vertical bands, in Fig. 6 (left) to select a data subset which corresponding on the map to regions with high vegetation. Clicking the **A** button and placing the cursor on the middle of one axis opens an angle, with vertex on the mid-range of the previous (left) axis, whose range is controlled by the arrow movements on the right axis. Actually this "rule" (i.e. relation among some parameters) for finding vegetation can be refined by twicking a couple of more parameters. This raises the topic of rule finding in general, *Classification*, which is taken up in Sect. 3.

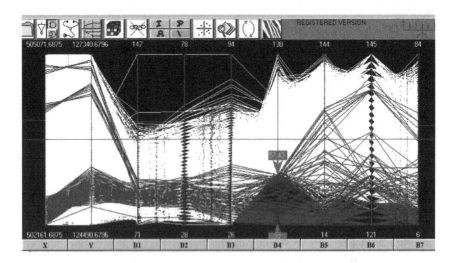

Fig. 3. Finding water regions. The contrast due to density differences around the lower values of $B4$ is the *visual cue* prompts this query.

Fig. 4. (Left) the lake – result of query shown in Fig. 3. On the right is just the lake's edge. It is the result of query shown in Fig. 5.

The *angle* and *pinch* queries are motivated by the ℓ *line* → *point* $\bar{\ell}$ duality

$$\ell : x_2 = mx_1 + b \leftrightarrow \bar{\ell} = \left(\frac{d}{1-m}, \frac{b}{1-m}\right) \tag{3}$$

in ‖-coords illustrated in Fig. 7 where the inter-axes distance is d. As seen from its x-coordinate, the point $\bar{\ell}$ lies between the parallel axes when the line's slope $m < 0$, to the right of the $\bar{X}_2$ axis for $0 < m < 1$ and left of $\bar{X}_1$ for $m > 1$. Lines with $m = 1$ are mapped to the *direction* with slope b/d in the on the xy-plane; with d the inter-axes distance and b the constant (intercept) in the equation of ℓ. This points out that dualities properly reside in the *Projective*, the *directions* being the *ideal points*, rather than the Euclidean plane. For sets of points having a "general" direction with negative slope, i.e. are "negatively

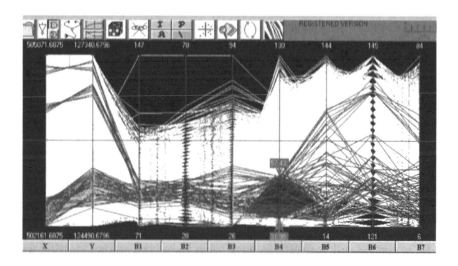

Fig. 5. Query finding the water's edge.

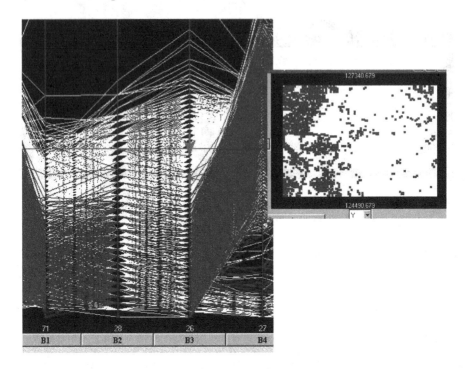

Fig. 6. Finding regions with vegetation. Here the *angle query* is used (left) between $B2, B3$ axes. Note the arrow-heads on the $B3$ axis which specify the angle-range for the selected lines.

correlated", the lines representing them in ∥ cross each other roughly in between the axes and they can be *selected with the pinch query*. For positively correlated sets of points their corresponding lines cross outside the axes and can be *selected with the angle query*. All this exemplifies the need to understand some of the basic geometry so as to work effectively with the queries and of course designing them properly. The three atomic queries having been introduced there remains to learn how they can be combined to construct complex queries.

Prior to that, Fig. 6 (left) begs the question: "what if the $B2$ and $B3$ axes were *not* adjacent"? Then the pattern and hence their pairwise relation would be missed. Hence the axes-permutation used for the exploration is important. In particular what is the minimum number of permutations among N-axes containing the *adjacencies* for all pairs of axes? It turns out [10]: M permutations are needed for even $N = 2M$ and $M + 1$ for odd $N = 2M + 1$. It is fun to see why. Label the N vertices of a graph with the index of the variables X_i, $i = 1, \ldots, N$ as shown in Fig. 8 for $N = 6$. An edge joining vertex **i** with **j** signifies that the axes indexed by **i** , **j** are adjacent. The graph on the left is a *Hamilton path* for it contains all the vertices. Such paths have been studied starting with Euler in the 18th century with modern applications to the "travelling salesman" problem and elsewhere ([9] pp. 66, [3] pp. 12). The graph corresponds to the

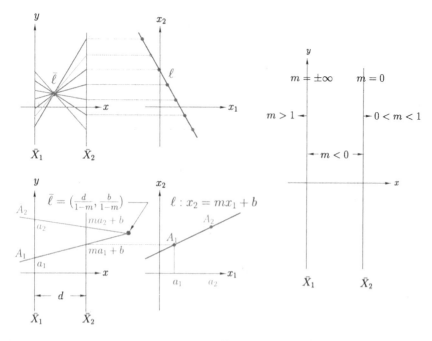

Fig. 7. Parallel coordinates induce a *point $\bar{\ell} \leftrightarrow \ell$ line* duality (left). (Right) the horizontal position of the point $\bar{\ell}$ representing the line ℓ is determined only by the line's slope m. The vertical line $\ell : x_1 = a_1$ is represented by the point $\bar{\ell}$ at the value a_1 on the $\bar{X}_1$ axis.

axes index permutation **126354**. On the right, the union with the additional two Hamiltonian paths, starting at vertices **2** and **3**, forms the complete graph which contains all possible edges. Hence the 3 permutations **126354** , **231465**, **342516** contain all possible adjacent pairs; just try it. The remaining permutations are obtained from the first by successively adding *1 mod 6* , and this

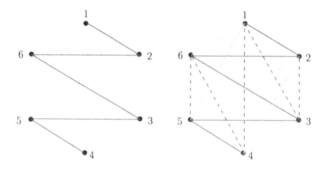

Fig. 8. (Left) first Hamiltonian path on vertices **1**, . . . , **6**. It corresponds to the (axes) index permutation **126354**. (Right) the complete graph as the union of the 3 distinct Hamiltonian paths starting successively at the vertices **1, 2, 3**.

works for general N [16]. As of this writing the authoritative reference on axes permutations is by C. Hurley and W. Olford [10]. Before leaving this interesting subject we pose the *triad permutation* problem. Namely, what is the minimum number of permuations needed to obtain all possible adjacent *triples* of axes?

Returning to EDA, the icon with the *Rubik's Cube* on *Parallax*'s toolbar activates a *permutation editor* which automatically generates the Hamiltonian permutations (abbr. *HP*). After scrutinizing the dataset display the recommended next step is to run through the $O(N/2)$ *HP*. This is how all nice adjacencies such as the one in Fig. 6 are discovered. Then using the editor, patch your own custom-made permutation containing all the parts you like in the *HP*. With this preprocessing cost, referred to earlier in list item 2 of the introduction, the user sets her own best permutation to work with. Of course, there is nothing to prevent one from including axes several times in different positions and experimenting with different permuations in the course of the exploration.

2.3   Compound Queries – Financial Data

To be explored next is the financial dataset shown in Fig. 9, the goal being to discover relations useful for investments and trading. The datat for the years 1986 (second tick on the 3rd axes) and 1992 are selected and compared. In 1986 the **Yen** had the greater volatility among the 3 currencies, interests varied in the mid-range, gold had a price gap while **SP**500 was uniformly low. By comparison in 1992, the **Y**en was stable while the **S**terling was very volatile (possibly due to Soros' speculation), interests and gold price were low and the **SP**500 was uniformly high. Two **I**nterval queries are combined with the *OR* boolean operator (i.e. Union) to obtain this picture. We continue

– **"looking for the gold" by checking out patterns that caught our attention.**

The data for 1986 is isolated in Fig. 10 and the lower range in the gold price gap is selected. Gold prices were low until the 2nd week in August when they jumped and stayed higher. The exploration was carried out in the presence of four financial experts who carefully recorded the relation between low **Yen**, high **3MTB** rates and low **G**old prices By the way, *low* **Yen** rate of exchange means the Yen has high value relative to the US $.

There are two bunches of crossing lines between 6th and 7th axes in Fig. 9 which together comprise more than 80 % of the dataset. This and recalling the previous discussion on the *line* ← *point* mapping in Fig. 7 points out the strong negative correlation between **Yen** and **3MTB** rates. The smaller cluster in Fig. 11 is selected. Moving from the top range of any of the two axes, with the **I** query, and lowering the range causes the other variable's range to rise and is a nice way to show negative correlation interactively. For the contrarians among us, we check also for positive correlation Fig. 12. We find that it exists when **G**old prices are low to mid-range as happened for a period in the 90's. This is a free investment tip for bucking the main trend shown in Fig. 11. It is also a nice opportunity for showing the *inversion* feature activated by the icon with 2 cyclical arrows. A variable is selected and the min/max values on that axes are inverted. Diverging lines (as for + correlation) now intersect Fig. 13 making it easier visually to spot the crossing and hence the correlation. Actually, the recommendation is to work with the **A** query experimenting with various angle ranges using the inversion to check out or confirm special clusters. When stuck don't just stand there but

– **vary one of the variables watching for interesting variations in the other variables.**

Doing this on the **Yen** axis, Fig. 14, we strike another gold connection. The (rough) intersection of a bunch of lines joining **Yen** to the **D**mark corresponds, by the duality, to their rate of exchange. When the rate of exchange changes so does the intersection **and the price of Gold!** That is movements in currency exchange rates and the price range of **G**old go together. Are there any indications that are associated with the high range of **G**old? The top price range is selected, Fig. 15, and prompted by the result of the previous query we check out the exchange rate between **S**terling and **D**mark (or **Yen**) and the resul is stunning: a perfect straight line. The slope *is* the rate of exchange which is constant when **G**old tops out. The relation between **S**terling and **D**mark is checked for different

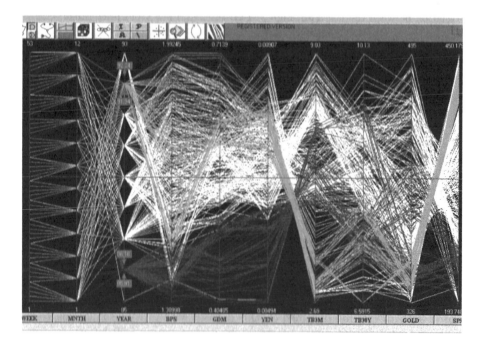

Fig. 9. Financial data. Quotes by **W**eek-on Monday, Month, **Y**ear first 3 axes fix the date; Sterling, Dmark, **Yen** rates per $ 4th, 5th, 6th axes; **3M**TB, **30Y**TB interest rates in %, 7th, 8th axes; Gold in $/ounce, 9th, SP500 index values on 10th axes.

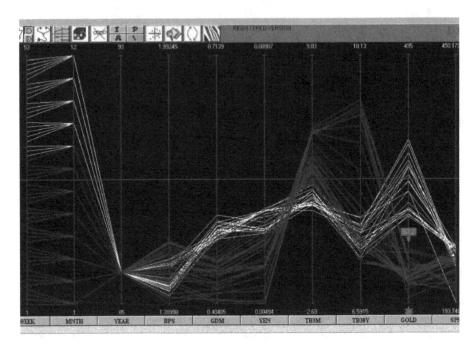

Fig. 10. Gold prices In 1986. Gold prices jumped in the 2nd week of August. Note the correlation between the low **Yen**, high **3MTB** rates and low **Gold** price range.

price ranges of **Gold**, Fig. 16, and the only regularity found is the one straight-line above. Aside from the trading guideline it establishes, it suggests "behind-the-scenes manipulation of the **Gold** market" ... we could have said that but we won't. We perish this thought and proceed with the boolean complement, Fig. 17 of an **I** (or any other) query. Not finding anything we select a narrow but dense range on the **Yen**, Fig. 18 and notice an interesting relation between **D**mark, interest rates and **Gold**.

There is an exploratory step akin to "multidimensional contouring" which we fondly call **Zebra** activated by the last icon button on the right with the appropriate skin-color. A variable axis is selected, the **SP**500 axis in Fig. 19, and divided into a number (user specified) intervals (here it is 4) and colored differently. This shows the connections (influence) of the intervals with the remaining variables which here is richly structured especially for the highest range. So what does it take for the **SP**500 to rise? This is a good question and helps introduce Parallax's classifier. The result, shown in Fig. 20 confirms the investment community's experience that low **3MTB** and **Gold** predict high **SP**500. A comparison with the results obtained on this dataset with other visualization tools would be instructive though unfortunately not available. Still let us consider such an analysis done by the scatterplot matrix. There are 10 variables (axes) which requires 45 pairwise scatterplots each, even with a large monitor screen being no larger than about $2.5 \times 2.5\,cm^2$ square. Varying 1, 2 or more variables

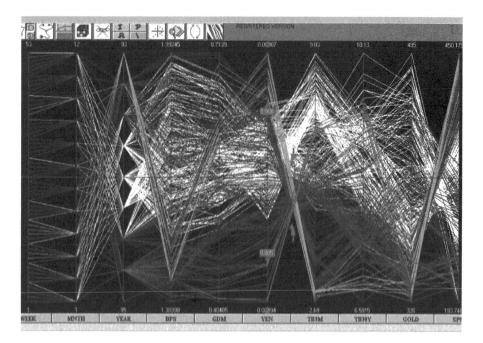

Fig. 11. Negative correlation. The crossing lines between the 6th and 7th axes in Fig. 9 show strong negative correlation between **Yen** and **3M**TB rates. One cluster is selected with the **Pinch** query and combined with the high and low ranges on the **Yen** axis. Data for the years 1986 and 1992 are selected.

in tandem and observing the effects *simultaneously* over **all** the variables in the 45 squares may be possible but quite challenging. By contrast, the effects of varying **Dmark**, *conditionally* for stable **Yen**, are easily seen on the two interest rates, **Gold** as well as the remaining variables in *one* Fig. 18. This example illustrates the difficulties due to high *Representational Complexity* (see Sect. 1.2 item # 2) which is $O(N^2)$ for the scatterplot matrix but $O(N)$ for ‖-coords and made even clearer with the next dataset.

2.4 Hundreds of Variables

An important question frequently asked is "how many variables can be handled with ‖-coords?" The largest dataset that I have effectively worked with had about 800 variables and 10,000 data entries. With various techniques developed over the years and the automatic classifier discussed in the next section much larger datasets can be handled. Still the relevant admonition is

– **be sceptical about the quality of datasets with large number of variables.**

When hundreds or more variables are involved, it is unlikely that there are many people around who have a good feel for what is happening as confirmed by my experience. A case in point is the dataset shown in Fig. 21 consisting of instrumentation measurements for a complex process. The immediate observation is that lots of instruments recorded 0 for the duration something which was unnoticed. Another curiosity was the repetitive patterns on the right. It turns that several variables were measured in more than one location using different names. When the dataset was cleaned-up of the superfluous information it was reduced to about 90 variables as shown in Fig. 22 and eventually to about 30 which contained the information of real interest. By my tracking the phenomenon of repetitive measurements is widespread with at least 10 % of the variables, occurring in large datasets, being duplicates or near duplicates, possibly due to instrument non-uniformities, as suggested in the 2 variable scatterplot2 in Fig. 22. Here the repetitive observations were easily detected due to the fortuitous variable permutation in the display. Since repetitive measurements occur frequently it may be worth adding to the software an automated feature to detect and exhibit the suspect variables. This brief exposure is just an indication that large (in dimension – i.e. in number of variables) datasets can still be gainfully explored in ‖-coords.

There follows a different example of EDA on a process control dataset s [11] where compound queries turned out to be very useful and where we learn to add, to the list of exploration guidelines, arguably the most important one:

– **test the assumptions and especially the "I am really sure of"s.**

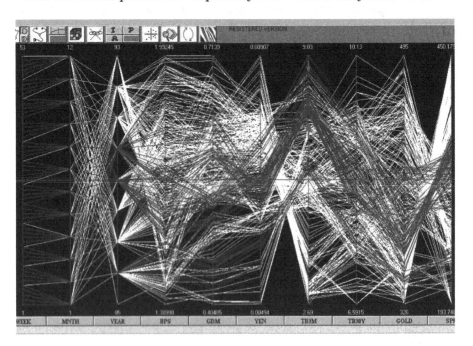

Fig. 12. Positive correlation. A positively correlated where the **Yen** and **3M**TB rates move *together when* **G**old *prices are low to mid-range.*

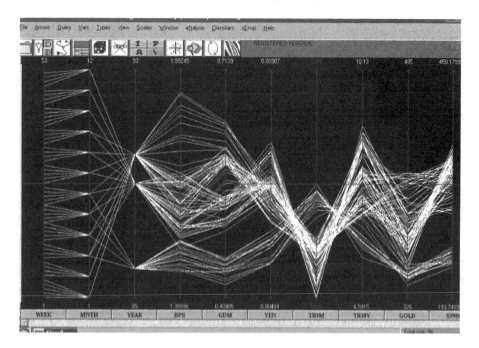

Fig. 13. Inverting the **3M**TB axis. Now the lines between the **Yen-3M**TB and **3M**TB-**30M**TB axes in Fig. 12 cross.

2.5   Production of VLSI (Chips)

The dataset, displayed in Fig. 23, consists of production data of several batches of a specific VLSI (computer chip) with measurements of 16 parameters involved in the process. The parameters are denoted by $X1$, $X2$, ... , $X16$. The *yield*, as the % (percent) of useful chips produced in the batch, is denoted by $X1$, and $X2$ is a measure of the *quality* (given in terms of speed performance) of the batch. Ten different categories of *defects* are monitored and the variables' scales of $X3$ through $X12$ are inverted so that 0 (zero) amount appears at the top and increasing amounts appear proportionately lower. The remaining $X13$ through $X16$ denote some physical parameters.

Since the goal here is to raise the yield, $X1$, while maintaining high quality, $X2$, we have a case of multi-objective optimization due to the presence of more than one objective. The production specialists believed that it was the presence of defects which prevented high yields and qualities. So their purpose in life was to keep on pushing – at considerable cost and effort – for zero defects.

With this in mind the result of our first query is shown in Fig. 24 where the batches having the highest $X1$ and $X2$ have been isolated. This in an attempt to obtain clues; and two real good ones came forth. Notice the resulting range of $X15$ where there is a significant separation into two clusters. As it turns out, this gap yielded important insight into the physics of the problem. The other clue is almost hidden. A careful comparison – and here interactivity of the software is

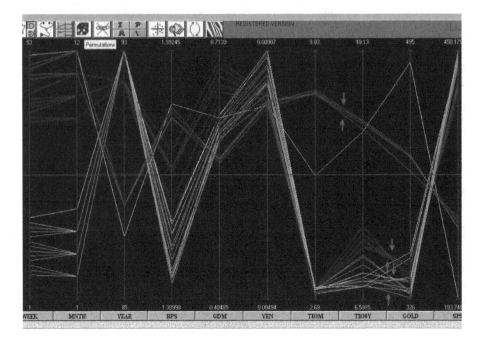

Fig. 14. Variations in currency exchange rates. Variations in the rate of exchange of the currencies correlate with movements in the price of **G**old.

essential – between Figs. 23 and 24 shows that some batches which were high in $X3$ (i.e. due to the inverted scale low in that defect) were not included in the selected subset. That casts some doubt into the belief that zero defects are the panacea and motivates the next query where we search for batches having zero defects in at least 9 (excluding $X3$ where we saw that there are problems) out of the 10 categories. The result is shown in Fig. 25 and is a shocker. There are 9 such batches and all of them have poor yields and for the most part also low quality! That this was not questioned and discovered earlier is surprising. We scrutinize the original picture Fig. 23 for visual cues relevant to our objectives and our findings so far. And ... there is one staring us in the face! Among the 10 defects $X3$ through $X12$ whatever $X6$ is, it's graph is very different than the others. It shows that the process is much more sensitive to variations in $X6$ than the others. For this reason, we chose to treat $X6$ differently and remove its zero defect constraint. This query (not shown) showed that the very best batch (i.e. highest yield with very high quality) does not have zeros (or the lowest values) for $X3$ and $X6$; a most heretical finding. It was confirmed by the next query which isolated the cluster of batches with the top yields (note the gap in $X1$ between them and the remaining batches). These are shown in Fig. 26 and they confirm that small amounts (the ranges can be clearly delimited) of $X3$ and $X6$ type defects are essential for high yields and quality.

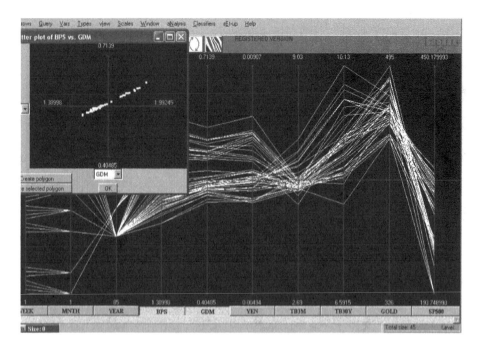

Fig. 15. High **Gold**. Note the perfect straight line in the Sterling vs. **D**mark plot. The slope is the *rate of exchange* between them and which remains constant when **Gold** prices peak.

Returning to the subset of data which best satisfied the objectives, Fig. 24 in order to explore the gap in the range of $X15$, we found that the cluster with the high range of $X15$ gives the lowest (of the high) yields $X1$, and worse it does not give consistently high quality $X2$, whereas the cluster corresponding to the lower range has the higher qualities and the full range of the high yield. It is evident that the small ranges of $X3$, $X6$ close to (but not equal to) zero, together with the short (lower) range of $X15$ provide necessary conditions for obtaining high yields and quality. This is also indicated in Fig. 26. By a stroke of good luck these 3 can also be checked early in the process avoiding the need of "throwing good money after bad"(i.e. by continuing the production of a batch whose values of $X3$, $X6$ and $X15$ are not in the small "good" ranges we have found).

These findings were significant and differed from those found with other methods for statistical process control [1]. This approach has been successfully used in a wide variety of applications from the manufacture of printed circuit boards, PVC and manganese production, financial data, determining skill profiles (i.e. as in drivers, pilots), etc.

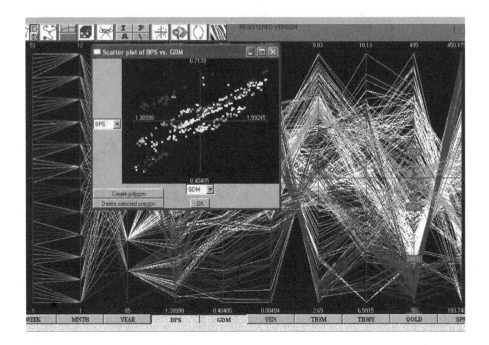

Fig. 16. Two price ranges of **G**old. The associated **S**terling vs. **D**mark plots show no regularity.

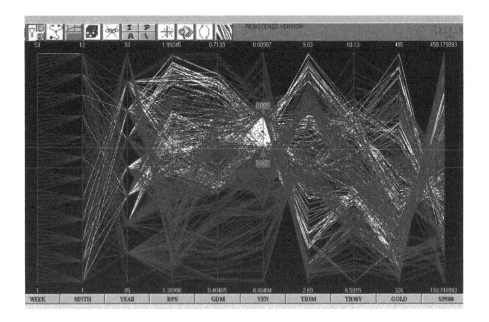

Fig. 17. The complement of an **I** query.

3   Classification

Though it is fun to undertake this kind of exploration, the level of skill and patience required tends to discourage some users. It is not surprising then that the most persistent requests and admonitions have been for tools which, at least partially, automate the knowledge discovery process [13]. Classification is a basic task in data analysis and pattern recognition and an algorithm accomplishing it is called a **Classifier** [5, 18, 19]. The input is a dataset P and a designated subset S. The output is a characterization, a set of conditions or rules, to distinguish elements of S from all other members of P the "global" dataset. The output may also be that there is insufficient information to provide the desired distinction.

With parallel coordinates a dataset P with N variables is transformed into a set of points in N-dimensional space. In this setting, the designated subset S can be described by means of a hypersurface which encloses just the points of S. In practical situations the strict enclosure requirement is dropped and some points of S may be omitted ("false negatives"), and some points of $P - S$ are allowed ("false positives") in the hypersurface. The description of such a hypersurface is equivalent to the rule for identifying, within some acceptable error, the elements of S. Casting the problem in a geometrical setting leads us to *visualize* how such may work. This entails:

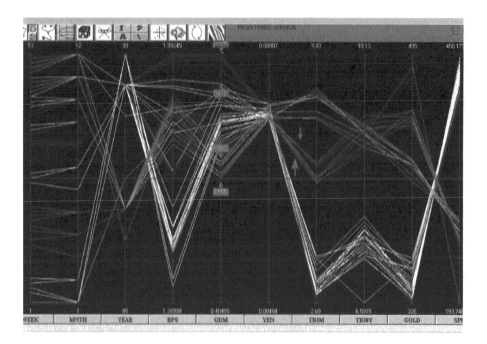

Fig. 18. Yen stable. For the **Yen** trading in a narrow range, high **Dmark** goes with low **3MTB** rates, low **Dmark** goes with high **3MTB** rates, while mid **3MTB** rates go with high **Gold**.

1. use of an efficient "wrapping" (a convex-hull approximation) algorithm to enclose the points of S in a hypersurface S_1 containing S and in general also some points of $P - S$; so $S \subset S_1$.
2. the points in $(P - S) \cap S_1$ are isolated and the wrapping algorithm is applied to enclose them, usually also enclosing some points of S_1, producing a new hypersurface S_2 with $S \supset (S_1 - S_2)$,
3. the points in S not included in $S_1 - S_2$ are next marked for input to the wrapping algorithm, a new hypersurface S_3 is produced containing these points as well as some other points in $P - (S_1 - S_2)$ resulting in $S \subset (S_1 - S_2) \cup S_3$,
4. the process is repeated alternatively producing upper and lower containment bounds for S; termination occurs when an error criterion (which can be user specified) is satisfied or when convergence is not achieved. After termination is obtained two error measures are available to estimate the rule's precision:
   - *Train & Test.* A portion of the dataset (usually 2/3) selected at random is used to derive the classification rule, which is then tested on the remaining 1/3 of the data.
   - Cross-Correlation.

It can and does happen that the process does not converge when P does not contain sufficient information to characterize S. It may also happen that S is so

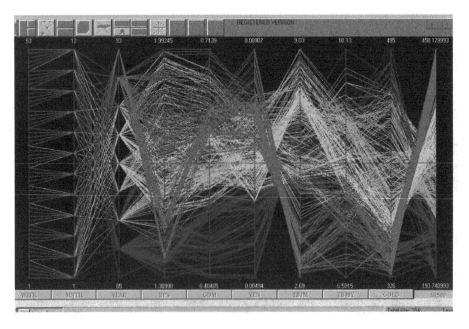

Fig. 19. The *zebra* query. It partitions and colors the segments of values differently. A variable, here the **SP**500 axis, is divided it into equal (here 4) intervals. This quickly reveals interelationships. Note especially those for the highest **SP**500 range and see next figure.

"porous" (i.e. sponge-like) that an inordinate number of iterations are required. On convergence, say at step $2n$, the description of S is provided as:

$$S \approx (S_1 - S_2) \cup (S_3 - S_4) \cup ... \cup (S_{2n-1} - S_{2n}) \qquad (4)$$

this being the terminating expression resulting from the algorithm which we call **Nested Cavities** (abbr. **NC**).

The user can select a subset of the available variables and restrict the rule generation to these variables. In certain applications, as in process control, not all variables can be controlled and hence it is useful to have a rule involving only the accessible (i.e. controllable) variables. An important additional benefit, is that the minimal set of variables needed to state the rule is found and ordered according to their predictive value. These variables may be considered as the best *features* to identify the selected subset. The algorithm is display independent there is no inherent limitation as to the size and number of variables in the dataset. Summarizing for **NC**,

– an approximate convex-hull boundary for each cavity is obtained,
– utilizing properties of the representation of multidimensional objects in $\|$-coords, a very low polynomial worst case complexity of $O(N^2|P|^2)$ in the number of variables N and dataset size $|P|$ is obtained; it is worth contrasting this with the often unknown, or unstated, or very high (even exponential) complexity of other classifiers,

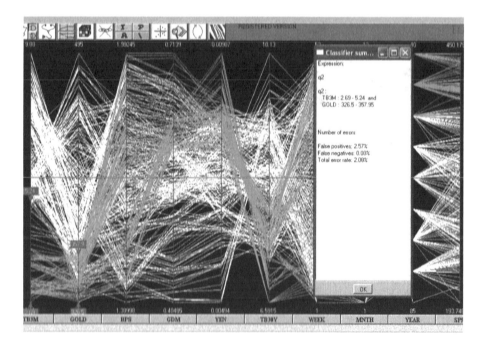

Fig. 20. The rule for high SP500. Both **3MTB** (the "short-bond" as it is called) and Gold are low and in this order of importance.

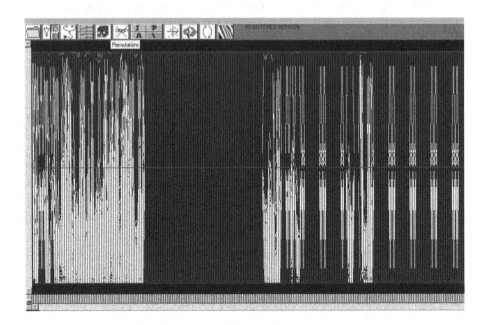

Fig. 21. Manufacturing process measurements – 400 variables.

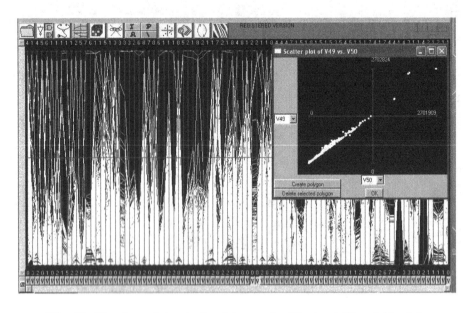

Fig. 22. The above dataset after "clean-up" with about 90 variables left.

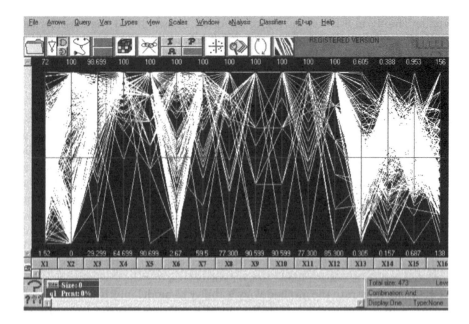

Fig. 23. Dataset – VLSI production with 16 parameters

- an intriguing prospect, due to the low complexity, is that the rule can be
  derived in near real-time making the classifier adaptive to changing conditions,
- the minimal subset of variables needed for classification is found,
- the rule is given explicitly in terms of conditions on these variables, i.e.
  included and excluded intervals, and provides "a picture" showing the complex distributions with regions where there are data and "holes" with no data
  providing important insights to the domain experts.

A *neural-pulse dataset* has interesting and unusual features. There are two
classes of neurons whose outputs to stimuli are to be distinguished. They consist
of 32 different pulses measured in a monkey's brain (poor thing!). There are
600 samples with 32 variables (the pulses)[3]. Various classification methods were
unable to obtain a rule. With **NC** convergence is obtained requiring only 9 of
the 32 parameters for the classification rule for class # 1. The resulting ordering
shows a striking separation. In Fig. 28 the first pair of variables x_1, x_2 in the
original order is plotted on the left. On the right the best pair x_{11}, x_{14}, as chosen
by the classifier's ordering speaks for itself. By the way, the discovery of this
manually would require constructing a scatterplot matrix with 496 pairs, then
carefully inspecting and comparing the individual plots. The implementation
provides all the next best sections, some of which are shown in Fig. 29, to aid
the visualization of the rule. The dataset consists of two "pretzel-like" clusters

---

[3] I am grateful to Prof. R. Coiffman and his group at the CS & Math. Depts at Yale
University for giving me this dataset.

wrapping closely in 8-D one enclosing the other; showing that the classifier can actually "carve" highly complex regions with the cavity shown. One can understand why separation of clusters by hyperplanes or nearest-neighbor techniques can fail badly for such datasets. The rule has 4 % error some of which are shown in Fig. 29.

The rules are explicit, "visualizable", optimally ordering the minimal set of variables needed to state the rule without loss of information. There are variations which apply in some situations where the **NC** classifier fails, such as the presence of several large "holes" (see [13]). Further, keeping in mind that the classification rule is the result of several iterations suggests heuristics for dealing with the pesky problem of *over-fitting*. The iterations can be stopped just where the corrections in Eq. (4) become very small, i.e. the S_i consist of a small number of points. The number of iterations is user defined and the resulting rule yields an error in the test stage more stable under variations in the number of points of the test set. In addition, the user can exclude variables from being used in the description of the rule; those ordered last are the ones providing the smaller corrections and hence more liable to over-correct.

Finally we illustrate the methodology's ability to model multivariate relations in terms of hypersurfaces – just as we model a relation between two variables by a planar region. Then by using the interior point algorithm, as shown in Fig. 37 of the next section, with the model we can do trade-off analyses, discover sensitivities, understand the impact of constraints, and in some cases do optimization. For this purpose we shall use a dataset consisting of the outputs of

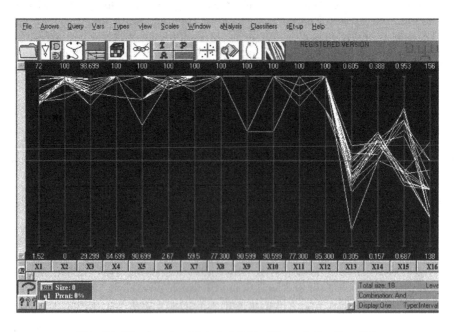

Fig. 24. The batches high in Yield, $X1$, and Quality, $X2$.

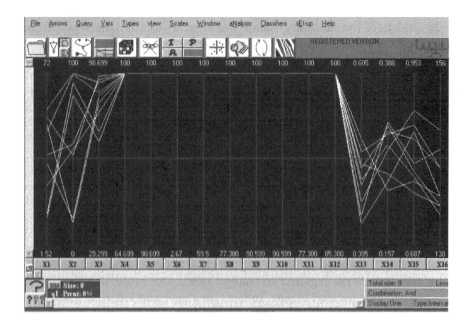

Fig. 25. The batches with zero in 9 out of ten defect types.

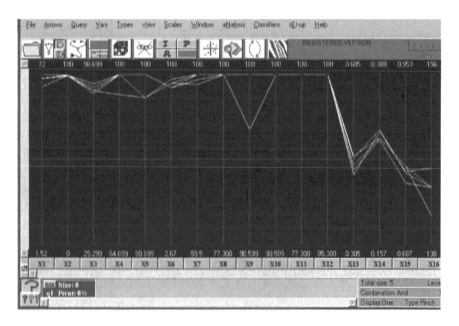

Fig. 26. The batches with the highest Yields. They do not have the lowest defects of type $X3$ and $X6$.

various economic sectors and other expenditures of a particular (and real) country. It consists of the monetary values over several years for the **Ag**ricultural, **Fish**ing, and **Min**ing sector outputs, **Man**ufacturing and **Con**struction industries, together with **Gov**ernment, Miscellaneous spending and resulting GNP; eight variables altogether. We will not take up the full ramifications of constructing a model from data. Rather, we want to illustrate how ∥-coords may be used as a modeling tool. Using the Least Squares technique we "fit" a function to this dataset and are not concerned at this stage whether the choice is "good" or not. The function obtained bounds a region in $\mathbb{R}^8$ and is represented by the upper and lower curves shown in Fig. 30.

The picture is in effect a simple visual model of the country's economy, incorporating its capabilities, limitations and interelationships among the sectors. A point interior to the region, satisfies all the constraints simultaneously, and therefore represents (i.e. the 8-tuple of values) a feasible economic policy for that country. Using the interior point algorithm we can construct such points. It can be done interactively by sequentially choosing values of the variables and we see the result of one such choice in Fig. 30. Once a value of the first variable is chosen (in this case the **Ag**ricultural output) within its range, the dimensionality of the region is reduced by one. In fact, the upper and lower curves between the 2nd and 3rd axes correspond to the resulting 7-dimensional hypersurface and show the available range of the second variable **Fish**ing reduced by the constraint. This can be seen (but not shown here) for the rest of the variables. That is, due to the relationship between the 8 variables, a constraint on one of them

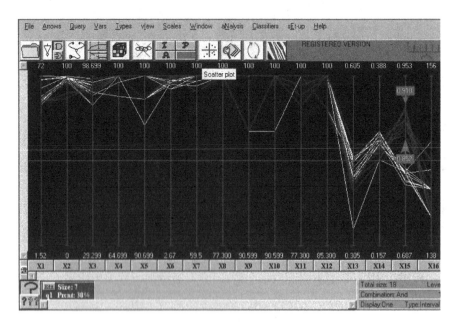

Fig. 27. Only the lower range of $X\,15$ is associated with the highest yields and quality.

impacts all the remaining ones and restricts their range. The display allows us to experiment and actually see the impact of such decisions downstream. By interactively varying the chosen value for the first variable we found, that it not possible to have a policy that favors **Agr**iculture without also favoring **Fish**ing and vice versa (Fig. 31).

Proceeding, a very high value from the available range of **Fish**ing is chosen and it corresponds to very low values of the **Min**ing sector. By contrast in Fig. 30 we see that a low value in **Fish**ing yields high values for the **Min**ing sector. This inverse correlation was examined and it was found that the country in question has a large number of migrating semi-skilled workers. When the fishing industry is doing well most of them are attracted to it leaving few available to work in the mines and vice versa. The comparison between the two figures shows the competition for the same resource between **Min**ing and **Fish**ing. It is especially instructive to discover this interactively. The construction of the interior point proceeds in the same way. In the next section in the discussion on surfaces this construction is shown for higher dimensional hypersurfaces.

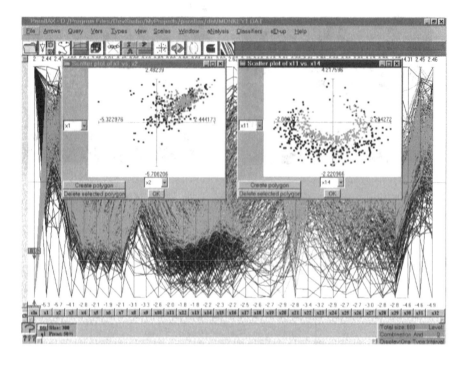

Fig. 28. The neural-pulses dataset with 32 parameters and two categories. Dataset is shown in the background. On the left plot are the first two parameters in the original order. The classifier found the 9 parameters needed to state the rule with 4 % error and ordered them according to their predictive value. The best two parameters are plotted on the right showing the separation achieved.

4   Parallel Coordinates – the Bare Essentials

The following short review of ∥-coords together Fig. 7 and the discussion on duality provide the essential background on ∥-coords to make this section self-contained. The detailed development of Parallel Coordinates is contained in [12].

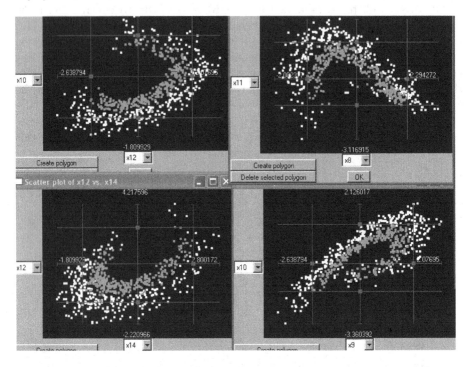

Fig. 29. Neural dataset classification. Further cross-sections of the hypersurface corresponding to the classification rule.

4.1   Lines

An N-dimensional line ℓ can be described by the $N - 1$ linear equations:

$$
\ell \; : \quad
\begin{cases}
\ell_{1,2} & : \quad x_2 = m_2 x_1 + b_2 \\
\ell_{2,3} & : \quad x_3 = m_3 x_2 + b_3 \\
\quad \cdots \\
\ell_{i-1,i} & : \quad x_i = m_i x_{i-1} + b_i \\
\quad \cdots \\
\ell_{N-1,N} & : \quad x_N = m_N x_{N-1} + b_N,
\end{cases}
\tag{5}
$$

each with a pair of adjacently indexed variables. In the $x_{i-1}x_i$-plane the relation labeled $\ell_{i-1,i}, N = 2, \ldots, N$ is a line, and by the *line* $\leftrightarrow$ *point* duality, Eq. (3), it can be represented by the point

$$\bar{\ell}_{i-1,i} = (\frac{1}{(1-m_i)} + (i-2), \frac{b_i}{(1-m_i)})$$ (6)

Here the inter-axes distance is 1 so that $i-2$ is distance between the y (or $\bar{X}_1$) and $\bar{X}_{i-1}$ axes. Actually any $N-1$ independent equations like

$$\ell_{i,j} : x_i = m_{i,j}x_j + b_{i,j},$$ (7)

can equivalently specify the line ℓ, for Eq. (7) is the projection of ℓ on the x_ix_j 2-D plane and $N-1$ such independent projections completely describe ℓ. There is a beautiful and very important relationship illustrated in (left) Fig. 32. For a line ℓ in 3-D the three points $\bar{\ell}_{12}, \bar{\ell}_{13}, \bar{\ell}_{23}$ are collinear, this line is denoted by $\bar{L}$, and any two points represent ℓ. It is easy to see that a polygonal line on all the $N-1$ points, given by Eq. (6) or their equivalent, represents a point on the line ℓ. Conversely, two points determine a line ℓ. Starting with the two polygonal lines representing the points, the $N-1$ intersections of their $\bar{X}_{i-1}, \bar{X}_i$ portions are the $\bar{\ell}_{i-1,i}$ points for the line ℓ. A line interval in 10-D and several of its points is seen on the (right) Fig. 32. By the way, the indexing of the points $\bar{\ell}$ is essential.

4.2   Planes & Hyperplanes

While a line can be determined from its projections, a plane even in 3-D can not. A new approach is called for [4]. Rather than discerning a p-dimensional object from its points, it is described in terms of its (p-1)-dimensional subsets constructed from the points. Let's see how this works. In Fig. 33 (left) polygonal

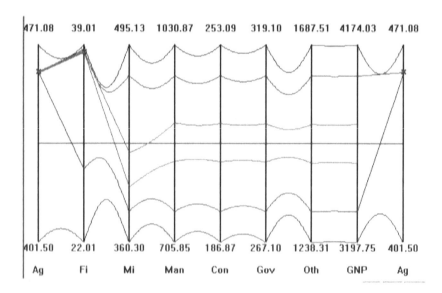

Fig. 30. Model of a country's economy. Choosing high **A**gricultural and high **F**ishing output **forces** low **M**ining output.

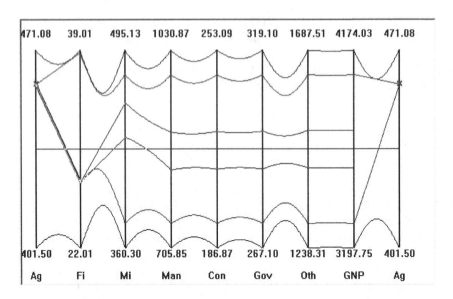

471.08	39.01	495.13	1030.87	253.09	319.10	1687.51	4174.03	471.08
401.50	22.01	360.30	705.85	186.87	267.10	1238.31	3197.75	401.50
Ag	Fi	Mi	Man	Con	Gov	Oth	GNP	Ag

Fig. 31. Competition for labor between the **Fish**ing & **Mi**ning sectors

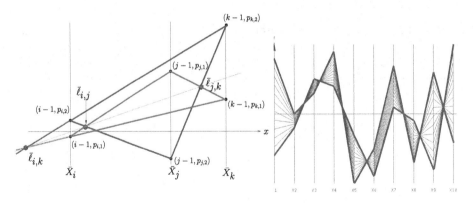

Fig. 32. Properties of multidimensional lines. (Left) the 3 points $\bar{\ell}_{i,j}, \bar{\ell}_{j,k}, \bar{\ell}_{i,k}$ are collinear for $i \neq j \neq k$. (Right) a line interval in 10-D.

lines representing a set of coplanar points in 3-D are seen. From this picture even the most persistent pattern-seeker can **not** detect any clues hinting at a relation among the three variables much less a linear one. The plane has dimension $p = 1$ so we look at *lines* (having dimension $p - 1 = 1$) on the plane constructed so that each pair of polygonal lines the lines $\bar{L}$ of the 3 point collinearity shown in Fig. 32 (left) are obtained. The result, shown on the right, is stunning. All the $\bar{L}$ lines intersect at a point which turns out to be characteristic of coplanarity but not enough to specify the plane. Translating the first axis $\bar{X}_1$ to the position $\bar{X}_{1'}$, one unit to the right of the $\bar{X}_3$ axis and repeating the construction, based on the axes triple $\bar{X}_2, \bar{X}_3, \bar{X}_{1'}$, yields a second point shown in Fig. 34 (left). For a plane described by:

$$\pi : c_1 x_1 + c_2 x_2 + c_3 x_3 = c_0, \tag{8}$$

the two points, in the order they are constructed, are respectively

$$\bar{\pi}_{123} = \left(\frac{c_2 + 2c_3}{S}, \ \frac{c_0}{S} \right), \ \bar{\pi}_{1'23} = \left(\frac{3c_1 + c_2 + 2c_3}{S}, \frac{c_0}{S} \right), \tag{9}$$

for $S = c_1 + c_2 + c_3$. Three subscripts correspond to the 3 variables appearing in the plane's equation and the axes triple used for their construction, and distinguish them from the points with two subscripts representing lines. The 2nd and 3rd axes can also be consecutively translated, as indicated in Fig. 33 (left), repeating the construction to generate two more points denoted by $\bar{\pi}_{1'2'3}, \bar{\pi}_{1'2'3'}$. These points can also be found otherwise in an easier way. The gist of all this is shown in Fig. 34 (right). The distance between successive points is $3c_i$. The equation of the plane π can actually be read from the picture!

In general, a hyperplane in N-dimensions is represented uniquely by $N-1$ points each with N indices. There is an algorithm which constructs these points *recursively*, raising the dimensionality by one at each step, as is done here starting from points (0-dimensional) constructing lines (1-dimensional). By the way, all the nice higher dimensional projective dualities like *point* ↔ *hyperplane, rotation* ↔ *translation* etc. hold. Further, a multidimensional object, represented in ‖-coords, can still be recognized after it has been acted on by projective transformation (i.e. translation, rotation, scaling and perspective). The recursive construction and its properties are at the heart of the ‖-coords visualization.

Challenge: Visualizing Families of Proximate Planes. Returning to 3-D, it turns out that for points as in Fig. 33 which are "nearly" coplanar (i.e. have small errors) the construction produces a pattern very similar to that in Fig. 34

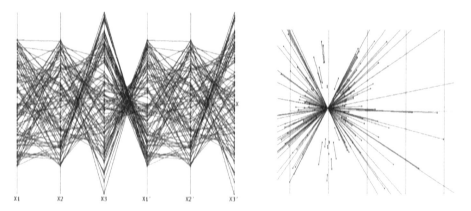

Fig. 33. Coplanarity. (Left)The polygonal lines on the first 3 axes represent a set of coplanar points in 3-D. (Right) Coplanarity! Forming lines on the plane, with the 3 point collinearity, the resulting lines intersect at point.

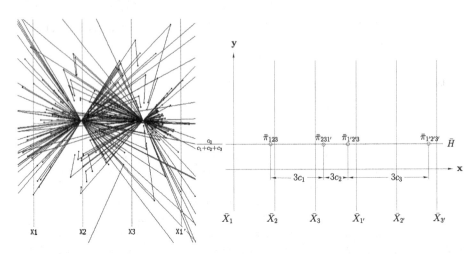

Fig. 34. Plane representation. (Left) the two points where the lines intersect uniquely determine a plane π in 3-D. (Right) from four points, similarly constructed by consecutive axes translation, the coefficients of $\pi : c_1x_1 + c_2x_2 + c_3x_3 = c_0$ can be read from the picture!

(left). A little experiment is in order. Let us return to the family of *proximate* (i.e. close) planes generated by

$$\Pi = \{\pi : c_1x_1 + c_2x_2 + c_3x_3 = c_0, \quad c_i \in [c_i^-, c_i^+], \quad i = 0, 1, 2, 3\}, \qquad (10)$$

randomly chosing values of the c_i within the allowed intervals to determine several planes $\pi \in \Pi$, keeping at first $c_0 = 1$, and plotting the two points $\bar{\pi}_{123}, \bar{\pi}_{1'23}$ as shown in Fig. 35 (left). Not only is closeness apparent but more significantly the distribution of the points is not chaotic. The outline of two hexagonal patterns can be discerned. The family of "close" planes is visualizable but also the variations in several directions. It is possible to see, estimate and compare errors or proximity [14].

It can be proved that in 3-D the set of pairs of points representing the family of proximate planes form two convex hexagons when $c_0 = 1$ with an example is shown in Fig. 35 (right), and are contained in octagons each with two vertical edges for varying c_0. In general, a family of proximate hyperplanes in N-D is represented by $N-1$ convex $2N$-agons when $c_0 = 1$ or $2(N+1)$-agons for c_0 varying. These polygonal regions can be constructed with $O(N)$ computational complexity. Choosing a point in one of the polygonal regions, an algorithm matches the possible remaining $N-2$ points, one each from the remaining convex polygons, which represent and identify hyperplanes in the family by $N-1$ points.

We pose the thesis that visualization is not about seeing lots of things but rather discovering **relations** among them. While the display of randomly sampled *points* from a family of proximate hyperplanes is utterly chaotic (the mess in Fig. 33 (right) from points in just *one* plane), their **proximate coplanarity relation** corresponds to a clear and compact pattern. With ∥-coords we can focus and *concentrate* the relational information rather than wallowing in the

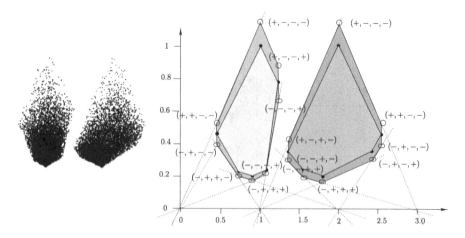

Fig. 35. A family of close planes. (Left) pair of point clusters representing close planes. (Right) the hexagonal regions (interior) are the regions containing the points $\bar{\pi}_{123}$ (left) and $\bar{\pi}_{1'23}$ for the family of planes with $c_0 = 1$ and $c_1 \in [1/3, 1.5], c_2 \in [1/3, 2.5], c_3 \in [1/3, 1]$. For c_0 varying, here $c_0 \in [.85, 1.15]$, the regions (exterior) are octagonal with two vertical edges.

details, ergo the remark "without loss of information" when referring to ‖-coords. This is the methodology's real strength and where he future lies. Here then is a visualization challenge: how else can proximate coplanarity be detected and seen?

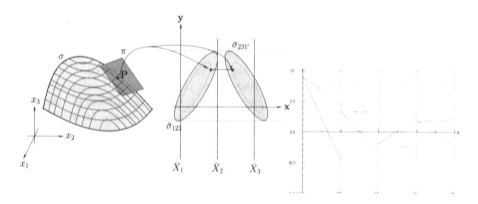

Fig. 36. Surface representation. (Left) A smooth surface σ is represented by two planar regions $\bar{\sigma}_{123}, \bar{\sigma}_{231'}$ consisting of pairs of points representing its tangent planes. (Right) One of the two hyperbolic regions representing a sphere in 3-D.

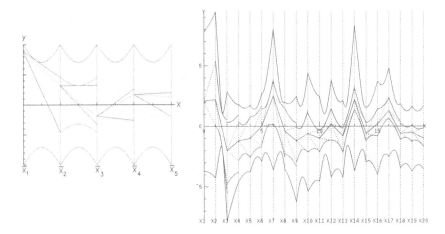

Fig. 37. Interior point construction. (Left) a sphere in 5-D showing the construction of an interior point (polygonal line). (Right) the general interior point (polygonal line) construction algorithm shown for a convex hypersurface in 20-D.

4.3   Nonlinear Multivariate Relations – Hypersurfaces

A relation among 2 real variables is represented geometrically by a unique region in 2-D. Analogously, a relation between N variables corresponds to a hypersurface in N-D, hence the need to say something about the representation of hypersurfaces in ‖-coords. A smooth surface in 3-D (and also N-D) can be described as the envelope of all its tangent planes. This is the basis for the representation shown in Fig. 36 (left). Every point of the surface is mapped into the two points representing its *tangent plane at the point*. This generates 2 planar regions and for N-D there are $N - 1$ such regions. These regions are *linked*, just as the polygons above, to provide the proper $N - 1$ points representing each tangent hyperplane and from which the hypersurface can be reconstructed. Classes of surfaces can be immediately distinguished from their ‖-coords display (see the section on surfaces for extensive treatment). For developable surfaces the regions consists of boundary curves only with no interior points, regions for ruled surfaces have grids consisting of straight lines, quadric surfaces have regions with conic boundaries these are some examples (Figs. 40 and 41).

There is a simpler but inexact surface representation which is quite useful when used judiciously. The polygonal lines representing points on the boundary are plotted and their envelope "represents" the surface; the "" are a reminder that this is not a *unique* representation. In Fig. 37 (left) are the upper and lower envelopes for a sphere in 5-D consisting of 4 overlapping hyperbolae which must be distinguished from those in Fig. 36 (right), which is exact and, interestingly enough are also hyperbolae, the curves determined by points representing the sphere's *tangent planes*. Retaining the exact surface description (i.e. its equation) internally, interior points can be constructed and displayed as shown for the 5-D sphere in Fig. 37 (left). On the right the same construction is shown but for a

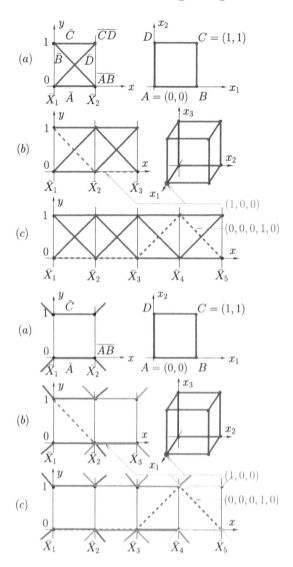

Fig. 38. Square (a), cube (b) and hypercube (c) in 5-D on the top represented by their vertices and on the bottom by the tangent planes. Note the hyperbola-like (with 2 asymptotes) regions showing that the object is convex.

more complex 20-dimensional convex hypersurface ("model"). The intermediate curves (upper and lower) also provide valuable information and previews of coming attractions. They indicate a neighborhood of the point (represented by the polygonal line) and provide a feel for the local curvature. Note the narrow strips around $X13, X14, X15$ (as compared to the surrounding ones), indicating that at this state these are the critical variables where the point is bumping the

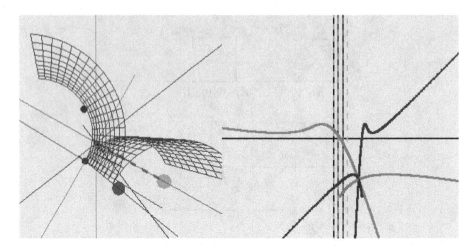

Fig. 39. Developable surfaces are represented by curves. Note the two dualities *cusp* ↔ *inflection point* and *bitangent plane* ↔ *crossing point*. Three such curves represent the corresponding hypersurface in 4-D and so on.

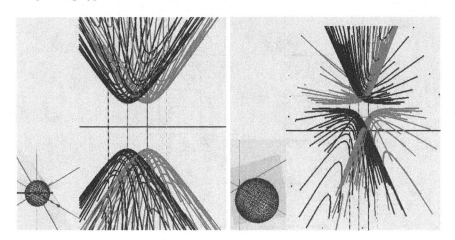

Fig. 40. Representation of a sphere centered at the origin (left) and after a translation along the x_1 axis (right) causing the two hyperbolas to rotate in opposite directions. Note the *rotation* ↔ *translation* duality. In N-D a sphere is represented by $(N - 1)$ such hyperbolic regions — pattern repeats as for hypercube above.

boundary. A theorem guarantees that a polygonal line which is in-between all the intermediate curves/envelopes represents an interior point of the hypersurface and all interior points can be found in this way. If the polygonal line is tangent at anyone of the intermediate curves then it represents a boundary point, while if it crosses anyone of the intermediate curves it represents an exterior point. The later enables us to see, in an application, the first variable for which the construction failed and what is needed to make corrections. By varying the choice

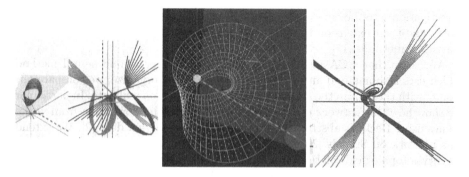

Fig. 41. Möbius strip and its representation for two orientations. The two cusps on the left show that it corresponds to an "inflection-point in 3-D" – see the duality in Fig. 39. The curves tending to infinity in the same direction upwards and downwards show that it is closed.

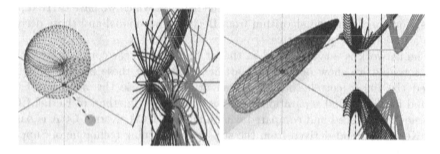

Fig. 42. Representation of a surface with 2 "dimples" (depressions with cusp) which are mapped into "swirls" and are **all** visible. By contrast, in the perspective (left) one dimple is hidden. On the right is a convex surface represented by hyperbola-like (having two assymptotes) regions.

of value over the available range of the variable interactively, sensitive regions (where small changes produce large changes downstream) and other properties of the model can be easily discovered. Once the construction of a point is completed it is possible to vary the values of each variable and see how this effects the remaining variables. So one can do trade-off analysis in this way and provide a powerful tool for, Decision Support, Process Control and other applications. As new data becomes available the model can be updated with decisions being made based on the most recent information. This algorithm is used in the earlier example on a model for a country's economy shown in Figs. 30 and 31.

5   A Neural Implementation of Canonical Correlation Analysis

This section reports an investigation into the well-known parallel between artificial neural networks and statistics; specifically it will present the results of an investigation into the development of Canonical Correlation Analysis neural networks and their extension to more complex forms not available to the standard statistical methods [17].

We show that CCA may be implemented in a very simple neural method which uses Hebbian and anti-Hebbian learning to self-organize; we compare our results with those from the standard statistical method on real data. In particular we show how this network can be used to extract depth information from visual scenes using Becker's abstraction of random dot stereogram data [2]. We extend this network on Stone's [20] random dot stereogram data so that the network preserves topological relationships within the correlations.

We then report in the list of reference that the basic CCA network can be derived from a probabilistic perspective and that, by making simple assumptions about Becker's models, we can derive a family of CCA-type networks all of which share similar Hebbian and anti-Hebbian rules. Extensive comparative studies have been performed on members of this family; we show that the network derived from Becker's second model and the network which we have derived previously out-performs the algorithm from Becker's first model and those derived from a probabilistic perspective.

The network is now extended so that it finds non-linear correlations within data sets and we show on real and artificial data how these correlations may exceed the linear correlations. We then attempt to use the nonlinear CCA method for the "blind separation of sources". The new method of Kernel CCA [7] is now introduced and compared with nonlinear CCA. Kernel CCA is based on a Kernel method derived from the supervised learning technique of Support Vector Machines and extensive simulations with different types of kernels are discussed. We show that both of these methods find correlations in excess of those found by linear methods but this finding comes with a caveat: the correlations may be functions of the methods rather than inherent in the data set and also some more recent research from my own work for instance, "Two Forms of Immediate Reward Reinforcement Learning for Exploratory Data Analysis" [24], "The Sphere-Concatenate Method for Gaussian Process Canonical Correlation Analysis" [15], etc.

5.1 The Canonical Correlation Network (The Network – Derivation/Theory)

Canonical Correlation Analysis [17] is used when we have two data sets which we believe have some underlying correlation. Consider two sets of input data, from which we draw iid samples to form a pair of input vectors, $\mathbf{x}_1$ and $\mathbf{x}_2$. Then in classical CCA, we attempt to find the linear combination of the variables which gives us maximum correlation between the combinations. Let

$$y_1 = \mathbf{w}_1 \mathbf{x}_1 = \sum_j w_{1j} x_{1j} \tag{11}$$

$$y_2 = \mathbf{w}_2 \mathbf{x}_2 = \sum_j w_{2j} x_{2j} \tag{12}$$

Then we wish to find those values of $\mathbf{w}_1$ and $\mathbf{w}_2$ which maximise the correlation between y_1 and y_2. Whereas Principal Components Analysis and Factor Analysis deals with the interrelationships within a set of variables, CCA deals with the relationships between two sets of variables. If the relation between y_1 and y_2 is believed to be causal, we may view the process as one of finding the best predictor of the set $\mathbf{x}_2$ by the set $\mathbf{x}_1$ and similarly of finding the most predictable criterion in the set $\mathbf{x}_2$ from the $\mathbf{x}_1$ data set.

We wish to maximize the correlation $E(y_1 y_2)$ where $E()$ denotes the expectation which will be taken over the joint distribution of $\mathbf{x}_1$ and $\mathbf{x}_2$. We may regard this problem as that of maximizing the function $g_1(\mathbf{w}_1|\mathbf{w}_2) = E(y_1 y_2)$ which is defined to be a function of the weights, $\mathbf{w}_1$ given the other set of parameters, $\mathbf{w}_2$. This is an unconstrained maximization problem which has no finite solution and so we must constrain the maximization.

Typically in CCA, we add the constraint that $E(y_1^2 = 1)$ and similarly with y_2 when we maximize $g_2(\mathbf{w}_2|\mathbf{w}_1)$. Using the method of Lagrange multipliers, this yields the constrained optimization functions,

$$J_1 = E(y_1 y_2) + \frac{1}{2}\lambda_1(1 - y_1^2) \text{ and}$$

$$J_2 = E(y_1 y_2) + \frac{1}{2}\lambda_2(1 - y_2^2)$$

We may equivalently use

$$J = E(y_1 y_2) + \frac{1}{2}\lambda_1(1 - y_1^2) + \frac{1}{2}\lambda_2(1 - y_2^2)$$

We therefore extend our neural implementation of CCA by maximising the correlation between outputs when such outputs are a nonlinear function of the inputs. We investigate a particular case of maximisation of $E(\mathbf{y_1 y_2})$ when the values y_i are a nonlinear function of the inputs, $\mathbf{x}_i$.

For the nonlinear optimization, we use e.g. $y_3 = \sum_j w_{3j} f_3(v_{3j} x_{1j}) = \mathbf{w}_3 \mathbf{f}_3$ where the function $f_3()$ is applied on an element-wise basis to give the vector $\mathbf{f}_3$.

The equivalent optimization function for the nonlinear problem is then

$$J_3(\mathbf{w}_3|\mathbf{w}_4) = E(y_3 y_4) + \frac{1}{2}\lambda_3(1 - y_3^2)$$

$$J_4(\mathbf{w}_4|\mathbf{w}_3) = E(y_4 y_3) + \frac{1}{2}\lambda_4(1 - y_4^2)$$

5.2   Experimental Results

We report simulations on both real and artificial data sets of increasing complexity. We begin with data sets in which there is a linear correlation and we demonstrate the effectiveness of the network on Becker's [2] random dot stereogram data. We then extend the method in two ways not possible with standard statistical techniques:

1. We maximise correlations between more than two input data sets.
2. We consider maximising correlations where such correlations may be on non-linear projections of the data.

Real Data. Our experiment uses a data set reported in [17], p. 290; it comprises 88 students' marks on 5 module exams. The exam results can be partitioned into two data sets: two exams were given as close book exams (C) while the other three were opened book exams (O). We thus split the five variables (exam marks) into two sets-the closed-book exams (x_{11}, x_{12}) and the opened-book exams (x_{21}, x_{22}, x_{23}). One possible quantity of interest here is how highly a student's ability on closed-book exams is correlated with his ability on open-book exams. (or vice versa) (Table 1).

Table 1. The converged weights from the neural network are compared with the values reported from a standard statistical technique [17].

Standard statistics maximum correlation	0.6962		
$\mathbf{w}_1$	0.0260	0.0518	
$\mathbf{w}_2$	0.0824	0.0081	0.0035
Neural network maximum correlation	0.6630		
$\mathbf{w}_1$	0.0264	0.0526	
$\mathbf{w}_2$	0.0829	0.0098	0.0041

Random Dot Stereograms. It has been suggested [2] that one of the goals of sensory information processing may be the extraction of common information between different sensors or across sensory modalities. Table 2.

Table 2. The converged weights clearly show that the first pair of neurons has learned a right shift while the second pair has learned a left shift.

$\mathbf{w}_1$	−0.002	**1.110**	0.007	−0.009
$\mathbf{w}_2$	0.002	0.025	**0.973**	0.020
$\mathbf{w}_3$	−0.014	0.026	**1.111**	−0.002
$\mathbf{w}_4$	0.013	**0.984**	0.003	−0.007

Stone's Stereogram Data. In the following experiments, we have used an artificial data set previously used by Stone [20]. This data simulates a moving surface with a slowly varying depth (Fig. 43).

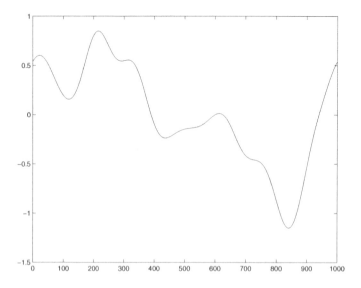

Fig. 43. Example of an array of 1000 disparity values generated using the method described above.

6   Future

Searching for *patterns* in a ‖-coords display is what skillful exploration is about. If there are multivariate relations in the dataset the patterns *are there* though they may be covered by the overlapping polygonal lines and that is not all. Our vision is not multidimensional. We do not perceive a room which is 3-dimensional from its points which are 0-dimensional, but from the 2-dimensional planes which enclose and define it. The recursive construction algorithm does exactly that for the visualization of p-dimensional objects from their $p - 1$-dimensional subsets; one dimension less. We advocate including this algorithm within our armory of interactive analysis tools. Whatever p-dimensional relations exist are revealed by the pattern from the representation of the tangent hypeplanes of the corresponding hypersurface. The polygonal lines are completely discarded for the *relation is concentrated in the pattern*: Linear relations into points, proximate coplanarity into convex polygons, quadrics into conics and so on. Note further, again with reference to Figs. 32 and 33, that relational information resides at the *crossings*. What can be achieved for the representation of complex relations by patterns is exemplified by the pictures in Fig. 38 through Fig. 42. These are state of the art results showing what is achievable and how easily it generalizes to N-D. Can one imagine a higher dimensional convex surface or various kinds of non-convexities much less the *non-orientable* the Möbius strip. It is possible to do such a process on a dataset though at present it is computationally slow. The challenge is to speed up the algorithm for real-time response and thus break the gridlock of multidimensional visualization. There will still be work and fun for the multidimensional detectives visually separating and classifying the *no longer hidden* regions identifying complex multivariate relations which evaded us until now.

References

1. Bassett, E.W.: IBM's IBM fix. Ind. Comput. **14**(41), 23–25 (1995)
2. Becker, S., Hinton, G.: A self-organizing neural network that discovers surfaces in random-dot stereograms. Nature (Lond.) **355**, 161–163 (1992)
3. Bollobas, B.: Graph Theory. Springer, New York (1979)
4. Eickemeyer, J.: Visualizing p-flats in N-space using parallel coordinates. Ph.D. thesis, Dept. Comp. Sc., UCLA (1992)
5. Fayad, G., Piatesky-Shapiro, U.M., Smyth, P., Uthurusamy, R.: Advances in Knowledge Discovery and Data Mining. AAAI/MIT Press, Cambridge Mass. (1996)
6. Friendly, M., et al.: Milestones in Thematic Cartography (2005). www.math.yorku. ca/scs/SCS/Gallery/milestones/
7. Fyfe, C., Lai, P.L.: ICA using kernel canonical correlation analysis. In: ICA 2000 (2000)
8. Gennings, C., Dawson, K.S., Carter, W.H., Myers, R.H.: Interpreting plots of a multidimensional dose-response surface in parallel coordinates. Biometrics **46**, 719–35 (1990)
9. Harary, F.: Graph Theory. Addison-Wesley, Reading (1969)
10. Hurley, C.B., Oldford, R.W.: Pairwise display of high-dimensional information via eulerian tours and hamiltonian decompositions. J. Comput. Graph. Stat. **19**(4), 861–886 (2010)
11. Inselberg, A.: Visual data mining with parallel coordinates. Comput. Stat. **13**(1), 47–64 (1998)
12. Inselberg, A.: Parallel Coordinates: VISUAL Multidimensional Geometry and its Applications. Springer, New York (2009)
13. Inselberg, A., Avidan, T.: The automated multidimensional detective. In: Proceedings of IEEE Information Visualization 1999, pp. 112–119. IEEE Comp. Soc., Los Alamitos (1999)
14. Inselberg, A., Lai, P.L.: Visualizing families of close planes, 66. In: Proceedings of the 5th Asian Conference on Statistics, Hong Kong (2005)
15. Lai, P.L., Leen, G., Fyfe, C.: The sphere-concatenate method for gaussian process canonical correlation analysis. In: Oja, E., Kollias, S.D., Stafylopatis, A., Duch, W. (eds.) ICANN 2006. LNCS, vol. 4132, pp. 302–310. Springer, Heidelberg (2006)
16. Lucas, D.E.: Recréations Mathematiques, vol. II. Gauthier Villars, Paris (1892)
17. Mardia, K.V., Kent, J.T., Bibby, J.M.: Multivariate Analysis. Academic Press, London (1979)
18. Mitchell, T.M.: Machine Learning. McGraw-Hill, New York (1997)
19. Quinlan, J.R.: C4.5: Programs for Machine Learning. Morgan Kaufman, San Francisco (1993)
20. Stone, J.: Learning perpetually salient visual parameters using spationtemporal smoothness constraints. Neural Comput. **8**(7), 1463–1492 (1996)
21. Tufte, E.R.: The Visual Display of Quantitative Information. Graphic Press, Connecticut (1983)
22. Tufte, E.R.: Envisioning Information. Graphic Press, Connecticut (1990)
23. Tufte, E.R.: Visual Explanation. Graphic Press, Connecticut (1996)
24. Ying, W., Fyfe, C., Lai, P.L.: Two forms of immediate reward reinforcement learning for exploratory data analysis. Neural Netw. **21**(6), 847–855 (2008)

Web as a Corpus: Going Beyond the n-gram

Preslav Nakov[✉]

Qatar Computing Research Institute,
Tornado Tower, Floor 10, P.O.box 5825, Doha, Qatar
pnakov@qf.org.qa

Abstract. The 60-year-old dream of computational linguistics is to make computers capable of communicating with humans in natural language. This has proven hard, and thus research has focused on sub-problems. Even so, the field was stuck with manual rules until the early 90s, when computers became powerful enough to enable the rise of statistical approaches. Eventually, this shifted the main research attention to machine learning from text corpora, thus triggering a revolution in the field.

Today, the Web is the biggest available corpus, providing access to quadrillions of words; and, in corpus-based natural language processing, size does matter. Unfortunately, while there has been substantial research on the Web as a corpus, it has typically been restricted to using page hit counts as an estimate for n-gram word frequencies; this has led some researchers to conclude that the Web should be only used as a baseline. We show that much better results are possible for structural ambiguity problems, when going beyond the n-gram.

Keywords: Web as a corpus · Surface features · Paraphrases · Noun compound bracketing · Prepositional phrase attachment · Noun phrase coordination · Syntactic parsing

1 Introduction

The 60-year-old dream of computational linguistics is to make computers capable of communicating with humans in natural language. This has proven hard, and thus research has focused on sub-problems. Even so, the field was stuck with manual rules until the early 90s, when computers became powerful enough to enable the rise of statistical approaches. Eventually, this shifted the main research attention to machine learning from text corpora, thus triggering a revolution in the field.

Today, the Web is the biggest available corpus, providing access to quadrillions of words; and, in corpus-based natural language processing, size does matter. Unfortunately, while there has been substantial research on the Web as a corpus, it has typically been restricted to using page hit counts as an estimate for n-gram word frequencies.

Joint work with Marti Hearst.

© Springer International Publishing Switzerland 2015
P. Braslavski et al. (eds.): RuSSIR 2014, CCIS 505, pp. 185–228, 2015.
DOI: 10.1007/978-3-319-25485-2_5

In 2001, Banko and Brill (2001) advocated for the use of very large text collections as an alternative to sophisticated algorithms and hand-built resources. They demonstrated the idea on a lexical disambiguation problem for which labeled examples are available "for free". The problem was to choose which of 2-3 commonly confused words (e.g., {*principle, principal*}) were appropriate for a given context. The labeled data was "free" because the authors could safely assume that in the carefully edited text in their training set the words were used correctly. They have shown that even using a very simple algorithm, the results continue to improve log-linearly with more training data, even out to a billion words. Thus, they concluded that getting more data may be a better idea than fine-tuning algorithms on small training datasets. Today, the obvious source of very large data is the Web.

The research interest in using the Web as a corpus started around the year 2000, and by 2003 there was enough momentum to trigger a special issue of the *Computational Linguistics* journal on this topic (Kilgariff and Grefenstette 2003). This was followed by a number of workshops, most notably, the Web as Corpus (WAC) workshop, which had its 9th edition in 2014, and the establishment of a Special Interest Group on the Web as a Corpus with the Association for Computational Linguistics: ACL SIGWAC[1]. The Web has been used as a corpus for a variety of NLP tasks, e.g.,

- machine translation (Grefenstette 1998; Resnik 1999a; Cao and Li 2002; Way and Gough 2003; Nakov 2008a),
- question answering (Dumais et al. 2002; Soricut and Brill 2004),
- word sense disambiguation (Mihalcea and Moldovan 1999; Rigau et al. 2002; Santamaría et al. 2003; Zahariev 2004),
- spelling correction (Keller and Lapata 2003; Bergsma et al. 2010),
- semantic relation extraction (Chklovski and Pantel 2004; Idan Szpektor and Coppola 2004; Shinzato and Torisawa 2004),
- noun compound interpretation (Nakov and Hearst 2006; Nakov and Hearst 2008; Nakov 2008c; Nakov and Kozareva 2011; Nakov and Hearst 2013),
- anaphora resolution (Modjeska et al. 2003),
- language modeling (Zhu and Rosenfeld 2001; Keller and Lapata 2003; Brants et al. 2007),
- query segmentation (Bergsma and Wang 2007),
- prepositional phrase attachment (Volk 2001; Calvo and Gelbukh 2003; Nakov and Hearst 2005c),
- noun compound bracketing (Nakov 2007; Nakov 2008b; Butnariu and Veale 2008; Kim and Nakov 2011),
- noun compound coordination (Nakov and Hearst 2005c),
- full syntactic parsing (Bansal and Klein 2011),
- etc.

Despite the variability of applications, the most popular use of the Web as a corpus has been as a means to obtain page hit counts, which are then used as estimates for n-gram word frequencies. Keller and Lapata (2003) demonstrated high

---

[1] http://www.sigwac.org.uk/

correlation between page hits and corpus bigram frequencies, as well as between page hits and plausibility judgments. They proposed using Web counts as a baseline unsupervised method for many NLP tasks and experimented with eight NLP problems (machine translation candidate selection, spelling correction, adjective ordering, article generation, noun compound bracketing, noun compound interpretation, countability detection and prepositional phrase attachment), and showed that variations on *n*-gram counts often perform nearly as well as more elaborate methods (Lapata and Keller 2005).

Below we show that the Web has the potential for more than just a baseline. Using various Web-derived surface features, in addition to paraphrases and *n*-gram counts, we demonstrate state-of-the-art results on the task of noun compound bracketing (Nakov and Hearst 2005a). We further show very strong results for prepositional phrase attachment and for noun phrase coordination (Nakov and Hearst 2005c).

2   Noun Compound Bracketing

2.1   The Problem

An important but understudied language analysis problem is that of noun compound bracketing, which is generally viewed as a necessary step towards noun compound (NC) interpretation. Consider the following contrastive pair of noun compounds:

(1) *liver cell antibody*
(2) *liver cell line*

In example (1) an *antibody* targets a *liver cell*, while (2) refers to a *cell line* which is derived from the *liver*. In order to make these semantic distinctions accurately, it can be useful to begin with the correct grouping of terms, since choosing a particular syntactic structure limits the options left for semantics.[2] Although equivalent at the part of speech (POS) level, these two noun compounds have different syntactic trees. The distinction can be represented as a binary tree or, equivalently, as a binary bracketing:

(1b) [[*liver cell*] *antibody*]   (left bracketing)
(2b) [*liver*   [*cell line*]] (right bracketing)

The best known early work on automated unsupervised NC bracketing is that of Lauer (1995) who introduces the probabilistic dependency model for the syntactic disambiguation of NCs and argues against the adjacency model, proposed by Marcus (1980), Pustejovsky *et al.* (1993) and Resnik (1993). Lauer collects *n*-gram statistics from Grolier's encyclopedia, which contains about eight million words. In order to overcome data sparseness problems, he estimated probabilities over conceptual categories in a taxonomy (Roget's thesaurus) rather than for individual words.

---

[2] See (Nakov 2013) for an overview on the syntax and semantics of noun compounds. See also the Nakov & Hearst (2013)

Lauer evaluated his models on a set of 244 unambiguous NCs derived from the same encyclopedia (inter-annotator agreement 81.50 %) and achieved 77.50 % for the dependency model above (baseline 66.80 %). Adding POS and further tuning allowed him to achieve the state-of-the-art result of 80.70 %.

Subsequently, Lapata and Keller (2004) proposed using Web counts as a baseline for many NLP tasks. They applied this idea to six NLP tasks, including the syntactic and semantic disambiguation of NCs following Lauer (1995), and showed that variations on bigram counts perform nearly as well as more elaborate methods. They did not use taxonomies and worked with the word n-grams directly, achieving 78.68 % with a much simpler version of the dependency model.

Girju et al. (2005) proposed a *supervised* model (decision tree) for NC bracketing *in context*, based on five semantic features (requiring the correct WordNet sense to be given): the top three WordNet semantic classes for each noun, derivationally related forms and whether the noun is a nominalization. The algorithm achieved 83.10 % accuracy.

Below we describe a highly accurate unsupervised method for making bracketing decisions for noun compounds. We improve on the current standard approach of using bigram estimates to compute adjacency and dependency scores introducing a new set of surface features for querying Web search engines which prove highly effective. We also experiment with paraphrases for improving prediction statistics.

2.2   Models and Features

Adjacency and Dependency Models. In related work, a distinction is often made between what is called the *dependency model* and the *adjacency model*. The main idea is as follows. For a given 3-word NC $w_1w_2w_3$, there are two reasons it may take on right bracketing, $[w_1[w_2w_3]]$. Either (a) w_2w_3 is a compound (modified by w_1), or (b) w_1 and w_2 independently modify w_3. This distinction can be seen in the examples *home health care* (*health care* is a compound modified by *home*) versus *adult male rat* (*adult* and *male* independently modify *rat*).

The adjacency model checks (a), whether w_2w_3 is a compound (i.e., how strongly w_2 modifies w_3 as opposed to w_1w_2 being a compound) to decide whether or not to predict a right bracketing. The dependency model checks (b), whether w_1 modify w_3 (as opposed to w_1 modifying w_2).

Left bracketing is a bit different since there is only modificational choice for a 3-word NC. If w_1 modifies w_2, this implies that w_1w_2 is a compound which in turn modifies w_3, as in *law enforcement agent*.

Thus the usefulness of the adjacency model vs. the dependency model can depend in part on the mix of left and right bracketing. Below we show that the dependency model works better than the adjaceny model, confirming other results in the literature.

Using Frequencies. The most straightforward way to compute adjacency and dependency scores is to simply count the corresponding frequencies. Lapata and

Keller (2004) achieved their best accuracy (78.68 %) with the dependency model and the simple symmetric score $\#(w_i, w_j)$.[3]

Computing Probabilities. Lauer (1995) assumes that adjacency and dependency should be computed via probabilities. Since they are relatively simple to compute, we investigate them in our experiments.

Consider the dependency model, as introduced above, and the NC $w_1 w_2 w_3$. Let $\Pr(w_i \rightarrow w_j | w_j)$ be the probability that the word w_i precedes w_j. Assuming that the distinct head-modifier relations are independent, we obtain

$$\Pr(\text{right}) = \Pr(w_1 \rightarrow w_3 | w_3)\Pr(w_2 \rightarrow w_3 | w_3)$$
$$\Pr(\text{left}) = \Pr(w_1 \rightarrow w_2 | w_2)\Pr(w_2 \rightarrow w_3 | w_3)$$

In order to choose the more likely structure, we can drop the shared factor and compare $\Pr(w_1 \rightarrow w_3 | w_3)$ to $\Pr(w_1 \rightarrow w_2 | w_2)$.

The alternative adjacency model compares the probability $\Pr(w_2 \rightarrow w_3 | w_3)$ to $\Pr(w_1 \rightarrow w_2 | w_2)$, i.e., the association strength between the last two words vs. that between the first two. If the former is bigger than the latter, the model predicts right.

The probability $\Pr(w_1 \rightarrow w_2 | w_2)$ can be estimated as $\#(w_1, w_2)/\#(w_2)$, where $\#(w_1, w_2)$ and $\#(w_2)$ are the corresponding bigram and unigram frequencies. They can be approximated as the number of pages returned by a search engine in response to queries for the exact phrase "$w_1 \ w_2$" and for the word w_2. In our experiments below, we smoothed[4] each of these frequencies by adding 0.5 to avoid problems caused by nonexistent n-grams.

Unless some particular probabilistic interpretation is needed,[5] there is no reason for us to use $\Pr(w_i \rightarrow w_j | w_j)$ rather than $\Pr(w_j \rightarrow w_i | w_i)$, $i < j$. This is confirmed by the adjacency model experiments in (Lapata and Keller 2004) on Lauer's NC set. Their results show that both ways of computing the probabilities make sense: using AltaVista queries, the former achieves a higher accuracy (70.49 % vs. 68.85 %), but the latter is better on the British National Corpus (65.57 % vs. 63.11 %).

Other Measures of Association. In both the adjacency and the dependency models, the probability $\Pr(w_i \rightarrow w_j | w_j)$ can be replaced by some (possibly

---

[3] This score worked best on training, when Keller and Lapata were doing model selection. On testing, Pr (with the dependency model) worked better and achieved accuracy of 80.32 %, but this result was ignored, as Pr did worse on training.

[4] Zero counts sometimes happen for $\#(w_1, w_3)$, but are rare for unigrams and bigrams on the Web, and there is no need for a more sophisticated smoothing.

[5] For example, as used by Lauer to introduce a prior for left-right bracketing preference. The best Lauer model does not work with words directly, but uses a taxonomy and further needs a probabilistic interpretation, so that the hidden taxonomy variables can be summed out. Because of that summation, the term $\Pr(w_2 \rightarrow w_3 | w_3)$ does not cancel in his dependency model.

symmetric) measure of association between w_i and w_j, such as *Chi squared* (χ^2). To calculate $\chi^2(w_i, w_j)$, we need the following:

(A) $\#(w_i, w_j)$;
(B) $\#(w_i, \overline{w_j})$, the number of bigrams in which the first word is w_i, followed by a word other than w_j;
(C) $\#(\overline{w_i}, w_j)$, the number of bigrams, ending in w_j, whose first word is other than w_i;
(D) $\#(\overline{w_i}, \overline{w_j})$, the number of bigrams in which the first word is not w_i and the second is not w_j.

They are combined in the following formula:

$$\chi^2 = \frac{N(AD - BC)^2}{(A+C)(B+D)(A+B)(C+D)} \tag{1}$$

In the above equation, $N = A + B + C + D$ is the total number of bigrams, $B = \#(w_i) - \#(w_i, w_j)$ and $C = \#(w_j) - \#(w_i, w_j)$. While it is hard to estimate D directly, we can calculate it as $D = N - A - B - C$. Finally, we estimate N as the total number of indexed bigrams on the Web. In our experiments, we estimated N as 8 trillion, assuming Google indexes about 8 billion pages and each contains about 1,000 words on average.

Other measures of word association are possible, such as *mutual information* (MI), which we can use with the dependency and the adjacency models, similarly to #, χ^2 or Pr. However, in our experiments, χ^2 worked better than other methods; this is not surprising, as χ^2 is known to outperform MI as a measure of association (Yang and Pedersen 1997).

Web-Derived Surface Features. Authors sometimes (consciously or not) disambiguate the NCs they write by using surface-level markers to suggest the correct structure. We have found that exploiting these markers, when they occur, can prove to be very helpful for making bracketing predictions. The enormous size of Web search engine indexes facilitates finding such markers frequently enough to make them useful.

One very productive feature is the *dash* (hyphen). Starting with the term *cell cycle analysis*, if we can find a version of it in which a dash occurs between the first two words, *cell-cycle*, which suggests a left bracketing for the full NC. Similarly, the dash in *donor T-cell* favors a right bracketing. The righthand dashes are less reliable though, as their scope is ambiguous. In *fiber optics-system*, the hyphen indicates that the noun compound *fiber optics* modifies *system*. There are also cases with multiple hyphens, as in *t-cell-depletion*, which are unusable.

The genitive ending, or *possessive* marker is another useful indicator. The phrase *brain's stem cells* suggests a right bracketing for *brain stem cells*, while *brain stem's cells* favors a left bracketing.[6]

---

[6] Features can also occur combined, e.g., *brain's stem-cells*.

Another highly reliable source is related to internal *capitalization*. For example *Plasmodium vivax Malaria* suggests left bracketing, while *brain Stem cells* would favor a right one. We disabled this feature on Roman digits and single-letter words to prevent problems with terms like *vitamin D deficiency*, where the capitalization is just a convention as opposed to a special mark to make the reader think that the last two terms should go together.

We can also make use of embedded *slashes*, e.g., in *leukemia/lymphoma cell*, the slash predicts a right bracketing since the first word is an alternative and thus it cannot be modify the second one.

In some cases, we can find instances of the NC in which one or more words are enclosed in parentheses, e.g., *growth factor (beta)* or *(growth factor) beta*, both of which indicate a left structure, or *(brain) stem cells*, which suggests a right bracketing.

Even a comma, a dot or a colon (or any special character) can act as indicators. For example, "*health care, provider*" or "*lung cancer: patients*" are weak predictors of a left bracketing, showing that the author chose to keep two of the words together, separating out the third one.

We can also exploit dashes to words external to the target noun compound, as in *mouse-brain stem cells*, which is a weak indicator of right bracketing.

Unfortunately, Web search engines ignore punctuation characters, thus preventing querying directly for terms containing hyphens, brackets, apostrophes, etc. We collect them indirectly by issuing queries with the NC as an exact phrase and then post-processing the resulting summaries, looking for the surface features of interest. Search engines typically allow the user to explore up to 1000 results. We collect all results and summary texts that are available for the target NC and then search for the surface patterns using regular expressions over the text. Each match increases the score for left or right bracketing, depending on which the pattern favors.

While some of the above features are clearly more reliable than others, we do not try to weigh them. For a given NC, we post-process the returned Web summaries, then we find the number of left-predicting surface feature instances (regardless of their type) and compare it to the number of right-predicting ones to make a bracketing decision.[7]

Some features can be obtained by using the overall counts returned by the search engine. As these counts are derived from the entire Web, as opposed to a set of up to 1,000 summaries, they are of different magnitude, and we did not want to simply add them to the surface features above. They appear as independent models in Tables 1 and 2.

First, in some cases, we can query for *possessive markers* directly: although search engines drop the apostrophe, they keep the *s*, so we can query for "*brain's*" (but not for "*brains'* "). We then compare the number of times the possessive marker appeared on the second vs. the first word, to make a bracketing decision.

Abbreviations are another important feature. For example, finding on the Web the variant "*tumor necrosis factor (NF)*" suggests a right bracketing, while

---

[7] This appears as *Surface features (sum)* in Tables 1 and 2.

"tumor necrosis (TN) factor" would favor left. We would like to issue exact phrase queries for the two patterns and see which one is more frequent. Unfortunately, search engines drop the brackets and ignore the capitalization, so we issue queries with the parentheses removed, as in *"tumor necrosis factor nf"*. This yields highly accurate results, although errors occur when the abbreviation is an existing word (e.g., *me*), a Roman digit (e.g., *IV*), a state (e.g., *CA*), etc.

Another reliable feature is *concatenation*. Consider the NC *health care reform*, which is left-bracketed. Now, consider the bigram *"health care"*. Google estimates 80,900,000 pages for it as an exact term. Now, if we try the word *healthcare*, we get 80,500,000 hits. At the same time, *carereform* returns just 109. This suggests that authors sometimes concatenate words that act as compounds. We find below that comparing the frequency of the concatenation of the left bigram to that of the right (adjacency model for concatenations) often yields accurate results. We also tried the dependency model for concatenations, as well as the concatenations of two words in the context of the third one (i.e., compare frequencies of *"healthcare reform"* and *"health carereform"*).

We also used Google's support for "*", which allows a single word wildcard, to see how often two of the words are present but separated from the third by some other word(s). This implicitly tries to capture paraphrases involving the two sub-concepts making up the whole. For example, we compared the frequency of *"health care * reform"* to that of *"health * care reform"*. We also used two and three stars and switched the word group order (indicated with *rev.* in Tables 1 and 2), e.g., *"care reform * * health"*. We also tried a simple *reorder* without inserting any stars, i.e., we compared the frequency of *"reform health care"* to the frequency of *"care reform health"*. For example, when analyzing *myosin heavy chain*, we see that *heavy chain myosin* is very frequent, which provides evidence against grouping *heavy* and *chain* together as they can commute.

Further, we tried to look inside the *internal inflection variability*. The idea is that if *"tyrosine kinase activation"* is left-bracketed, then the first two words probably make a whole, and thus the second word can be found inflected elsewhere, but the first word cannot, e.g., *"tyrosine kinases activation"*. Alternatively, if we find different internal inflections of the first word, this would favor a right bracketing.

Finally, we tried switching the word order of the first two words. If they independently modify the third one (which implies a right bracketing), then we could expect to see also a form with the first two words switched, e.g., if we are given *"adult male rat"*, we would also expect *"male adult rat"*.

Paraphrases. Warren (1978) proposed that the semantics of the relations between words in a noun compound are often made overt by a paraphrase. Example of a *prepositional paraphrase*: an author describing the concept of *brain stem cells* may choose to expand it as *stem cells in the brain*. This contrast can be helpful for syntactic bracketing, suggesting that the full NC takes on right bracketing, since *stem* and *cells* are kept together in the expanded version. However, this NC is ambiguous, and can also be paraphrased as *cells from the brain stem*, implying a left bracketing.

Of course, not all noun compounds can be paraphrased with a preposition. For some, it is possible to use a *copula paraphrase*, e.g., *skyscraper office building* can be paraphrased as *office building that/which is a skyscraper*, which suggests right bracketing. Another option is to use a *verbal paraphrase*, e.g., *arthritis migraine pain* can be paraphrased as *pain associated with arthritis migraine*, suggesting left bracketing.

Other researchers have used prepositional paraphrases as a proxy for determining the semantic relations that hold between nouns in a compound (Lauer 1995; Keller and Lapata 2003; Girju *et al.* 2005). Since most NCs have a prepositional paraphrase, Lauer builds a model trying to choose between the most likely candidate prepositions: *of, for, in, at, on, from, with* and *about* (excluding *like* which is mentioned by Warren). This could be problematic though, since as a study by Downing (1977) shows, when no context is provided, people often come up with incompatible interpretations.

In contrast, we use paraphrases in order to make syntactic bracketing assignments. Instead of trying to manually decide the correct paraphrases, we can issue queries using paraphrase patterns and find out how often each occurs in the corpus. We then add up the number of hits predicting a left versus a right bracketing and compare the counts.

Unfortunately, search engines lack linguistic annotations, making general verbal paraphrases too expensive. Instead we used a small set of hand-chosen paraphrases: *associated with, caused by, contained in, derived from, focusing on, found in, involved in, located at/in, made of, performed by, preventing, related to* and *used by/in/for*. It is however feasible to generate queries predicting left/right bracketing with/without a determiner for every preposition.[8] For the copula paraphrases, we combine two verb forms *is* and *was*, and three complementizers *that, which* and *who*. These are optionally combined with a preposition or a verb form, e.g., *themes that are used in science fiction*.

2.3 Experiments

We experimented with *Lauer's dataset* (Lauer 1995), which is the benchmark dataset for the task of NC bracketing.

For comparison purposes, we further experimented with a *Biomedical dataset* (Nakov and Hearst 2005a) using a domain-specific text corpus with suitable linguistic annotations instead of the Web. We used the Layered Query Language and architecture (Nakov *et al.* 2005b; Nakov *et al.* 2005a) in order to acquire *n*-gram and paraphrase frequency statistics. Our corpus consists of about 1.4 million MEDLINE abstracts, each one being about 300 words long on the average, which means about 420 million indexed words in total. Suppose *Google* indexes about eight billion pages; if we assume that each one contains about 500 words on the average, this yields about four trillion indexed words, which is about a million times bigger than our corpus. Still, the subset of MEDLINE we

---

[8] In addition to the articles (*a, an, the*), we also used quantifiers (e.g., *some, every*) and pronouns (e.g., *this, his*).

use is about four times bigger than the 100 million word *BNC* used by Lapata and Keller (2004). It is also more than fifty times bigger than the eight million word *Grolier's encyclopedia* used by Lauer (1995).

In our experiments, we collected the *n*-grams, surface features, and paraphrase counts by issuing exact phrase queries against a search engine, limiting the pages to English and requesting filtering of similar results.[9] For each NC, we generated all possible word inflections (e.g., *tumor* and *tumors*) as well as alternative word variants (e.g., *tumor* and *tumour*). For the biomedical dataset, they were automatically obtained from the UMLS Specialist lexicon.[10] For Lauer's dataset, we used Carroll's morphological tools.[11] For bigrams, we inflected only the second word. Similarly, for a prepositional paraphrase, we generated all possible inflected forms for the two parts, before and after the preposition.

2.4   Results and Discussion

The results are shown in Tables 1 and 2. As NCs are left-bracketed at least 2/3rds of the time (Lauer 1995), a straightforward baseline is to always assign a left bracketing. Tables 1 and 2 suggest that the surface features perform best. The paraphrases are equally good on the biomedical dataset, but on Lauer's set their performance is lower and is comparable to that of the dependency model.

The dependency model clearly outperforms the adjacency one (as other researchers have found) on Lauer's set, but not on the biomedical set, where it is equally good. χ^2 barely outperforms #, but on the biomedical set χ^2 is a clear winner (by about 1.5 %) on both dependency and adjacency models.

The frequencies (#) outperform or at least rival the probabilities on both sets and for both models. This is not surprising, given the previous results by Lapata and Keller (2004). Frequencies also outperform Pr on the biomedical set. This may be due to the abundance of single-letter words in that set (because of terms like *T cell*, *B cell*, *vitamin D* etc.; similar problems are caused by Roman digits like *ii*, *iii* etc.), whose Web frequencies are rather unreliable, as they are used by Pr but not by frequencies. Single-letter words cause potential problems for the paraphrases as well, by returning too many false positives, but they work very well with concatenations and dashes, e.g., *T cell* is often written as *Tcell*.

As Table 4 shows, most of the surface features that we predicted to be right-bracketing actually indicated left. Overall, the surface features were very good at predicting left bracketing, but unreliable for right-bracketed examples. This is probably in part due to the fact that they look for adjacent words, i.e., they act as a kind of adjacency model.

---

[9] In our experiments, we used MSN Search (now Bing) statistics for the *n*-grams and the paraphrases (unless the pattern contained a "*"), and Google for the surface features. MSN always returned exact numbers, while Google and Yahoo rounded their page hits, which generally leads to lower accuracy (Yahoo was better than Google for these estimates).

[10] http://www.nlm.nih.gov/pubs/factsheets/umlslex.html.

[11] http://www.cogs.susx.ac.uk/lab/nlp/carroll/morph.html.

Table 1. NC bracketing, Lauer dataset. Shown are numbers for correct ($\sqrt{}$), incorrect ($\times$), and no prediction ($\emptyset$), followed by accuracy (Acc, calculated over $\sqrt{}$ and $\times$ only) and coverage (C, % examples with prediction). We use "$\rightarrow$" for back-off to another model in case of $\emptyset$.

Model	$\sqrt{}$	$\times$	$\emptyset$	Accuracy (%)	Coverage (%)
# adjacency	183	61	0	75.00	100.00
Pr adjacency	180	64	0	73.77	100.00
MI adjacency	182	62	0	74.59	100.00
χ^2 adjacency	184	60	0	**75.41**	100.00
# dependency	193	50	1	79.42	99.59
Pr dependency	194	50	0	79.51	100.00
MI dependency	194	50	0	79.51	100.00
χ^2 dependency	195	50	0	**79.92**	100.00
# adjacency (*)	152	41	51	78.76	79.10
# adjacency (**)	162	43	39	79.02	84.02
# adjacency (***)	150	51	43	74.63	82.38
# adjacency (*, rev.)	163	48	33	77.25	86.47
# adjacency (**, rev.)	165	51	28	76.39	88.52
# adjacency (***, rev.)	156	57	31	73.24	87.30
Concatenation adj	175	48	21	78.48	91.39
Concatenation dep.	167	41	36	**80.29**	85.25
Concatenation triples	76	3	165	**96.20**	32.38
Inflection Variability	69	36	139	65.71	43.03
Swap first two words	66	38	140	63.46	42.62
Reorder	112	40	92	73.68	62.30
Abbreviations	21	3	220	**87.50**	9.84
Possessives	32	4	208	**88.89**	14.75
Paraphrases	174	38	32	**82.08**	86.89
Surface features (sum)	183	31	30	**85.51**	87.70
Majority vote	210	22	12	90.52	95.08
Majority vote → left	*218*	*26*	*0*	*89.34*	*100.00*
Baseline (choose left)	163	81	0	66.80	100.00

We obtained our best overall results by combining the most reliable models, marked in bold in Tables 1, 2 and 4. As they have independent errors, we used a majority vote combination.

Table 3 compares our results to those of Lauer (1995) and of Lapata and Keller (2004). It is important to note though, that our results are *directly* comparable to those of Lauer, while the Keller and Lapata's are not, since they used half of the Lauer set for development and the other half for testing.[12] Following Lauer, we used everything for testing. Lapata and Keller also used the AltaVista search engine, which no longer exists in its earlier form. The table does not contain the results of Girju *et al.* (2005), who achieved 83.10 % accuracy, but used a

---

[12] In fact, the differences are negligible; their system achieved very similar result on the half split as well as on the whole set (personal communication).

Table 2. NC bracketing, Biomedical dataset.

Model	$\checkmark$	$\times$	$\emptyset$	Accuracy (%)	Coverage (%)
# adjacency	374	56	0	86.98	100.00
Pr adjacency	353	77	0	82.09	100.00
MI adjacency	372	58	0	86.51	100.00
χ^2 **adjacency**	379	51	0	**88.14**	100.00
# dependency	374	56	0	86.98	100.00
Pr dependency	369	61	0	85.81	100.00
MI dependency	369	61	0	85.81	100.00
χ^2 **dependency**	380	50	0	**88.37**	100.00
# adjacency (*)	373	57	0	86.74	100.00
# adjacency (**)	358	72	0	83.26	100.00
# adjacency (***)	334	88	8	79.15	98.14
# adjacency (*, rev.)	370	59	1	86.25	99.77
# adjacency (**, rev.)	367	62	1	85.55	99.77
# adjacency (***, rev.)	351	79	0	81.63	100.00
Concatenation adj	370	47	13	88.73	96.98
Concatenation dep.	366	43	21	**89.49**	95.12
Concatenation triple	238	37	155	**86.55**	63.95
Inflection Variability	198	49	183	80.16	57.44
Swap first two words	90	18	322	83.33	25.12
Reorder	320	78	32	80.40	92.56
Abbreviations	133	23	274	**85.25**	36.27
Possessives	48	7	375	**87.27**	12.79
Paraphrases	383	44	3	**89.70**	99.30
Surface features (sum)	382	48	0	**88.84**	100.00
Majority vote	403	17	10	95.95	97.67
Majority vote → right	*410*	*20*	*0*	*95.35*	*100.00*
Baseline (choose left)	361	69	0	83.95	100.00

supervised algorithm and targeted bracketing *in context*. They further "shuffled" the Lauer's set, mixing it with additional data, thus making their results even harder to compare to these in the table.

The results for the *Biomedical dataset* are shown in Table 5. In addition to probabilities (Pr), we also use counts (#) and χ^2 (with the dependency and the adjacency models). The prepositional paraphrases are much more accurate: 93.3 % (with 83.62 % coverage). By combining the paraphrases with the χ^2 models in a majority vote, and by assigning the undecided cases to right-bracketing, we achieve 92.24 % accuracy, which is slightly worse than 95.35 % we achieved using the Web. This difference is not statistically significant,[13] which suggests that in some cases a big domain-specific corpus with suitable linguistic annotations could be a possible alternative to using the Web. This is not true, however, for general domain compounds: for example, our subset of MEDLINE can

---

[13] Note however that here we experiment with 232 of the 430 examples.

Table 3. NC bracketing, comparison to previous unsupervised results on
Lauer's set. The results of Keller and Lapata are on half of Lauer's set and thus are
only indirectly comparable (note the different baseline).

Model	Accuracy
LEFT (baseline)	66.80
Lauer adjacency	68.90
Lauer dependency	77.50
Our χ^2 dependency	*79.92*
Lauer tuned	80.70
"Upper bound" (humans - Lauer)	81.50
Our majority vote → left	*89.34*
Keller and Lapata: LEFT (baseline)	63.93
Keller and Lapata: best BNC	68.03
Keller and Lapata: best AltaVista	78.68

Table 4. NC bracketing, surface features analysis (%s), for the biomedical set.

Example	Predicts	Accuracy	Coverage
brain-stem cells	left	**88.22**	**92.79**
brain stem's cells	left	**91.43**	**16.28**
(brain stem) cells	left	96.55	6.74
brain stem (cells)	left	100.00	1.63
brain stem, cells	left	96.13	42.09
brain stem: cells	left	97.53	18.84
brain stem cells-death	left	80.69	60.23
brain stem cells/tissues	left	83.59	45.35
brain stem Cells	left	**90.32**	**36.04**
brain stem/cells	left	100.00	7.21
brain. stem cells	left	97.58	38.37
brain stem-cells	*right*	*25.35*	*50.47*
brain's stem cells	right	**55.88**	**7.90**
(brain) stem cells	*right*	*46.67*	*3.49*
brain (stem cells)	*right*	*0.00*	*0.23*
brain, stem cells	right	54.84	14.42
brain: stem cells	*right*	*44.44*	*6.28*
rat-brain stem cells	*right*	*17.97*	*68.60*
neural/brain stem cells	*right*	*16.36*	*51.16*
brain Stem cells	*right*	*24.69*	*18.84*
brain/stem cells	right	53.33	3.49
brain stem. cells	*right*	*39.34*	*14.19*

Table 5. NC bracketing, results on the *Biomedical dataset* using 1.4M MED-LINE abstracts. For each model, the number of correctly classified, wrongly classified, and non-classified examples is shown, followed by accuracy and coverage (in %).

Model	$\sqrt{}$	$\times$	$\emptyset$	Accuracy	Coverage
# adjacency	196	36	0	84.48±5.22	100.00
Pr adjacency	173	59	0	74.57±5.97	100.00
χ^2 adjacency	200	32	0	86.21±5.03	100.00
# dependency	195	37	0	84.05±5.26	100.00
Pr dependency	193	39	0	83.19±5.34	100.00
χ^2 dependency	196	36	0	84.48±5.22	100.00
PrepPar	181	13	38	93.30±4.42	83.62
PP+χ^2adj+χ^2dep	207	13	12	94.09±3.94	94.83
PP+χ^2adj+χ^2dep→right	214	18	0	**92.24±4.17**	100.00
Baseline (choose left)	193	39	0	83.19±5.34	100.00

provide prepositional paraphrases for only 23 of the 244 examples in *Lauer's dataset* (i.e., for less than 10 %), and for 12 of them the predictions are wrong (i.e., the accuracy is below 50 %).

3   Prepositional Phrase Attachment

3.1   The Problem

A long-standing challenge for syntactic parsers is the attachment decision for prepositional phrases. In a configuration where a verb takes a noun complement that is followed by a prepositional phrase (PP), the problem arises of whether the PP attaches to the noun or to the verb. Consider the following contrastive pair of sentences:

(1) *Peter spent millions of dollars.*   (noun)
(2) *Peter spent time with his family.* (verb)

In the first example, the PP *millions of dollars* attaches to the noun *millions*, while in the second the PP *with his family* attaches to the verb *spent*.

Past work on PP-attachment has often cast these associations as the quadruple (v, n_1, p, n_2), where v is the verb, n_1 is the head of the direct object, p is the preposition (the head of the PP) and n_2 is the head of the NP inside the PP. For example, the quadruple for (2) is *(spent, time, with, family)*.

Early work on PP-attachment ambiguity resolution relied on syntactic, e.g., "minimal attachment" and "right association", and pragmatic considerations. Most recent work can be divided into supervised and unsupervised approaches. Supervised approaches tend to make use of semantic classes or thesauri in order to deal with data sparseness problems. Brill and Resnik (1994) used the supervised transformation-based learning method and lexical and conceptual classes derived from WordNet, achieving 82 % accuracy on 500 randomly selected examples. Ratnaparkhi *et al.* (1994) created a benchmark dataset of 27,937 quadruples (v, n_1, p, n_2), extracted from the Wall Street Journal. They found the human performance on this task to be 88 %.[14] Using this dataset, they trained a maximum entropy model and a binary hierarchy of word classes derived by mutual information, achieving 81.6 % accuracy. Collins and Brooks (1995) used a supervised back-off model to achieve 84.5 % accuracy on the Ratnaparkhi test set. Stetina and Makoto (1997) used a supervised method with a decision tree and Word-Net classes to achieve 88.1 % accuracy on the same test set. Toutanova *et al.* (2004) used a supervised method that makes use of morphological and syntactic analysis and WordNet synsets, yielding 87.5 % accuracy.

In the unsupervised approaches, the attachment decision depends largely on co-occurrence statistics drawn from text collections. The pioneering work in this area was that of Hindle and Rooth (1993). Using a partially parsed corpus, they calculated and compared lexical associations over subsets of the tuple (v, n_1, p), ignoring n_2, and achieved 80 % accuracy at 80 % coverage.

More recently, Ratnaparkhi (1998) developed an unsupervised method that collects statistics from text annotated with part-of-speech tags and morphological base forms. An extraction heuristic is used to identify unambiguous attachment decisions, for example, the algorithm can assume a noun attachment if there is no verb within k words to the left of the preposition in a given sentence, among other conditions. This extraction heuristic uncovered 910 K unique tuples of the form (v, p, n_2) and (n_1, p, n_2), although the results are very noisy, suggesting the correct attachment only about 69 % of the time. The tuples are used as training data for classifiers, the best of which achieves 81.9 % accuracy on the Ratnaparkhi test set. Pantel and Lin (2000) described an unsupervised method that uses a collocation database, a thesaurus, a dependency parser, and a large corpus (125M words), achieving 84.3 % accuracy on the Ratnaparkhi test set. Using simple combinations of Web-derived n-grams, Lapata and Keller (2005) achieved lower results, in the low 70s.

Using a different collection consisting of German PP-attachment decisions, Volk (2000) used the Web to obtain n-gram counts. He compared $\Pr(p|n_1)$ to $\Pr(p|v)$, where $\Pr(p|x) = \#(x, p)/\#(x)$. Here x can be n_1 or v. The bigram frequencies $\#(x, p)$ were obtained using the AltaVista NEAR operator. The method was able to make a decision on 58 % of the examples with 75 % accuracy (baseline 63 %). Volk (2001) then improved on these results by comparing $\Pr(p, n_2|n_1)$ to $\Pr(p, n_2|v)$. Using inflected forms, he achieved 75 % accuracy and 85 % coverage.

Calvo and Gelbukh (2003) experimented with a variation of this, using exact phrases instead of the NEAR operator. For example, to disambiguate *"Veo al gato con un telescopio."*, they compared frequencies for phrases such as "ver con telescopio" and "gato con telescopio". They tested this idea on 181 randomly chosen Spanish disambiguation examples, achieving 91.97 % accuracy and 89.5 % coverage.

---

[14] When presented with a whole sentence, average humans score 93 %.

3.2 Models and Features

n-gram Models. We used two co-occurrence models:

(*i*) $\Pr(p|n_1)$ vs. $\Pr(p|v)$
(*ii*) $\Pr(p, n_2|n_1)$ vs. $\Pr(p, n_2|v)$.

Each of these was computed in two different ways: using Pr (probabilities) and # (frequencies). We estimated the n-gram counts using exact phrase queries (with inflections, derived from WordNet 2.0) using the MSN Search Engine. We also allowed for determiners, where appropriate, e.g., between the preposition and the noun when querying for $\#(p, n_2)$. We added up the frequencies for all possible variations. Web frequencies were reliable enough and did not need smoothing for (*i*), but for (*ii*), smoothing using the technique described in Hindle and Rooth (1993) led to better coverage. We also tried back-off from (*ii*) to (*i*), as well as back-off plus smoothing, but did not find improvements over smoothing alone. We found n-gram counts to be unreliable when pronouns appear in the test set rather than nouns, and disabled them in these cases. Such examples can still be handled by paraphrases or surface features (see below).

Web-Derived Surface Features. We used various surface features as we did for NC bracketing. For example, *John opened the door with a key* is a difficult verb attachment example because doors, keys, and opening are all semantically related. To determine if this should be a verb or a noun attachment, we search for cues that indicate which of these terms tend to associate most closely. If we see parentheses used as follows:

> *"open the door (with a key)"*

this suggests a verb attachment, since the parentheses signal that "with a key" acts as its own unit. Similarly, hyphens, colons, capitalization, and other punctuation can help signal disambiguation decisions. For *John ate spaghetti with sauce*, if we see

> *"eat: spaghetti with sauce"*

this suggests a noun attachment.

Table 6 illustrates a wide variety of surface features, along with the attachment decisions they are assumed to suggest (we ignored events with a frequency of 1). The surface features for PP-attachment have low coverage: for most of the examples, we could not extract any surface features.

Paraphrases. We further paraphrased the relation of interest, checking whether it can be found in its alternative form, which could suggest an attachment decision. We used the following patterns along with their associated attachment predictions:

(1) $v\ n_2\ n_1$ (noun)
(2) $v\ p\ n_2\ n_1$ (verb)
(3) $p\ n_2\ *\ v\ n_1$ (verb)
(4) $n_1\ p\ n_2\ v$ (noun)
(5) $v\ pronoun\ p\ n_2$ (verb)
(6) $be\ n_1\ p\ n_2$ (noun)

The idea behind Pattern (1) is to determine if "$n_1\ p\ n_2$" can be expressed as a noun compound; if this happens sufficiently often, we can predict a noun attachment. For example, $meet/v\ demands/n_1\ from/p\ customers/n_2$ becomes $meet/v\ the\ customers/n_2\ demands/n_1$.

Note that the pattern could wrongly target ditransitive verbs, e.g., it could turn $gave/v\ an\ apple/n_1\ to/p\ him/n_2$ into $gave/v\ him/n_2\ an\ apple/n_1$. To prevent this, we do not allow a determiner before n_1, but we do require one before n_2. In addition, we disallow the pattern if the preposition is to and we require both n_1 and n_2 to be nouns (as opposed to numbers, percents, pronouns, determiners, etc.).

Table 6. PP-attachment surface features. Accuracy and coverage shown are across all examples, not just the door example shown.

Example	Predicts	Accuracy (%)	Coverage (%)
open Door with a key	noun	100.00	0.13
(open) door with a key	noun	66.67	0.28
open (door with a key)	noun	71.43	0.97
open - door with a key	noun	69.70	1.52
open / door with a key	noun	60.00	0.46
open, door with a key	noun	65.77	5.11
open: door with a key	noun	64.71	1.57
open; door with a key	noun	60.00	0.23
open. door with a key	noun	64.13	4.24
open? door with a key	noun	83.33	0.55
open! door with a key	noun	66.67	0.14
open door With a Key	verb	0.00	0.00
(open door) with a key	verb	50.00	0.09
open door (with a key)	verb	73.58	2.44
open door - with a key	verb	68.18	2.03
open door / with a key	verb	100.00	0.14
open door, with a key	verb	58.44	7.09
open door: with a key	verb	70.59	0.78
open door; with a key	verb	75.00	0.18
open door. with a key	verb	60.77	5.99
open door! with a key	verb	100.00	0.18

Pattern (2) predicts a verb attachment. It presupposes that "p n_2" is an indirect object of the verb v and tries to switch it with the direct object n_1, e.g., had/v a $program/n_1$ in/p $place/n_2$ → had/v in/p $place/n_2$ a $program/n_1$. We require n_1 to be preceded by a determiner (to prevent "n_2 n_1" from forming a noun compound).

Pattern (3) looks for appositions, where the PP has moved in front of the verb, e.g., to/p him/n_2 I $gave/v$ an $apple/n_1$. The symbol * indicates a wildcard position where we allow up to three intervening words.

Pattern (4) looks for appositions, where the PP has moved in front of the verb together with n_1. It would transform $shaken/v$ $confidence/n_1$ in/p $markets/n_2$ into $confidence/n_1$ in/p $markets/n_2$ $shaken/v$.

Pattern (5) is motivated by the observation that if n_1 is a pronoun, this suggests a verb attachment (Hindle and Rooth 1993); a separate feature checks if n_1 is a pronoun. The pattern substitutes n_1 with him or her, e.g., it will convert put/v a $client/n_1$ at/p $odds/n_2$ into put/v him at/p $odds/n_2$.

Pattern (6) is motivated by the observation that the verb to be is typically used with a noun attachment; a separate feature checks whether v is a form of the verb to be. This pattern substitutes v with is and are, e.g., it could transform eat/v $spaghetti/n_1$ $with/p$ $sauce/n_2$ into is $spaghetti/n_1$ $with/p$ $sauce/n_2$.

These patterns all allow for determiners where appropriate, unless explicitly stated otherwise. For a given example, a prediction is made if at least one instance of the pattern has been found.

3.3   Evaluation

For the evaluation, we used the test part (3,097 examples) of the benchmark dataset by Ratnaparkhi et $al.$ (1994). We used all 3,097 test examples in order to make our results directly comparable.

Unfortunately, there are numerous errors in the test set.[15] There are 149 examples in which a bare determiner is labeled as n_1 or n_2 rather than the actual head noun. Supervised algorithms can deal with this problem by learning from the training set that "the" can act as a noun in this collection, but unsupervised algorithms cannot do so.

Moreover, there are around 230 examples in which the nouns contain special symbols such as %, slash, &,', which are lost when querying against a search engine. This poses a problem for our algorithm, but this is not a problem with the test set itself.

The results are shown in Table 7. Following Ratnaparkhi (1998), we predict a noun attachment if the preposition is of (a very reliable heuristic). The table shows the performance for each feature in isolation (excluding examples whose preposition is of). The surface features are represented by a single score in Table 7: for a given example, we sum up separately the number of noun- and verb-attachment pattern matches, and we assign the attachment with the larger number of matches.

---

[15] Ratnaparkhi (1998) noted that the test set contains errors, but did not correct them.

Table 7. PP-attachment results, in %.

Model	Accuracy (%)	Coverage (%)
Baseline (noun attach)	41.82	100.00
$\#(x,p)$	58.91	83.97
$\Pr(p\|x)$	66.81	83.97
$\Pr(p\|x)$ smoothed	**66.81**	83.97
$\#(x,p,n_2)$	65.78	81.02
$\Pr(p,n_2\|x)$	68.34	81.62
$\Pr(p,n_2\|x)$ smoothed	**68.46**	83.97
(1) "$v\ n_2\ n_1$"	**59.29**	22.06
(2) "$p\ n_2\ v\ n_1$"	**57.79**	71.58
(3) "$n_1\ *\ p\ n_2\ v$"	**65.78**	20.73
(4) "$v\ p\ n_2\ n_1$"	**81.05**	8.75
(5) "$v\ pronoun\ p\ n_2$"	**75.30**	30.40
(6) "$be\ n_1\ p\ n_2$"	**63.65**	30.54
n_1 is $pronoun$	**98.48**	3.04
v is $to\ be$	**79.23**	9.53
Surface features (summed)	**73.13**	9.26
Maj. vote, of $\rightarrow$ *noun*	85.01±1.21	91.77
Maj. vote, of $\rightarrow$ *noun, N/A* $\rightarrow$ *verb*	**83.63±1.30**	**100.00**

We combined the bold rows of Table 7 in a majority vote (assigning noun attachment to all *of* instances), obtaining 85.01 % accuracy and 91.77 % coverage. To get 100 % coverage, we assigned all undecided cases to *verb*, since the majority of the remaining non-*of* instances attach to the verb, which yielded 83.63 % accuracy. We show 0.95-level confidence intervals for the accuracy, computed by a general method based on constant Chi-square boundaries (Fleiss 1981).

A test for statistical significance reveals that our results are as strong as those of the leading unsupervised approach on this collection (Pantel and Lin 2000). Unlike that work, we do not require a collocation database, a thesaurus, a dependency parser, nor a large domain-dependent text corpus, which makes our approach easier to implement and to extend to other languages.

4 Coordination

4.1 The Problem

Coordinating conjunctions such as *and*, *or*, *but*, etc., pose major challenges to parsers and their proper handling is essential for the understanding of the sentence. Consider the following somewhat "cooked" example:

*The Department of Chronic Diseases **and** Health Promotion leads **and** strengthens global efforts to prevent **and** control chronic diseases **or** disabilities **and** to promote health **and** quality of life.*

Conjunctions can link two words, two constituents, e.g., NPs, two clauses or even two sentences. Thus, the first challenge is to identify the boundaries of the conjuncts of each coordination. The next problem comes from the interaction of the coordinations with other constituents that attach to its conjuncts (most often prepositional phrases). In the example above, we need to decide between two structures: *[health and [quality of life]]* and *[[health and quality] of life]*. Semantically, we also need to determine whether the *or* in *chronic diseases or disabilities* really means *or* or is used as an *and* (Agarwal and Boggess 1992). Finally, we need to choose between a *non-elided* and an *elided* reading:

[[chronic diseases] or disabilities] vs. [chronic [diseases or disabilities]]

Below we focus on a special case of the latter problem: noun compound coordination. Consider the NC *car and truck production*. What it really means is *car production and truck production*. However, due to the principle of economy of expression, the first instance of *production* has been compressed out by means of ellipsis. In contrast, in *president and chief executive*, *president* is coordinated with *chief executive*. There is also an all-way coordination, where the conjunct is part of the whole, as in *Securities and Exchange Commission*.

More formally, we consider configurations of the kind n_1 c n_2 h, where n_1 and n_2 are nouns, c is a coordination (*and* or *or*) and h is the head noun.[16] The task is to decide whether there is ellipsis or not, independently of the local context. Syntactically, this can be expressed by the following two bracketings: $[[n_1 \ c \ n_2] \ h]$ vs. $[n_1 \ c \ [n_2 \ h]]$. In order to make the task more realistic (from a parser's perspective), we ignore the option of all-way coordination and we try to predict the bracketing in Penn Treebank (Marcus *et al.* 1994) for configurations of this kind. The Penn Treebank brackets NCs with ellipsis as, e.g.,

(NP car/NN and/CC truck/NN production/NN)

and without ellipsis as

(NP (NP president/NN) and/CC (NP chief/NN executive/NN))

The NPs with ellipsis are flat, while the others contain internal NPs. The all-way coordinations can appear bracketed either way and make the task harder.

Coordination ambiguity is under-explored, despite being one of the three major sources of structural ambiguity (together with prepositional phrase attachment and noun compound bracketing), and belonging to the class of ambiguities for which the number of analyses is the number of binary trees over the corresponding nodes (Church and Patil 1982), and despite the fact that conjunctions are among the most frequent words.

---

[16] The configurations of the kind $n \ h_1 \ c \ h_2$ (e.g., *company/n cars/h_1 and/c trucks/h_2*) can be handled in a similar way.

Rus *et al.* (2002) presented a deterministic rule-based approach for bracketing *in context* of coordinated NCs of the kind $n_1 \, c \, n_2 \, h$, as a necessary step towards logical form derivation. Their algorithm used POS tagging, syntactic parses, semantic senses of the nouns (manually annotated), lookups in a semantic network (WordNet) and the type of the coordination conjunction to make a 3-way classification: ellipsis, no ellipsis and all-way coordination. Using a back-off sequence of 3 different heuristics, they achieved 83.52 % accuracy (baseline 61.52 %) on a set of 298 examples. When 3 additional context-dependent heuristics and 224 additional examples with local contexts were added, the precision jumped to 87.42 % (baseline 52.35 %), with 71.05 % coverage.

Resnik (1999b) worked with the following two patterns: n_1 *and* $n_2 \, n_3$ and $n_1 \, n_2$ *and* $n_3 \, n_4$, e.g., *[food/n_1 [handling/n_2 and/c storage/n_3] procedures/n_4]*. While there are two options for the former (all-way coordinations are not allowed), there are 5 valid bracketings for the latter. Following Kurohashi and Nagao (1992), Resnik made decisions based on similarity of form (i.e., number agreement: Acc=53 %, Cov=90.6 %), similarity of meaning (Acc=66 %, Cov=71.2 %) and conceptual association (Acc=75.0 %, Cov=69.3 %). Using a decision tree to combine the three information sources, he achieved 80 % accuracy (baseline 66 %) at 100 % coverage for the 3-noun coordinations. For the 4-noun coordinations, the accuracy was 81.6 % (baseline 44.9 %), 85.4 % coverage.

Chantree *et al.* (2005) covered a large set of ambiguity types, not limited to nouns. They allowed the head word to be a noun, a verb or an adjective, and the modifier to be an adjective, a preposition, an adverb, etc. They extracted distributional information from the British National Corpus and distributional similarities between words, similarly to (Resnik 1999b). In two different experiments, they achieved Acc=88.2 %, Cov=38.5 % and Acc=80.8 %, Cov=53.8 % (baseline Acc=75 %).

Goldberg (1999) resolved the *attachment of ambiguous coordinate phrases* of the kind $n_1 \, p \, n_2 \, c \, n_3$, e.g., *box/$n_1$ of/p chocolates/n_2 and/c roses/n_3*. Using an adaptation of the algorithm proposed by Ratnaparkhi (1998) for PP-attachment, she achieved Acc=72 % (baseline: 64 %) for Cov=100.00 %.

Agarwal and Boggess (1992) focused on the *identification of the conjuncts of coordinate conjunctions*. Using POS and case labels in a deterministic algorithm, they achieved Acc=81.6 %. Kurohashi and Nagao (1992) worked on the same problem for Japanese. Their algorithm looked for similar word sequences among with sentence simplification, achieving 81.3 % accuracy.

4.2   Models and Features

n-gram Models. We used the following n-gram models:

(*i*) $\#(n_1, h)$ vs. $\#(n_2, h)$
(*ii*) $\#(n_1, h)$ vs. $\#(n_1, c, n_2)$

Model (*i*) compares how likely it is that n_1 modifies h, as opposed to n_2 modifying h. Model (*ii*) checks which association is stronger: between n_1 and h,

Table 8. Coordination surface features. Accuracy and coverage shown are across all examples, not just the *buy and sell orders* shown.

Example	Predicts	Accuracy (%)	Coverage (%)
(buy) and sell orders	NO ellipsis	33.33	1.40
buy (and sell orders)	NO ellipsis	70.00	4.67
buy: and sell orders	NO ellipsis	0.00	0.00
buy; and sell orders	NO ellipsis	66.67	2.80
buy. and sell orders	NO ellipsis	68.57	8.18
buy[...] and sell orders	NO ellipsis	49.00	46.73
buy- and sell orders	ellipsis	77.27	5.14
buy and sell / orders	ellipsis	50.54	21.73
(buy and sell) orders	ellipsis	92.31	3.04
buy and sell (orders)	ellipsis	90.91	2.57
buy and sell, orders	ellipsis	92.86	13.08
buy and sell: orders	ellipsis	93.75	3.74
buy and sell; orders	ellipsis	100.00	1.87
buy and sell. orders	ellipsis	93.33	7.01
buy and sell[...] orders	ellipsis	85.19	18.93

or between n_1 and n_2. Regardless of whether the coordination is *or* or *and*, we query for both and we add up the corresponding counts.

Web-Derived Surface Features. The set of surface features is similar to the one we used for PP-attachment. These are brackets, slash, comma, colon, semicolon, dot, question mark, exclamation mark, and any character. There are two additional ellipsis-predicting features: a dash after n_1 and a slash after n_2, see Table 8.

Paraphrases. We further used the following paraphrase patterns:

(1) $n_2 \ c \ n_1 \ h$ (ellipsis)
(2) $n_2 \ h \ c \ n_1$ (NO ellipsis)
(3) $n_1 \ h \ c \ n_2 \ h$ (ellipsis)
(4) $n_2 \ h \ c \ n_1 \ h$ (ellipsis)

If matched frequently enough, each of these patterns predicts the coordination decision indicated in parentheses. If found only infrequently or not found at all, the opposite decision is made. Pattern (1) switches the places of n_1 and n_2 in the coordinated NC. For example, *bar and pie graph* would be transformed to *pie and bar graph*, founding which on the Web would favor ellipsis. Pattern (2) moves n_2 and h together to the left of the coordination conjunction, and places n_1 to the right. If this happens frequently enough, there is no ellipsis. Pattern

Table 9. Coordination results, in percentages.

Model	Accuracy (%)	Coverage (%)
Baseline: ellipsis	56.54	100.00
(n_1, h) vs. (n_2, h)	**80.33**	28.50
(n_1, h) vs. (n_1, c, n_2)	61.14	45.09
(n_2, c, n_1, h)	**88.33**	14.02
(n_2, h, c, n_1)	**76.60**	21.96
(n_1, h, c, n_2, h)	**75.00**	6.54
(n_2, h, c, n_1, h)	**78.67**	17.52
Heuristic 1	**75.00**	0.93
Heuristic 4	64.29	6.54
Heuristic 5	61.54	12.15
Heuristic 6	**87.09**	7.24
Number agreement	**72.22**	46.26
Surface sum	**82.80**	21.73
Majority vote	83.82	80.84
Majority vote, N/A → no ellipsis	***80.61***	***100.00***

(3) inserts the elided head h after n_1 with the hope that if there is ellipsis, we will find the full phrase elsewhere in the data. Pattern (4) combines patterns (1) and (3); it not only inserts h after n_1, but also switches the places of n_1 and n_2.

As shown in Table 9, we further included four of the heuristics by Rus *et al.* (2002). Heuristic 1 predicts that there is no coordination when n_1 and n_2 are the same, e.g., *milk and milk products*. Heuristics 2 and 3 perform a lookup in WordNet and we did not use them. Heuristics 4, 5 and 6 exploit the local context, namely the adjectives modifying n_1 and/or n_2. Heuristic 4 predicts no ellipsis if both n_1 and n_2 are modified by adjectives. Heuristic 5 predicts ellipsis if the coordination is *or* and n_1 is modified by an adjective, but n_2 is not. Heuristic 6 predicts no ellipsis if n_1 is not modified by an adjective, but n_2 is. We used versions of heuristics 4, 5 and 6 that check for determiners rather than adjectives.

Finally, we included the number agreement feature (Resnik 1993): (a) if n_1 and n_2 match in number, but n_1 and h do not, predict ellipsis; (b) if n_1 and n_2 do not match in number, but n_1 and h do, predict no ellipsis; (c) otherwise leave undecided.

4.3 Evaluation

We evaluated the algorithms on a collection of 428 examples that we extracted from the Penn Treebank (Nakov and Hearst 2005c). On extraction, determiners and non-noun modifiers were allowed, but the program was only presented with the quadruple (n_1, c, n_2, h). As Table 9 shows, our overall performance of 80.61 % is on par with other approaches, whose best scores fall into the low 80's for accuracy; direct comparison is not possible, as the tasks and the datasets differ.

As Table 9 shows, n-gram model (i) performs well, but n-gram model (ii) performs poorly, probably because the (n_1, c, n_2) contains three words, as opposed to two for (n_1, h), and thus a priori is less likely to be observed.

The surface features are less effective for resolving coordinations. As Table 8 shows, they are very good predictors of ellipsis, but are less reliable when predicting NO ellipsis. We combined the bold rows of Table 9 in a majority vote, obtaining 83.82 % accuracy, 80.84 % coverage. We assigned all undecided cases to no ellipsis, which yielded 80.61 % accuracy.

5   On the Stability of Web Page Hit Estimates

5.1   Problems and Limitations

Web search engines provide a convenient way for researchers to obtain statistics over an enormous corpus, but using them for this purpose is not without drawbacks. We will discuss these drawbacks below; see (Nakov and Hearst 2005b: Nakov 2007; Kilgarriff 2007) for further discussion.

First, there are limitations on what kinds of queries can be issued, mainly because of the lack of linguistic annotation. For example, if we want to estimate the probability that *health* precedes *care* $\frac{\#("health\ care")}{\#(care)}$, we need the frequencies of *"health care"* and *care*, where both *health* and *care* are used as nouns. Unfortunately, a query for *care* will return not only noun uses but also many verb uses, while a query for *health care* would return results where *care* is almost always a noun. Even when both *health* and *care* are used as nouns and are adjacent, they may belong to different NPs, but sit next to each other only by chance. Furthermore, since search engines ignore punctuation characters, the two nouns may also come from different sentences.

Web search engines also prevent querying directly for terms containing hyphens or possessive markers such as *amino-acid sequence* and *protein synthesis' inhibition*. They also disallow querying for a term like *bronchoalveolar lavage (BAL) fluid*, which contains an internal parenthesized abbreviation. They also do not support queries that make use of generalized POS information such as

stem cells VERB PREP DET brain

in which the uppercase patterns stand for any verb, any preposition and any determiner, e.g., *stem cells derived from the brain*.

Furthermore, using page hits as a proxy for n-gram frequencies can produce some counter-intuitive results. Consider the bigrams w_1w_4, w_2w_4 and w_3w_4 and a page that contains each bigram exactly once. A search engine will contribute a page count of 1 for w_4 instead of a frequency of 3; thus the number of page hits for w_4 can be smaller than that for the sum of the bigrams that contain it. See (Keller and Lapata 2003) for more potential problems with page hits.

Another potential problem is instability of the n-gram counts. Today Web search engines are too complex to be run on a single machine, and instead the queries are served by hundreds, sometimes thousands of servers, which collaborate to produce the final result. Moreover, the Web is dynamic, since at any given time some pages disappear, some appear for the first time, and some change frequently. Thus search engines need to update their indexes frequently, and in fact the different engines compete on how "fresh" their indexes are. As a result, the number of page hits for a given query changes over time in unpredictable ways.

The indexes themselves are too big to be stored on a single machine and so are spread across multiple machines (Brin and Page 1998). For availability and efficiency reasons, there are also multiple copies of the same part of the index, and these are not always synchronized with one another since the different copies are updated at different times. As a result, if we issue the same query multiple times in rapid succession, we may connect to different physical machines and get different results. This is known as search engine "dancing".

From a research perspective, "dancing" and dynamics over time are potentially undesirable, as they preclude the exact replicability of any results obtained using search engines. At best, one could reproduce the same initial conditions, and expect similar outcomes.

Another potentially undesirable aspect of using Web search engines is that search engines often round their page hit estimates. This rounding is probably done because for most users' purposes exact counts are not necessary once the numbers get somewhat large, and computing the exact numbers is expensive if the index is distributed and continually changing. It might also indicate that under high load search engines sample from their indexes, rather than performing an exact computation. There have also been speculations on more nefarious reasons, e.g., see (Véronis 2005a; Véronis 2005c; Véronis 2005b).

It is unclear what the implications of these inconsistencies are on using the Web to obtain n-gram frequencies. If the estimates are close to accurate and consistent across queries, this should not have a big impact for most applications, since they only need the ratios of different n-grams.

Below we study the impact of rounding and inconsistencies in a suit of experiments organized around a real NLP task. We chose noun compound bracketing, which, while being a simple task, can be solved using several different methods which make use of n-grams of different lengths, as we have seen above.

5.2   Experiments and Results

We performed series of experiments comparing the accuracy of several of the above Web-based models for the problem of noun compound bracketing across four dimensions: (1) *search engine* (Google vs. Yahoo vs. MSN), (2) *time*, (3) *language filter* (English only vs. any), and (4) *inflected wordforms* usage.

In these experiments, we compared the results using the Chi square test for statistical significance as computed by (Lapata and Keller 2005). In nearly every case, we found that the differences were not statistically significant. The only exceptions are for *concatenation triple* in Tables 2 and 3 (marked with a *).

As above, we experimented with the dataset from (Lauer 1995), in order to produce results comparable to those of both Lauer and Keller and Lapata.

For all *n*-grams, we issued exact phrase queries within a single day. Unless otherwise stated, the queries were not inflected and no language filter was applied. We used a threshold of five for the difference between the left- and the right-predicting *n*-gram frequencies: we did not make a decision when the module of that difference was below that threshold. This slightly lowers the coverage, but potentially increases the accuracy.

Figures 1 and 2 show the variability over time for Google and for MSN Search respectively. (As Yahoo behaves similarly to Google, it is omitted here due to space limitations.) We chose time samples at varying time intervals in an attempt to capture index changes, in case they happen in the same fixed time intervals. For Google (see Fig. 1), we observe a low variability in the adjacency- and dependency-based models and a more sizable variability for the other models and features. The variability is especially high for *apostrophe* and *concatenation triple*: while in the first two time snapshots the accuracy of the apostrophes is much lower than in the last two, it is the reverse for concatenation.

MSN Search exhibits a more uniform behavior overall (see Fig. 2); however, while the variability in the adjacency- and dependency-based models is still a bit lower than that of the last five features, it is bigger than Google's. We think that this is due to rounding: because Google's counts are rounded, they change less over time, especially for very large counts. In contrast, these counts are exact for MSN Search, which makes its unigram and bigram counts more sensitive to variation. For the higher-order *n*-grams, both engines show higher variability: these counts are smaller, and so are more likely to be represented by exact numbers in Google, and they are also more sensitive to index updates for both search engines. However, the difference between the accuracy for May 4, 2005 and that for the other five dates is statistically significant for MSN Search only.

Figure 3 compares the three search engines at the same fixed time point. The biggest difference in accuracy is exhibited by *concatenation triple*; with MSN Search it achieves an accuracy of 92 %, which is better than the others' by 11 % (statistically significant). Other large variations (not statistically signif-

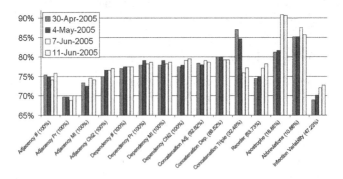

Fig. 1. Comparison over time for Google. Accuracy for any language, no inflections. Average coverage is shown in parentheses.

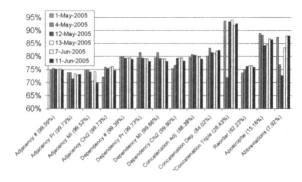

Fig. 2. Comparison over time for MSN Search. Accuracy for any language, no inflections. Average coverage is shown in parentheses.

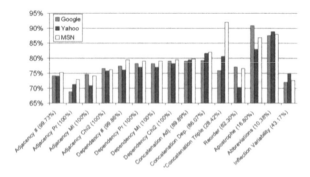

Fig. 3. Comparison by search engine. *Accuracy* (in %) for any language, no inflections. All results are for 6/6/2005. Average coverage is shown in parentheses.

icant) are seen for *apostrophe*, *reorder*, and to a lesser extent for the adjacency- and dependency-based models. As we expected, MSN Search looks best overall (especially on the unigram- and bigram-based models), which we attribute to the better accuracy of its *n*-gram estimates. Google is almost 5 % ahead of the others for *apostrophes* and *reorder*. Yahoo leads on *abbreviations* and *inflection variability*. The fact that different search engines exhibit strength on different kinds of queries and models shows the potential of combining them: in a majority vote combining some of the best models, we would choose *concatenation triple* from MSN Search and *apostrophe* from Google and *abbreviations* from Yahoo (together with *concatenation dependency*, χ^2 *dependency* and χ^2 *adjacency*). Figure 4 shows the corresponding coverage for some of the methods (it is about 100 % for the rest). We can see that Google exhibits slightly higher coverage, which suggests it might have a bigger index compared to Yahoo and MSN Search.

Figure 5 compares, on a fixed date (6/6/2005), for all the three search engines the impact of language filtering, meaning requiring only documents in English versus no restriction on language. The impact of the language filter on the accuracy seems minor and inconsistent for all three search engines: sometimes

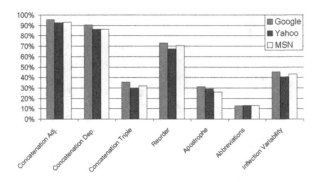

Fig. 4. Comparison by search engine. *Coverage* (in %) for any language, no inflections. All results are for 6/6/2005.

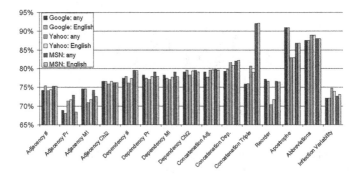

Fig. 5. Comparison by search engine: any language vs. English. *Accuracy* shown in %, no inflections. All results are for 6/6/2005.

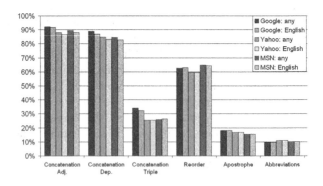

Fig. 6. Comparison by search engine: any language vs. English. *Coverage* shown in %, no inflections. All results are for 6/6/2005.

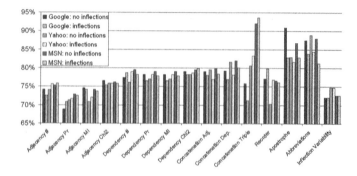

Fig. 7. Comparison by search engine: no inflections vs. using inflections. *Accuracy* shown in %, any language. All results are for 6/6/2005.

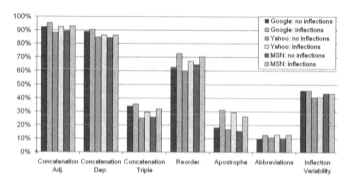

Fig. 8. Comparison by search engine: no inflections vs. using inflections. *Coverage* shown in %, any language. All results are for 6/6/2005.

the results are improved slightly and sometimes they are negatively impacted. Figure 6 compares the corresponding coverage for some of the models (the rest are omitted as the coverage for them is about 100 %). As we can see, using English only leads to a drop in coverage, as one would expect, but this drop is small.

Finally, Fig. 7 compares for the three search engines the impact of using inflections. When we estimate the frequency of a word, e.g., *tumor*, we also add up the frequencies of all possible variants, e.g., *tumors, tumour, tumours*. For bigrams, we inflect only the second word, and for n-grams only the last one. The results are again mixed, but the impact on accuracy is more significant compared to that of the language filter, especially on the high-order n-grams (of course, there is no impact on *inflection variability*). Figure 8 compares the corresponding coverage for some of the models (for the rest it is about 100 %). As one would expect, the coverage goes up when using inflection. The change for *apostrophe, reorder* and *concatenation triple* is again the biggest.

6 Measuring Semantic Similarity

The above problems were syntactic; here, we turn to semantic ones.

6.1 The Problem

Despite the tremendous amount of work on word similarity – see (Budanitsky and Hirst 2006) for an overview –, there is surprisingly little research on the important related problem of *relational similarity* – semantic similarity between pairs of words. Students who took the SAT test before 2005 or who are taking the GRE test nowadays are familiar with an instance of this problem – verbal analogy questions, which ask whether, e.g., the relationship between *ostrich* and *bird* is more similar to that between *lion* and *cat*, or rather between *primate* and *monkey*. These analogies are difficult, and the average test taker gives a correct answer 57 % of the time (Turney and Littman 2005).

Many NLP applications could benefit from solving relational similarity problems, including but not limited to question answering, information retrieval, machine translation, word sense disambiguation, and information extraction. For example, a relational search engine such as **TextRunner**, which serves queries like "find all X such that X causes wrinkles", asking for all entities that are in a particular relation with a given entity (Cafarella *et al.* 2006), needs to recognize that *laugh wrinkles* is an instance of **CAUSE-EFFECT**. See (Turney 2006; Nakov and Hearst 2008; Nakov and Kozareva 2011; Jurgens *et al.* 2012; Mikolov *et al.* 2013; Levy and Goldberg 2014) for further discussion on relational similarity, and (Girju *et al.* 2007; Girju *et al.* 2009; Hendrickx *et al.* 2010; Butnariu *et al.* 2010; Hendrickx *et al.* 2013; Nastase *et al.* 2013) for more detail on semantic relation extraction in general.

6.2 Model and Features

Given a pair of nouns, we try to characterize the semantic relation between them by leveraging the vast size of the Web to build linguistically-motivated lexically-specific features. We mine the Web for sentences containing the target nouns, and we extract the connecting verbs, prepositions, and coordinating conjunctions, which we use in a vector-space model to measure relational similarity.

The process of extraction starts with exact phrase queries issued against a Web search engine (*Google*) using the following patterns:

$$\text{"}infl_1 \text{ THAT } * \; infl_2\text{"}$$
$$\text{"}infl_2 \text{ THAT } * \; infl_1\text{"}$$
$$\text{"}infl_1 * infl_2\text{"}$$
$$\text{"}infl_2 * infl_1\text{"}$$

where: $infl_1$ and $infl_2$ are inflected variants of $noun_1$ and $noun_2$ generated using the *Java WordNet Library*[17]; THAT is a complementizer and can be *that*, *which*, or *who*; and * stands for 0 or more (up to 8) instances of *Google*'s star operator.

The first two patterns are subsumed by the last two and are used to obtain more sentences from the search engine since including, e.g., *that*, in the query changes the set of returned results and their ranking.

For each query, we collect the text snippets from the result set (up to 1,000 per query). We split them into sentences, and we filter out all incomplete ones and those that do not contain the target nouns. We further make sure that the word sequence following the second mentioned target noun is nonempty and contains at least one non-noun, thus ensuring the snippet includes the entire noun phrase: snippets representing incomplete sentences often end with a period anyway. We then perform POS tagging using the *Stanford POS tagger* (Toutanova *et al.* 2003) and shallow parsing with the *OpenNLP tools*,[18] and we extract the following types of features:

Verb: We extract a verb if the subject NP of that verb is headed by one of the target nouns (or an inflected form thereof), and its direct object NP is headed by the other target noun (or an inflected form). For example, the verb *include* will be extracted from "The *committee* <u>includes</u> many *members*." We also extract verbs from relative clauses, e.g., "This is a *committee* which <u>includes</u> many *members*." Verb particles are also recognized, e.g., "The *committee* must <u>rotate off</u> 1/3 of its *members*." We ignore modals and auxiliaries, but we retain the passive *be*. Finally, we lemmatize the main verb using *WordNet*'s morphological analyzer *Morphy* (Fellbaum 1998).

Verb+Preposition: If the subject NP of a verb is headed by one of the target nouns (or an inflected form thereof), and its indirect object is a PP containing an NP that is headed by the other target noun (or an inflected form thereof), we extract the verb and the preposition heading that PP, e.g., "The thesis advisory *committee* <u>consists of</u> three qualified *members*." As in the verb case, we extract verb+preposition from relative clauses, we include particles, we ignore modals and auxiliaries, and we lemmatize the verbs.

Preposition: If one of the target nouns is the head of an NP containing a PP with an internal NP headed by the other target noun (or an inflected form), we extract the preposition heading that PP, e.g., "The *members* <u>of</u> the *committee* held a meeting."

Coordinating Conjunction: If the two target nouns are the heads of coordinated NPs, we extract the coordinating conjunction.

In addition to the lexical part, for each extracted feature, we keep a direction. Therefore, the preposition *of* represents two different features in the following examples "*member* <u>of</u> the *committee*" and "*committee* <u>of</u> *members*". See Table 10 for examples.

---

[17] JWNL: http://jwordnet.sourceforge.net.
[18] OpenNLP: http://opennlp.sourceforge.net.

Table 10. The most frequent Web-derived features for *committee member*.
Here V stands for verb (possibly +preposition and/or +particle), P for preposition
and C for coordinating conjunction; $1 \rightarrow 2$ means *committee* precedes the feature and
member follows it; $2 \rightarrow 1$ means *member* precedes the feature and *committee* follows it.

Freq	Feature	POS	Direction
2205	of	P	$2 \rightarrow 1$
1923	be	V	$1 \rightarrow 2$
771	include	V	$1 \rightarrow 2$
382	serve on	V	$2 \rightarrow 1$
189	chair	V	$2 \rightarrow 1$
189	have	V	$1 \rightarrow 2$
169	consist of	V	$1 \rightarrow 2$
148	comprise	V	$1 \rightarrow 2$
106	sit on	V	$2 \rightarrow 1$
81	be chaired by	V	$1 \rightarrow 2$
78	appoint	V	$1 \rightarrow 2$
77	on	P	$2 \rightarrow 1$
66	and	C	$1 \rightarrow 2$
66	be elected	V	$1 \rightarrow 2$
58	replace	V	$1 \rightarrow 2$
48	lead	V	$2 \rightarrow 1$
47	be intended for	V	$1 \rightarrow 2$
45	join	V	$2 \rightarrow 1$
...	...	...	...
4	be signed up for	V	$2 \rightarrow 1$

We use the above-described features to calculate relational similarity, i.e.,
similarity between pairs of nouns. In order to downweigh very common features
such as *of*, we use TF.IDF-weighting:

$$w(x) = TF(x) \times \log\left(\frac{N}{DF(x)}\right) \quad (2)$$

In the above formula, $TF(x)$ is the number of times the feature x has been
extracted for the target noun pair, $DF(x)$ is the total number of training noun
pairs that have that feature, and N is the total number of training noun pairs.
Given two nouns and their TF.IDF-weighted frequency vectors A and B, we
calculate the similarity between them using the following generalized variant of
the Dice coefficient:

$$Dice(A, B) = \frac{2 \times \sum_{i=1}^{n} \min(a_i, b_i)}{\sum_{i=1}^{n} a_i + \sum_{i=1}^{n} b_i} \quad (3)$$

Other variants are also possible, e.g., Lin (1998).

6.3   Experiments and Evaluation

6.4   SAT Verbal Analogy

Following Turney (2006), we use *SAT verbal analogy* as a benchmark problem for relational similarity. We experiment with the 374 SAT questions collected by Turney and Littman (2005). Table 11 shows two sample questions: the top word pairs are called *stems*, the ones in italic are the *solutions*, and the remaining ones are *distractors*. Turney (2006) achieves 56 % accuracy on this dataset, which matches the average human performance of 57 %, and represents a significant improvement over the 20 % random-guessing baseline.

Table 11. SAT verbal analogy: sample questions. The stem is in **bold**, the correct answer is in *italic*, and the distractors are in plain text.

	Ostrich:Bird		Palatable:Toothsome
(a)	*lion:cat*	(a)	rancid:fragrant
(b)	goose:flock	(b)	chewy:textured
(c)	ewe:sheep	(c)	*coarse:rough*
(d)	cub:bear	(d)	solitude:company
(e)	primate:monkey	(e)	no choice

Note that the righthand side example in Table 11 is missing one distractor; so do 21 questions. The dataset also mixes different parts of speech: while *solitude* and *company* are nouns, all remaining words are adjectives. Other examples contain verbs and adverbs, and even relate pairs of different POS. This is problematic for our approach, which requires that both words be nouns[19]. After having filtered all examples containing non-nouns, we ended up with 184 questions, which we used in the evaluation.

Given a verbal analogy example, we build six feature vectors – one for each of the six word pairs. We then calculate the relational similarity between the stem of the analogy and each of the five candidates, and we choose the pair with the highest score; we make no prediction in case of a tie.

The evaluation results for leave-one-out cross-validation are shown in Table 12. We also show 95 %-confidence intervals for the accuracy. The last line in the table shows the performance of Turney's LRA when limited to the 184 noun-only examples. Our best model $v + p + c$ performs a bit better, 71.3 % vs. 67.4 %, but the difference is not statistically significant. However, this "inferred" accuracy could be misleading, and the LRA could have performed better if it was restricted to solve *noun-only* analogies, which seem easier than the general ones, as demonstrated by the significant increase in accuracy for LRA when limited to nouns: 67.4 % vs. 56 %.

---

[19] It can be extended to handle adjective-noun pairs as well, as demonstrated in Sect. 6.5 below.

Table 12. SAT verbal analogy: 184 noun-only examples. v stands for verb, p for preposition, and c for coordinating conjunction. For each model, the number of correct ($\checkmark$), wrong ($\times$), and nonclassified examples ($\varnothing$) is shown, followed by accuracy and coverage (in %).

Model	$\checkmark$	$\times$	$\varnothing$	Accuracy	Coverage
$v + p + c$	129	52	3	**71.3±7.0**	**98.4**
v	122	56	6	68.5±7.2	96.7
$v + p$	119	61	4	66.1±7.2	97.8
$v + c$	117	62	5	65.4±7.2	97.3
$p + c$	90	90	4	50.0±7.2	97.8
p	84	94	6	47.2±7.2	96.7
baseline	37	147	0	20.0±5.2	100.0
LRA	122	59	3	67.4±7.1	98.4

6.5 Head-Modifier Relations

Next, we experiment with the *Diverse* dataset of Barker and Szpakowicz (1998), which consists of 600 head-modifier pairs: noun-noun, adjective-noun and adverb-noun. Each example is annotated with one of 30 fine-grained relations, which are further grouped into the following 5 coarse-grained classes (the fine-grained relations are shown in parentheses): CAUSALITY (*cause, effect, purpose, detraction*), TEMPORALITY (*frequency, time_at, time_through*), SPATIAL (*direction, location, location_at, location_from*), PARTICIPANT (*agent, beneficiary, instrument, object, object_property, part, possessor, property, product, source, stative, whole*) and QUALITY (*container, content, equative, material, measure, topic, type*). For example, *exam anxiety* is classified as *effect* and therefore as CAUSALITY, and *blue book* is *property* and therefore also PARTICIPANT.

Some examples in the dataset are problematic for our method. First, in three cases, there are two modifiers, e.g., *infectious disease agent*, and we had to ignore the first one. Second, seven examples have an adverb modifier, e.g., *daily exercise*, and 262 examples have an adjective modifier, e.g., *tiny cloud*. We treat them as if the modifier was a noun, which works in many cases, since many adjectives and adverbs can be used predicatively, e.g., '*This exercise is performed daily.*' or '*This cloud looks very tiny.*'

For the evaluation, we created a feature vector for each head-modifier pair, and we performed leave-one-out cross-validation: we left one example for testing and we trained on the remaining 599 ones, repeating this procedure 600 times so that each example be used for testing. Following Turney and Littman (2005) we used a 1-nearest-neighbor classifier. We calculated the similarity between the feature vector of the testing example and each of the training examples' vectors. If there was a unique most similar training example, we predicted its class, and if there were ties, we chose the class predicted by the majority of the tied examples, if there was a majority.

Table 13. Head-modifier relations, 30 classes: evaluation on the *Diverse* dataset, micro-averaged (in %).

Model	✓	×	∅	Accuracy	Coverage
$v + p$	240	352	8	**40.5±3.9**	**98.7**
$v + p + c$	238	354	8	40.2±3.9	98.7
v	234	350	16	40.1±3.9	97.3
$v + c$	230	362	8	38.9±3.8	98.7
$p + c$	114	471	15	19.5±3.0	97.5
p	110	475	15	19.1±3.0	97.5
baseline	49	551	0	8.2±1.9	100.0
LRA	239	361	0	39.8±3.8	100.0

The results for the 30-class *Diverse* dataset are shown in Table 13. Our best model achieved 40.5 % accuracy, which is slightly better than LRA's 39.8 %, but the difference is not statistically significant.

Table 13 shows that the verbs are the most important features, yielding about 40 % accuracy regardless of whether used alone or in combination with prepositions and/or coordinating conjunctions; not using them results in 50 % drop in accuracy.

The reason coordinating conjunctions do not help is that head-modifier relations are typically expressed with verbal or prepositional paraphrases. Therefore, coordinating conjunctions only help with some infrequent relations like *equative*, e.g., finding *player and coach* on the Web suggests an equative relation for *player coach* (and for *coach player*).

As Table 12 shows, this is different for SAT verbal analogy, where verbs are still the most important feature type and the only whose presence/absence makes a statistical difference. However, this time coordinating conjunctions (with prepositions) do help a bit (the difference is not statistically significant) since SAT verbal analogy questions ask for a broader range of relations, e.g., antonymy, for which coordinating conjunctions like *but* are helpful.

6.6 Relations Between Nominals

We further experimented with the *SemEval'07* task 4 dataset (Girju *et al.* 2007), where each example consists of a sentence, a target semantic relation, two nominals to be judged on whether they are in that relation, manually annotated *WordNet* senses, and the Web query used to obtain the sentence:

```
"Among the contents of the <e1>vessel</e1> were a set of carpenter's
<e2>tools</e2>, several large storage jars, ceramic utensils, ropes and
remnants of food, as well as a heavy load of ballast stones."
WordNet(e1) = "vessel%1:06:00::",
WordNet(e2) = "tool%1:06:00::",
Content-Container(e2, e1) = "true",
Query = "contents of the * were a"
```

Table 14. Relations between nominals: evaluation on the *SemEval* dataset. Accuracy is macro-averaged (in %), up to 10 *Google* stars are used unless otherwise stated.

Model	Accuracy
$v + p + c + sent + query$ (type C)	**68.1±4.0**
v	67.9±4.0
$v + p + c$	67.8±4.0
$v + p + c + sent$ (type A)	**67.3±4.0**
$v + p$	66.9±4.0
$sent$ (sentence words only)	59.3±4.2
p	58.4±4.2
Baseline (majority class)	57.0±4.2
$v + p + c + sent + query$ (C), 8 stars	67.0±4.0
$v + p + c + sent$ (A), 8 stars	65.4±4.1
Best type C on *SemEval*	67.0±4.0
Best type A on *SemEval*	66.0±4.1

The following nonexhaustive and possibly overlapping relations are possible: Cause-Effect (e.g., *hormone-growth*), Instrument-Agency (e.g., *laser-printer*), Theme-Tool (e.g., *work-force*), Origin-Entity (e.g., *grain-alcohol*), Part-Whole (e.g., *leg-table*), Product-Producer (e.g., *honey-bee*), and Content-Container (e.g., *bananas-basket*). Each relation is considered in isolation; there are 140 training and at least 70 test examples per relation.

Given an example, we reduced the target entities e_1 and e_2 to single nouns by retaining their heads only. We then mined the Web for sentences containing these nouns, and we extracted the above-described feature types: verbs, prepositions and coordinating conjunctions. We further used the following problem-specific contextual feature types:

Sentence words: after stop words removal and stemming with the Porter (1980) stemmer;

Entity words: lemmata of the words in e_1 and e_2;

Query words: words part of the query string

Each feature type has a specific prefix, which prevents it from mixing with other feature types; the last feature type is used for type C only (see below).

The *SemEval* competition defined four types of systems, depending on whether the manually annotated *WordNet* senses and the *Google* query are used: A (WordNet=no, Query=no), B (WordNet=yes, Query=no), C (WordNet=no, Query=yes), and D (WordNet=yes, Query=yes). We experimented with types A and C only since we believe that having the manually annotated *WordNet* sense keys is an unrealistic assumption for a real-world application.

As before, we used a 1-nearest-neighbor classifier with TF.IDF-weighting, breaking ties by predicting the majority class on the training data. The evaluation results are shown in Table 14. We studied the effect of different subsets of features and of more *Google* star operators. As the table shows, using up to ten *Google* stars instead of up to eight (see Sect. 6.2) yields a slight improvement in accuracy for systems of both type A (65.4 % vs. 67.3 %) and type C (67.0 % vs. 68.1 %). Both results represent a statistically significant improvement over the majority class baseline and over using sentence words only, and a slight improvement over the best type A and type C systems on *SemEval'07*, which achieved 66 % and 67 % accuracy, respectively.[20]

6.7   Noun-Noun Compound Relations

The last dataset we experimented with was a subset of the 387 examples listed in the appendix of (Levi 1978). Levi's theory is one of the most important linguistic theories of the syntax and semantics of *complex nominals* – a general concept grouping together the partially overlapping classes of nominal compounds (e.g., *peanut butter*), nominalizations (e.g., *dream analysis*), and nonpredicate noun phrases (e.g., *electric shock*).

In Levi's theory, complex nominals can be derived from relative clauses by removing one of the following 12 abstract predicates: CAUSE$_1$ (e.g., *tear gas*), CAUSE$_2$ (e.g., *drug deaths*), HAVE$_1$ (e.g., *apple cake*), HAVE$_2$ (e.g., *lemon peel*), MAKE$_1$ (e.g., *silkworm*), MAKE$_2$ (e.g., *snowball*), USE (e.g., *steam iron*), BE (e.g., *soldier ant*), IN (e.g., *field mouse*), FOR (e.g., *horse doctor*), FROM (e.g., *olive oil*), and ABOUT (e.g., *price war*). In the resulting nominals, the modifier is typically the object of the predicate; when it is the subject, the predicate is marked with the index 2. The second derivational mechanism in the theory is nominalization; it produces nominals whose head is a nominalized verb.

Since we are interested in noun compounds only, we manually cleansed the set of 387 examples. We first excluded all concatenations (e.g., *silkworm*) and examples with adjectival modifiers (e.g., *electric shock*), thus obtaining 250 noun-noun compounds (*Levi-250* dataset). We further filtered out all nominalizations for which the dataset provides no abstract predicate (e.g., *city planner*), thus ending up with 214 examples (*Levi-214* dataset).

As in the previous experiments, for each of the 214 noun-noun compounds, we mined the Web for sentences containing both target nouns, from which we extracted paraphrasing verbs, prepositions and coordinating conjunctions. We then performed leave-one-out cross-validation experiments with a 1-nearest-neighbor classifier, trying to predict the correct predicate for the testing example. The results are shown in Table 15. As we can see, using prepositions alone yields about 33 % accuracy, which is a statistically significant improvement over the majority-class baseline. Overall, the most important features were the verbs: they yielded 45.8 % accuracy when used alone, and 50 % together with prepositions. Adding coordinating conjunctions helps a bit with verbs, but not with

---

[20] The best type B system on *SemEval* achieved 76.3 % accuracy using the manually-annotated *WordNet* senses in context for each example, which constitutes an additional data source, as opposed to an additional resource. The systems that used *WordNet* as a resource only, i.e., ignoring the manually annotated senses, were classified as type A or C. (Girju *et al.* 2007).

Table 15. Noun-noun compound relations, 12 classes: evaluation on *Levi-214* dataset. Shown are micro-averaged accuracy and coverage in %, followed by average number of features (ANF) and average sum of feature frequencies (ASF) per example. The righthand side reports the results when the query patterns involving THAT were not used. For comparison purposes, the top rows show the performance with the human-proposed verbs used as features.

Model	Accuracy	Coverage	ANF	ASF
Human: all v	78.4±6.0	99.5	34.3	70.9
Human: first v from each worker	72.3±6.4	99.5	11.6	25.5
$v + p + c$	50.0±6.7	99.1	216.6	1716.0
$v + p$	50.0±6.7	99.1	208.9	1427.9
$v + c$	46.7±6.6	99.1	187.8	1107.2
v	45.8±6.6	99.1	180.0	819.1
p	33.0±6.0	99.1	28.9	608.8
$p + c$	32.1±5.9	99.1	36.6	896.9
Baseline	19.6±4.8	100.0	–	–

prepositions. Note however that none of the differences between the different feature combinations involving verbs is statistically significant.

We also show the average number of distinct features and sum of feature counts per example: as we can see, there is a strong positive correlation between number of features and accuracy.

7 Summary

We have shown that simple unsupervised algorithms that make use of bigrams, surface features and paraphrases extracted from a very large corpus are effective for several structural ambiguity resolutions tasks, yielding results competitive with the best unsupervised results, and close to supervised results. The proposed method does not require labeled training data, nor lexicons nor ontologies. We think this is a promising direction for a wide range of NLP tasks. In future work, we intend to explore better-motivated evidence combination algorithms and to apply the approach to other NLP problems.

Moreover, we have shown that the variability over time and across search engines, as well as using language filters and morphologically inflected word-forms, does not significantly affect the results of an NLP application and thus does not greatly impact the interpretation of results obtained using Web-derived n-gram frequencies. In order to further bolster these results, we will need to perform similar studies for other NLP tasks, which make use of Web-derived n-gram estimates. We would also like to run similar experiments for languages other than English, where the language filter could be much more important, and where the impact of the inflection variability may differ, especially in the case of a morphologically rich language.

8   The Future

In the near future, we should keep an eye on the following resources and information sources:

Google Web 1 T 5-gram. A major issue with using the Web as a corpus is the need to issue a large number of queries against a search engine, which is not always feasible. Google has offered an alternative, namely to use a static collection of its Web n-grams, which they have released as a corpus: LDC2006T13. Subsequently, a version with syntactic annotations was prepared (Lin *et al.* 2010), but was not made publicly available.
Google Book N-grams. Google further released the Google Book N-grams,[21] which contain syntactic and part-of-speech annotations of books in multiple languages (Lin *et al.* 2012).
Microsoft Web N-grams Services. Unfortunately, the Google corpora were frozen in time, contained truncated counts and did not indicate which part of the webpage the n-gram is extracted from. The Microsoft Web N-grams Services address these issues by offering a Web interface to regularly updated, smoothed counts with clear origin (e.g., title vs. hyperlinks vs. body text) indication for the n-grams.[22] See (Wang *et al.* 2011) for an example showing why all this is so important.

Finally, we should note that the Web is not just text, but also images, video, etc. There is some interesting research on combining visual and textual information, e.g., for building bilingual dictionaries (Bergsma and Goebel 2011; Bergsma and Van Durme 2011).

Acknowledgements. This research was supported by NSF DBI-0317510, and a gift from Genentech.

References

Rajeev, A., Boggess, L.: A simple but useful approach to conjunct identification. In: Proceedings of ACL, pp. 15–21 (1992)

Michele, B., Brill, E.: Scaling to very very large corpora for natural language disambiguation. In: Proceedings of ACL (2001)

Bansal, M., Klein, D.: Web-scale features for full-scale parsing. In: Proceedings of the 49th Annual Meeting of the Association for Computational Linguistics: Human Language Technologies - vol.1, HLT 2011, pp. 693–702. PA, USA, Stroudsburg (2011)

Barker, K., Szpakowicz, S.: Semi-automatic recognition of noun modifier relationships. In: Proceedings of the 17th international conference on Computational linguistics, 96–102. Association for Computational Linguistics, Morristown, NJ, USA (1998)

Bergsma, S., Goebel, R.: Using visual information to predict lexical preference. In: Proceedings of the International Conference Recent Advances in Natural Language Processing 2011, pp. 399–405. RANLP 2011 Organising Committee, Hissar, Bulgaria (2011)

---

[21] http://books.google.com/ngrams.
[22] http://weblm.research.microsoft.com/.

Pitler, E., Lin, D.: Creating robust supervised classifiers via web-scale n-gram data. In: Proceedings of the 48th Annual Meeting of the Association for Computational Linguistics, pp. 865–874. Uppsala, Sweden (2010)

Van Durme, B.: Learning bilingual lexicons using the visual similarity of labeled web images. In: Proceedings of the Twenty-Second International Joint Conference on Artificial Intelligence -Volume Volume Three, IJCAI 2011, pp. 1764–1769. AAAI Press (2011)

Iris Wang, Q.: Learning noun phrase query segmentation. In: Proceedings of the 2007 Joint Conference on Empirical Methods in Natural Language Processing and Computational Natural Language Learning (EMNLP-CoNLL), pp. 819–826 (2007)

Brants, T., Popat, A.C., Peng, X., Och, F.J., Dean, J.: Large language models in machine translation. In: Proceedings of the 2007 Joint Conference on Empirical Methods in Natural Language Processing and Computational Natural Language Learning (EMNLP-CoNLL), pp. 858–867. Czech Republic, Prague (2007)

Brill, E., Resnik, P.: A rule-based approach to prepositional phrase attachment disambiguation. In: Proceedings of COLING (1994)

Brin, S., Page, L.: The anatomy of a large-scale hypertextual web search engine. Comput. Netw. **30**, 107–117 (1998)

Budanitsky, A., Hirst, G.: Evaluating wordnet-based measures of lexical semantic relatedness. Comput. Linguist. **32**, 13–47 (2006)

Butnariu, C., Kim, SN., Nakov, P., Séaghdha, D., Szpakowicz, S., Veale, T.: Noun compounds using paraphrasing verbs and prepositions. In: Proceedings of the 5th International Workshop on Semantic Evaluations (SemEval-2), Uppsala, Sweden, 11–16 July 2010, pp. 39–44 (2010)

Veale, T.: A concept-centered approach to noun-compound interpretation. In: Proceedings of the 22nd International Conference on Computational Linguistics (Coling 2008), pp. 81–88. Manchester, UK (2008)

Cafarella, M., Banko, M., Etzioni, O.: Technical Report 02 April 2006, University of Washington, Department of Computer Science and Engineering (2006)

Calvo, H., Gelbukh, A.: Improving prepositional phrase attachment disambiguation using the web as corpus. In: Sanfeliu, A., Ruiz-Shulcloper, J. (eds.) CIARP 2003. LNCS, vol. 2905, pp. 604–610. Springer, Heidelberg (2003)

Cao, Y., Li, H.: Base noun phrase translation using web data and the EM algorithm. In: COLING, pp. 127–133 (2002)

Chantree, F., Kilgarriff, A., De Roeck, A., Willis, A.: Using a distributional thesaurus to resolve coordination ambiguities. In: Technical Report 2005/02. The Open University, UK (2005)

Chklovski, T., Pantel, P.: Proceedings of the Conference on Empirical Methods in Natural Language Processing, pp. 33–40 (2004)

Church, K., Patil, R.: Coping with syntactic ambiguity or how to put the block in the box on the table. Am. J. Comput. Linguist. **8**, 139–149 (1982)

Collins, M., Brooks, J.: Prepositional phrase attachment through a backed-off model. In: Proceedings of EMNLP, pp. 27–38 (1995)

Downing, P.: On the creation and use of english compound nouns. Language **53**(4), 810–842 (1977)

Dumais, S., Banko, M., Brill, E., Lin, J., Andrew Ng.: Web question answering: Is more always better?. In: Proceedings of SIGIR, pp. 291–298 (2002)

Fellbaum, C.: Wordnet: An Electronic Lexical Database. MIT Press, Cambridge (1998)

Fleiss, J.L.: Statistical Methods for Rates and Proportions, 2nd edn. John Wiley & Sons Inc, New York (1981)

Girju, R., Moldovan, D., Tatu, M., Antohe, D.: On the semantics of noun compounds. Special Issue on Multiword Expressions **19**(4), 479–496 (2005)

Girju, R., Nakov, P., Nastase, Szpakowicz, S., Turney, P., Yuret. D.: Semeval-2007 task 04: classification of semantic relations between nominals. In: Proceedings of the Fourth International Workshop on Semantic Evaluations (SemEval-2007), pp. 13–18, Prague, Czech Republic (2007)

Nakov, P., Nastase, V., Szpakowicz, S., Turney, P., Yuret, D.: Language Resources and Evaluation **43**, 105–121 (2009)

Goldberg, M.: An unsupervised model for statistically determining coordinate phrase attachment. In: Proceedings of ACL, pp. 610–614 (1999)

Grefenstette, G.: The world wide web as a resourcefor example-based machine translation tasks. In: Proceedings of the ASLIB Conference on Translating and the Computer (1998)

Hendrickx, I., Kim, S.N., Kozareva, Z., Nakov, P., Séaghdha, D., Padó, S., Romano, M., Szpakowicz, S.: SemEval-2010 Task 8: Multi-way classification of semantic relations between pairs of nominals. In: Proceedings of the 5th International Workshop on Semantic Evaluations (SemEval-2), Uppsala, Sweden, 11– 16 July 2010, 33–38 (2010)

Weber, I.M.: Semantic Methods for Execution-level Business Process Modeling. LNBIP, vol. 40. Springer, Heidelberg (2009)

Hindle, D., Rooth, M.: Structural ambiguity and lexical relations. Comput. Linguist. **19**, 103–120 (1993)

Szpektor, I., Tanev, H., Dagan, I., Coppola, B.: Scaling web-based acquisition of entailment relations. In: Proceedings of the Conference on Empirical Methods in Natural Language Processing, pp. 401–48 (2004)

Weber, I.M.: Evaluation. Semantic Methods for Execution-level Business Process Modeling. LNBIP, vol. 40, pp. 203–225. Springer, Heidelberg (2009)

Keller, F., Lapata, M.: Using the Web to obtain frequencies for unseen bigrams. Comput. Linguist. **29**, 459–484 (2003)

Kilgariff, A., Grefenstette, G.: Introduction to the special issue on the web as corpus. Comput. Linguist. **29**, 333–347 (2003)

Kilgarriff, A.: Googleology is bad science. Comput. Linguist. **33**, 147–151 (2007)

Nam, K.S., Nakov, P.: Large-scale noun compound interpretation using bootstrapping and the web as a corpus. In: Proceedings of the 2011 Conference on Empirical Methods in Natural Language Processing, pp. 648–658. Edinburgh, Scotland, UK (2011)

Kurohashi, S., Nagao, M.: Dynamic programming method for analyzing conjunctive structures in Japanese. In: Proceedings of COLING, vol. 1 (1992)

Lapata, M., Keller, F.: The Web as a baseline: evaluating the performance of unsupervised Web-based models for a range of NLP tasks. In: Proceedings of HLT-NAACL, pp. 121–128, Boston (2004)

Keller, F.: Web-based models for natural language processing. ACM Trans. Speech Lang. Process. **2**(1), 1–31 (2005)

Lauer, M.: Designing statistical language learners: experiments on noun compounds. Department of Computing Macquarie University NSW 2109 Australia dissertation (1995)

Levi, J.: The syntax and semantics of complex nominals. Academic Press, New York (1978)

Levy, O., Goldberg, Y.: Linguistic regularities in sparse and explicit word representations. In: Proceedings of the Eighteenth Conference on Computational Natural Language Learning, 171–180 (2014)

Lin, D.: An information-theoretic definition of similarity. In: ICML 1998: Proceedings of the Fifteenth International Conference on Machine Learning, pp. 296–304. Morgan Kaufmann Publishers Inc San Francisco, CA, USA (1998)

Church, K., Ji, H., Sekine, S., Yarowsky, D., Bergsma, S., Patil, K., Pitler, E., Lathbury, R., Rao, V., Dalwani, K., Narsale, S.: New tools for web-scale n-grams. In: Proceedings of the Seventh International Conference on Language Resources and Evaluation (LREC 2010) Calzolari, N., (Conference Chair), Choukri, K., Maegaard, B., Mariani, J., Odijk, J., Piperidis, S., Rosner, M.,Tapias, D., Valletta, M.: European Language Resources Association (ELRA) (2010)

Lin, Y., Michel, J.-B., Lieberman, E.A., Orwant, J., Brockman, W., Petrov, S.: Syntactic annotations for the google books ngram corpus. In: Proceedings of the ACL 2012 System Demonstrations, pp. 169–174. Jeju Island, Korea (2012)

Marcus, M.: A Theory of Syntactic Recognition for Natural Language. MIT Press, Cambridge (1980)

Santorini, B., Marcinkiewicz, M.: Building a large annotated corpus of english: The PennTreebank. Comput. Linguist. **19**, 313–330 (1994)

Mihalcea, R., Moldovan, D.: A method for word sense disambiguation of unrestricted text. In: ACL, pp. 152–158 (1999)

Mikolov, Tomas, Yih, Wen-tau, Zweig, Geoffrey: Linguistic regularities in continuous space word representations.Proceedings of the 2013 Conference of the North American Chapter of the Association for Computational Linguistics: Human Language Technologies, pp. 746–751. Atlanta, Georgia (2013)

Modjeska, N., Markert, K. Nissim, M.: Using the web in machine learning for other-anaphora resolution. In: Proceedings of the 2003 Conference on Empirical Methods in Natural Language Processing, 176–183 (2003)

Nakov, P.: Using the web as an implicit training set: Application to noun compound syntax and semantics. EECS Department, University of California, Berkeley, UCB/EECS-2007-173 dissertation (2007)

Improved statistical machine translation using monolingual paraphrases. In: Proceedings of the European Conference on Artificial Intelligence, ECAI 2008, pp. 338–342. Patras, Greece (2008a)

Nakov, P.: Noun compound interpretation using paraphrasing verbs: feasibility study. In: Dochev, D., Pistore, M., Traverso, P. (eds.) AIMSA 2008. LNCS (LNAI), vol. 5253, pp. 103–117. Springer, Heidelberg (2008)

Paraphrasing verbs for noun compound interpretation. In: Proceedings of the LREC'08 Workshop: Towards a Shared Task for Multiword Expressions, MWE 2008, pp. 46–49. Marrakech, Morocco (2008c)

On the interpretation of noun compounds: Syntax, semantics, and entailment. Natural Lang. Eng. vol. 19, pp. 291–330 (2013)

Hearst, M.: Search engine statistics beyond the n-gram: Application to noun compound bracketing. In: Proceedings of CoNLL-2005, Ninth Conference on Computational Natural Language Learning (2005a)

Hearst, M.: A study of using search engine page hits as a proxy for n-gram frequencies. In: Proceedings of RANLP 2005, pp. 347–353. Borovets, Bulgaria (2005)

Hearst, M.: Using the web as an implicit training set: application to structural ambiguity resolution. In: HLT 2005: Proceedings of the conference on Human Language Technology and Empirical Methods in Natural Language Processing, pp. 835–842. Association for Computational Linguistics, Morristown, NJ, USA (2005c)

Hearst, M.: Solving relational similarity problems using the web as a corpus. In: Proceedings of the 46th Annual Meeting on Association for Computational Linguistics, ACL 2008, pp. 452–460. Columbus, OH (2008)

Nakov, P., Hearst, M.: Using verbs to characterize noun-noun relations. In: Euzenat, J., Domingue, J. (eds.) AIMSA 2006. LNCS (LNAI), vol. 4183, pp. 233–244. Springer, Heidelberg (2006)

Kozareva, Z.: Combining relational and attributional similarity for semantic relation classification. In: Proceedings of the International Conference Recent Advances in Natural Language Processing, RANLP 2011, pp. 323–330. Hissar, Bulgaria (2011)

Schwartz, A., Wolf, B., Hearst, M.: Scaling up BioNLP: application of a text annotation architecture to noun compound bracketing. In: Proceedings of SIG BioLINK (2005a)

Schwartz, A., Wolf, B., Hearst, M.: Proceedings of the ACL 2005 on interactive poster and demonstration sessions, pp. 65–68. Association for Computational Linguistics, Morristown, NJ, USA (2005b)

Nakov, P.I., Hearst, M.A.: Semantic interpretation of noun compounds using verbal and other paraphrases. ACM Trans. Speech Lang. Process. **10**, 1–51 (2013)

Nastase, V., Nakov, P., Séaghdha, D.Ó., Szpakowicz, S.: Semantic Relations Between Nominals: Synthesis Lectures on Human Language Technologies. Morgan & Claypool Publishers, San Rafael (2013)

Pantel, P., Lin, D.: An unsupervised approach to prepositional phrase attachment using contextually similar words. In: Proceedings of ACL (2000)

Porter, M.: An algorithm for suffix stripping. Program **14**, 130–137 (1980)

Pustejovsky, J., Anick, P., Bergler, S.: Lexical semantic techniques for corpus analysis. Comput. Linguist. **19**, 331–358 (1993)

Ratnaparkhi, A.: Statistical models for unsupervised prepositional phrase attachment. In: Proceedings of COLING-ACL vol. 2, pp. 1079–1085 (1998)

Reynar, J., Roukos, S.: A maximum entropy model for prepositional phrase attachment. In: Proceedings of the ARPA Workshop on Human Language Technology, pp. 250–255 (1994)

Resnik, P.: Selection and information: a class-based approach to lexical relationships. University of Pennsylvania, UMI Order No. GAX94-13894 dissertation (1993)

Mining the web for bilingual text. In: Proceedings of the 37th Annual Meeting of the Association for Computational Linguistics on Computational Linguistics, pp. 527–534. Association for Computational Linguistics, Morristown, NJ, USA (1999a)

Semantic similarity in a taxonomy: An information-based measure and its application to problems of ambiguity in natural language. In: JAIR 11, pp. 95–130 (1999b)

Rigau, G., Magnini, B., Agirre, E., Carroll, J.: Meaning: A roadmap to knowledge technologies. In: Proceedings of COLING Workshop on A Roadmap for Computational Linguistics (2002)

Rus, V., Moldovan, D., Bolohan, O.: Bracketing compound nouns for logic form derivation. In: Haller, S.M., Simmons, G. (eds.) FLAIRS Conference, pp. 198–202. AAAI Press (2002)

Santamaría, C., Gonzalo, J., Verdejo, F.: Automatic association of web directories with word senses. Comput. Linguist. **29**, 485–502 (2003)

Shinzato, K., Torisawa, K.: Acquiring hyponymy relations from web documents. In: Proceedings of HLT-NAACL, pp. 73–80 (2004)

Soricut, R., Brill, E.: Automatic question answering: Beyond the factoid. In: Proceedings of HLT-NAACL, pp. 57–64 (2004)

Stetina, J., Makoto.: Corpus based PP attachment ambiguity resolution with a semantic dictionary. In: Proceedings of WVLC, pp. 66–80 (1997)

Toutanova, K., Klein, D., Manning, C., Singer, Y.: Feature-rich part-of-speech tagging with a cyclic dependency network. In: Proceedings of HLT-NAACL 2003, pp. 252–259 (2003)

Toutanova, K., Manning, C.D., Andrew Y.Ng.: Learning random walk models for inducing word dependency distributions. In: Proceedings of ICML (2004)

Turney, P., Littman, M.: Corpus-based learning of analogies and semantic relations. Mach. Learn. J. **60**, 251–278 (2005)

Turney, P.D.: Similarity of semantic relations. Comput. Linguist. **32**, 379–416 (2006)

Véronis, J.: Web: Google adjusts its counts. Jean Veronis' blog: (2005a). http://aixtal.blogspot.com/2005/03/web-google-adjusts-its-counts.html

Web: MSN cheating too? Jean Veronis' blog: (2005b). http://aixtal.blogspot.com/2005/02/web-msn-cheating-too.html

Web: Yahoo doubles its counts! Jean Veronis' blog: (2005c). http://aixtal.blogspot.com/2005/03/web-yahoo-doubles-its-counts.html

Volk, M.: Scaling up. using the www to resolve PP attachment ambiguities. In: Proceedings of Konvens-2000. Sprachkommunikation (2000)

Exploiting the WWW as a corpus to resolve PP attachment ambiguities. In: Proceedings of Corpus Linguistics (2001)

Wang, K., Thrasher, C., Paul Hsu, B.-J.: Web scale NLP: A case study on url word breaking. In: Proceedings of the 20th International Conference on World Wide Web, WWW 2011, pp. 357–366. ACM, New York, NY, USA (2011)

Warren, B.: Semantic patterns of noun-noun compounds. In: Gothenburg Studies in English 41, Goteburg, Acta Universtatis Gothoburgensis (1978)

Way, A., Gough, N.: wEBMT: developing and validating an example-based machine translation system using the world wide web. Comput. Linguist. **29**, 421–457 (2003)

Yang, Y., Pedersen, J.: A comparative study on feature selection in text categorization. In: Proceedings of ICML1997, pp. 412–420 (1997)

Zahariev, M.: School of Computing Science, Simon Fraser University, USA dissertation (2004)

Zhu, X., Rosenfeld, R.: Improving trigram language modeling with the world wide web. In: Proceedings of ICASSP I, pp. 533–536 (2001)

Author Profiling and Plagiarism Detection

Paolo Rosso[(✉)]

Natural Language Engineering Lab, PRHLT Research Center,
Universitat Politècnica de València, Valencia, Spain
prosso@dsic.upv.es
http://www.dsic.upv.es/prosso

Abstract. In this paper we introduce the topics that we will cover in
the RuSSIR 2014 course on Author Profiling and Plagiarism Detection
(APPD). Author profiling distinguishes between classes of authors studying how language is shared by classes of people. This task helps in identifying profiling aspects such as gender, age, native language, or even
personality type. In case of the plagiarism detection task we are not
interested in studying how language is shared. On the contrary, given a
document we are interested in investigating if the writing style changes in
order to unveil text inconsistencies, i.e., unexpected irregularities through
the document such as changes in vocabulary, style and text complexity.
In fact, when it is not possible to retrieve the source document(s) where
plagiarism has been committed from, the intrinsic analysis of the suspicious document is the only way to find evidence of plagiarism. The
difficulty in retrieving the source of plagiarism could be due to the fact
that the documents are not available on the web or the plagiarised text
fragments were obfuscated via paraphrasing or translation (in case the
source document was in another language). In this overview, we also discuss the results of the shared tasks on author profiling (gender and age
identification) and plagiarism detection that we help to organise at the
PAN Lab on Uncovering Plagiarism, Authorship, and Social Software
Misuse (http://pan.webis.de).

1   Author Profiling: How Writing Style is Shared

Author profiling tries to determine an author's gender, age, native language,
personality type, etc. solely by analysing her texts. Profiling anonymous authors
is a problem of growing importance, both from forensic and marketing perspectives. From a forensic perspective it is important to identify the linguistic profile
of an author of a harassing text message or a potential online paedophile on
the basis of the analysis of his writing style in order, for instance, to unveil his
age [7,58]. From a marketing viewpoint, companies may be interested in knowing the demographics of their target group in order to achieve a better market
segmentation.

In this section we will introduce the reader to the profiling aspects of gender
and age identification, describing the shared task that was organised at PAN, and
briefly discussing the obtained results and the way the problem was addressed

© Springer International Publishing Switzerland 2015
P. Braslavski et al. (eds.): RuSSIR 2014, CCIS 505, pp. 229–250, 2015.
DOI: 10.1007/978-3-319-25485-2_6

by the participants. PAN was the first lab to offer author profiling as a shared task. At PAN 2013 [58] we aimed at identifying age and gender from a large corpus collected from social media. Most of the participants used combinations of style-based features such as frequency of punctuation marks, capital letters, quotations, and so on, together with POS tags and content-based features such as latent semantic analysis, bag-of-words, tfidf, dictionary-based words, topic-based words, and so on. The winner of the PAN 2013 task [29] used second order representations based on relationships between documents and profiles, whereas another well-performing approach, winner of the English subtask [35], used collocations. Following we summarise the evaluation of 10 author profilers that have been submitted to the shared task that was organised in 2014.

Evaluation Corpora. In the author profiling task at PAN 2013 [58] participants approached the task of identifying age and gender in a large corpus collected from social media, and age was annotated with three classes: 10 s (13–17), 20 s (23–27), and 30 s (33–47). At PAN 2014, we continued to study the gender and age aspects of the author profiling problem, however, four corpora of different genres were considered—social media, blogs, Twitter, and hotel reviews—both in English and Spanish. Moreover, we annotated age with the following continuous classes: 18–24; 25–34; 35–49; 50–64; and 65+.

The social media corpus was built by sampling parts of the PAN 2013 evaluation corpus. We selected only authors with an average number of words greater than 100 in their posts. We also reviewed manually the data in order to remove authors who appeared to be fake profiles such as bots. The blogs and Twitter corpora were manually collected and annotated by three annotators. The Twitter corpus was built in collaboration with RepLab,[1] where the main goal of author profiling in the context of reputation management on Twitter was to decide how influential a given user is in a domain of interest. For each blog, we provided up to 25 posts and for each twitter profile, we provided up to 1000 tweets. The hotel review corpus is derived from another corpus that was originally used for aspect-level rating prediction [69].[2] The original corpus was crawled from the hotel review site TripAdvisor[3] and manually checked for quality and compliance with the format requirements of PAN 2014.

Evaluation Results. In Table 1 joint identification accuracies for both gender and age prediction are shown per data set and averaged over all corpora, which also serves as ranking criterion. The baseline considered the 1000 most frequent character trigrams. In summary, simple content features, such as bag-of-words or word n-grams achieve best accuracies. Bag-of-words features are used by Liau and Vrizlynn [28], word n-grams are used by Maharjan et al. [31], and term vector models are used by Villena-Román and González-Cristóbal [67]. They achieved competitive performances on almost all corpora. Weren et al. [70] employed information retrieval features and Marquardt et al. [32] mixed content

---

Table 1. Author profiling: joint identification (gender and age) results in terms of accuracy.

Team	Overall	Social media		Blogs		Twitter		Reviews
		en	es	en	es	en	es	en
López-Monroy	**0.2895**	0.1902	0.2809	**0.3077**	**0.3214**	**0.3571**	0.3444	0.2247
Liau	0.2802	0.1952	**0.3357**	0.2692	0.2321	0.3506	0.3222	**0.2564**
Shrestha	0.2760	**0.2062**	0.2845	0.2308	0.2500	0.3052	**0.4333**	0.2223
Weren	0.2349	0.1914	0.2792	0.2949	0.1786	0.2013	0.2778	0.2211
Villena-Román	0.2315	0.1905	0.1961	**0.3077**	0.2321	0.2078	0.2667	0.2199
Marquardt	0.1998	0.1428	0.2102	0.1282	0.2679	0.1948	0.3111	0.1437
Baker	0.1677	0.1277	0.1678	0.1282	0.2321	0.1688	0.2111	0.1382
Baseline	*0.1404*	*0.0930*	*0.1820*	*0.0897*	*0.0536*	*0.1494*	*0.2333*	*0.1821*
Mechti	0.1067	0.1244	0.1060	0.0897	0.1786	0.0584	0.1444	0.0451
Castillo Juarez	0.0946	0.1445	0.1254	0.1795	0.0893	–	–	0.1236
Ashok	0.0834	0.1318	–	0.1282	–	0.1948	–	0.1291

and style features. Some readability measures were also used: Automated Readability index [20,32], Coleman-Liau index [20,32], Rix Readability index [20,32], Gunning Fog index [20], Flesch-Kinkaid [70]. The approach of López-Monroy *et al.* [30] obtained the best overall using a second order representation based on relationships between documents and profiles.

From the results of Table 1, it can be seen that: (*a*) the highest joint accuracies were achieved on Twitter data, and, (*b*) the smallest joint accuracies were achieved in English social media and hotel reviews. It is an open question why these differences can be observed, whereas possible explanations may be that people express themselves more spontaneously on Twitter compared to the other genres, whereas the low scores are due to the approaches' difficulty of predicting gender in social media and age in hotel reviews. A complete version of the report can be found in [59], where a more in-depth analysis of the obtained results as well as a survey of detection approaches are given.

2   Plagiarism Detection: How Writing Style Changes

Plagiarism is the re-use of someone else's prior ideas, processes, results, or words without explicitly acknowledging the original author and source [26]. A person that fails to provide its corresponding source is suspected of plagiarism. In the academic domain, some surveys estimate that around 30 % of student reports include plagiarism [2] and a more recent study increases this percentage to more than 40 % [10]. Indeed the amount of text available in electronic media nowadays has caused cases of plagiarism to increase. As a result, its manual detection has become infeasible. Models for automatic plagiarism detection are being developed as a countermeasure. Their main objective is assisting people in the task of detecting plagiarism—as a side effect, plagiarism is discouraged.

However, not always it is straightforward to retrieve the document(s) that have been the source of plagiarism because they may be not available or with a high level of paraphrasing or even in another language. In this section we describe basic stylistic analysis techniques in order to spot irregularities in the writing style of the suspicious document. As well we illustrate how performance of plagiarism detectors decreases in case of obfuscation of the source text via paraphrasing or translation.

Fig. 1. Identifying changes in writing style within a document.

2.1 Stylistic Analysis

When it is not possible to retrieve the document(s) plagiarism has been committed from, or because they are not available in the given collection (or even on the web) or due to the high level of obfuscation via paraphrasing or translation, the evidence of plagiarism has to be found in the document itself (intrinsic plagiarism detection). The aim is to spot changes in vocabulary, text complexity and writing style (see Fig. 1). In fact, the insertion of text fragments from a different author into the suspicious document causes style and complexity irregularities. The quantification can be made by measuring vocabulary richness (type/tokens ratio), basic statistics (average sentence length, average word length, etc.), n-gram profiles (character level statistics), and text readability measures (e.g. Gunning Fog, Flesch-Kinkaid, etc.) [36, 62]. We have already mentioned in the previous section how readability measures help in author profiling as well. In fact, complexity in texts and writing style change with the author's demographics (e.g. her age, gender or personality). The formula below refers to the Gunning Fog (GF) index which, in order to determine the complexity of a given text, takes into account its total number of sentences, words and complex words, where complex words are those words with three or more syllables [23]. The value resulting of this calculation can be interpreted as the number of years of formal education required to understand the document contents.

$$GF = 0.4(\frac{|words|}{|sentences|} + 100 * \frac{|complex - words|}{|words|}) \tag{1}$$

Typical values for GF of texts such as a comic, a Newsweek article, and scientific texts are: GF(comic)=6, GF(Newsweek)=10, GF(T1) = 15.2, and GF(T2) = 14.1. Let us analyse the three text fragments of the example below.

Example 1. In this work, we have carried out some research on the influence that mineral salts on the mood of people. For this research I have worked with 5 people who have taken water with different amount of mineral salts. Our theory is that the more minerals are in the water, the more moody people are. [...]

Mineral salts are inorganic molecules of easy ionization in presence of water; in living beings they appear by precipitation as well as dissolved mineral salts. [...] Dissolved mineral salts are always ionized. These salts have a structural function and pH regulating functions, of the osmotic pressure and and of biochemical reactions, in which specific ions are involved.

It seems to me that the results are good. [...]

Figure 2 illustrates statistics and measures used for stylistic analysis. Values for the first and third paragraphs (third column) are in back, whereas the values for the more formal second paragraph (second column) are in red.

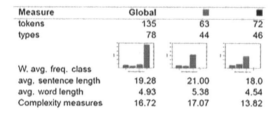

Measure	Global	■	■
tokens	135	63	72
types	78	44	46
W. avg. freq. class			
avg. sentence length	19.28	21.00	18.0
avg. word length	4.93	5.38	4.54
Complexity measures	16.72	17.07	13.82

Fig. 2. Stylisic analysis (Color figure online).

Although when the suspicious document is short an expert simply reading it could quite easily detect text inconsistencies and writing style changes, when the document is long (e.g. a thesis or a report) spotting irregularities in text is not always straightforward. Therefore, it is important to have tools that could help the experts (e.g. forensic linguists, teachers, etc.) in highlighting suspicious text fragments. For instance, Stylysis[4] is a tool whose aim is to provide the expert with a linguistic profile of the document on the basis of a stylistic analysis in order to determine whether or not there are text fragments of different writing styles. Stylysis analyses documents in English, Spanish or Catalan. The tool divides the text into fragments and for each of them, it calculates basic statistics, as well as vocabulary richness measure (function K proposed by Yule [72] and function R proposed by Honore [25]) and text readability measures (Gunning Fog index [23] and Flesch-Kincaid Readability test [11]).

---

[4] http://memex2.dsic.upv.es/StylisticAnalysis

2.2   Obfuscation via Paraphrasing

The linguistic phenomena underlying plagiarism have barely been analyzed. In [33] different kinds of plagiarism are identified: of ideas, of references, of authorship, word by word, and paraphrase plagiarism. In the first case, ideas, knowledge or theories from another person are claimed without proper citation. In plagiarism of references and authorship, citations and entire documents are included without any mention of their authors. Word by word plagiarism, also known as copy–paste or verbatim copy, consists of the exact copy of a text (fragment) from a source into the plagiarised document. Regarding paraphrase plagiarism, a different form expressing the same content is used.

For this purpose we will show how the plagiarism detectors that participated in the PAN shared task in 2010 decreased their performance on the subset of paraphrase plagiarism cases of the PAN-PC-10 corpus [48]. First, we briefly describe the evaluation measures that are employed in the shared task on plagiarism detection [45].

Evaluation Measures. As automatic plagiarism detection is identified as an information retrieval task, evaluation is usually carried out on the basis of recall and precision. Nevertheless, plagiarism detection aims at retrieving plagiarised–source fragments rather than documents. Given a suspicious document d_q and a collection of potential source documents D, the detector should retrieve: *a)* a specific text fragment $s_q \in d_q$, potential case of plagiarism; and *b)* a specific text fragment $s \in d$, the claimed source for s_q. Therefore, special versions of precision and recall have been proposed at PAN in order to fit in this framework. The plagiarized text fragments are treated as basic retrieval units, with $s_i \in S$ defining a query for which a plagiarism detection algorithm returns a result set $R_i \subseteq R$. The recall and precision of a plagiarism detection algorithm are defined as:

$$prec_{PDA}(S, R) = \frac{1}{|R|} \sum_{r \in R} \frac{|\bigcup_{s \in S} (s \sqcap r)|}{|r|} \text{ and} \tag{2}$$

$$recall_{PDA}(S, R) = \frac{1}{|S|} \sum_{s \in S} \frac{|\bigcup_{r \in R} (s \sqcap r)|}{|s|} \tag{3}$$

where $\sqcap$ computes the positionally overlapping characters. In both equations, S and R represent the entire set of actually plagiarized text fragments and detections, respectively.

Consider Fig. 3 for an illustrative example. $\{s_1, s_2, s_3\} \in S$ represent text sequences in the document that are known to be plagiarised. A given detector recognises the sequences $\{r_1, r_2, r_3, r_4, r_5\} \in R$ as plagiarised. Substituting the values in Eqs. 2 and 3:

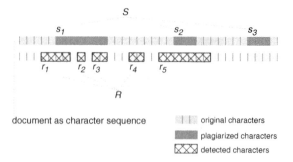

Fig. 3. A document as character sequence, including plagiarized sections S and detections R returned by a plagiarism detector.

$$precision_{PDA}(S, R) = \frac{1}{|R|} \cdot \left(\frac{|r_1 \sqcap s_1|}{|r_1|} + \frac{|r_2 \sqcap s_1|}{|r_2|} + \frac{|r_3 \sqcap s_1|}{|r_3|} + \frac{\cancel{|\emptyset|}^{\,0}}{\cancel{|r_4|}} + \frac{|r_5 \sqcap s_2|}{|r_5|} \right)$$

$$= \frac{1}{5} \cdot \left(\frac{2}{4} + \frac{1}{1} + \frac{2}{2} + \frac{3}{7} \right) = 0.5857 \text{ and}$$

$$recall_{PDA}(S, R) = \frac{1}{|S|} \cdot \left(\frac{|(s_1 \sqcap r_1) \bigcup (s_1 \sqcap r_2) \bigcup (s_1 \sqcap r_3)|}{|s_1|} + \frac{|s_2 \sqcap r_5|}{|s_2|} + \frac{\cancel{\emptyset}^{\,0}}{\cancel{|s_3|}} \right)$$

$$= \frac{1}{3} \cdot \left(\frac{5}{7} + \frac{3}{3} \right) = 0.5714$$

Once precision and recall are computed, they are combined into their harmonic mean (F_1-measure).

Evaluation Results. Figure 4(a) shows the evaluations computed by considering the entire PAN-PC-10 corpus. The best recall values are around 0.70, with very good values of precision, some of them above 0.90. The results, when considering only the simulated cases, that is, those generated by manual paraphrasing, are presented in Fig. 4(b). In most of the cases, the quality of the detections decreases dramatically compared to the results on the entire corpus, which also contains translated, verbatim and automatically modified plagiarism. The difficulty to detect paraphrase plagiarism cases in the PAN-PC-10 corpus was also stressed in [64]. Manually created cases seem to be much harder to detect than the other, artificially generated, cases (when the simulated cases in the PAN-PC-10 corpus were generated, volunteers had specific instructions to create rewritings with a high obfuscation degree). This can be appreciated when looking at the difference of capabilities of the plagiarism detector applied at the 2009 and 2010 shared tasks by [21, 22], practically the same implementation. At the first shared task, whose corpus included artificial cases only, its recall was of

0.66 while in the second one, with simulated (i.e., paraphrase plagiarism) cases, it decreased to 0.48.

Interestingly, the best performing plagiarism detectors on the paraphrase plagiarism corpus are not the ones that performed the best at the PAN-10 shared task. For instance, this is the case of [39] that apply greedy string tiling, which aims at detecting as long as possible identical text fragments. This approach outperforms the rest of detectors when dealing with cases with a high density of identical fragments (with paraphrase plagiarism cases in between).

The complete analysis of the results can be found in [6], where a paraphrase typology is employed in order to investigate further the relationship between paraphrasing and plagiarism. Figure 4(c) shows the evaluation results when considering only the cases included in the P4Psubset of the corpus that was annotated with the types of the paraphrase typology.[5]

2.3   Obfuscation via Translation

The detection of plagiarism is even more difficult when it concerns documents written in different languages. Cross-language (CL) plagiarism detection attempts to identify and extract automatically plagiarism among documents in different languages. Recently a survey was done on scholar practices and attitudes [3], also from a cross-language plagiarism perspective which manifests that CL plagiarism is a real problem: in fact, only 36.25 % of students think that translating a text fragment and including it into their report is plagiarism.

In recent years there have been a few approaches to cross-language similarity analysis that can be used for CL plagiarism detection [5]. A simple, yet effective approach is the cross-language character n-gram (CL-CNG) model [34]. Using character n-grams, it takes into account the syntax of documents, and offers remarkable performance for languages with syntactic similarities. Cross-language explicit semantic analysis (CL-ESA) [46,50] represents a document by its similarities to a collection of documents. These similarities in turn are computed with a monolingual retrieval model such as the vector space model. The cross-language alignment-based similarity analysis (CL-ASA) model [4,50] is instead based on statistical machine translation and combines probabilistic translations, using a statistical bilingual dictionary and similarity analysis. The cross-language conceptual thesaurus based similarity (CL-CTS) model [24] tries to measure the similarity between the documents in terms of shared concepts, using a conceptual thesaurus, and named entities among them.

Plagiarised fragments can be translated verbatim copies or may alter their structure to hide the copying, which is known as paraphrasing and is more difficult to detect. In order to improve the detection of paraphrase plagiarism, a model named cross-language knowledge graph analysis (CL-KBS) was introduced [15]. Its goal is to exploit explicit semantics for a better representation of the documents. CL-KGA provides a context model by generating knowledge graphs that expand and relate the original concepts from suspicious and source

---

[5] http://clic.ub.edu/corpus/en/paraphrases-en

(*a*) **overall (PAN-PC-10)**

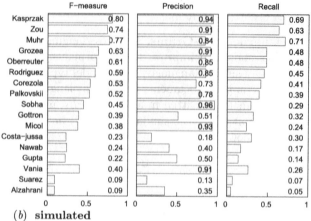

(*b*) **simulated**

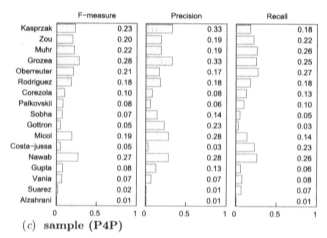

(*c*) **sample (P4P)**

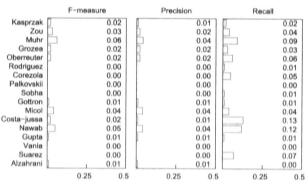

Fig. 4. Evaluation of the PAN-10 competition participants' plagiarism detectors. Figures show evaluations over: (*a*) entire PAN-PC-10 corpus (including artificial, translated, and simulated cases); (*b*) simulated cases only; (*c*) sample of simulated cases annotated on the basis of the paraphrases typology: the P4P corpus. Note the change of scale in (*c*).

text fragments. Finally, the similarity is measured in a semantic graph space. In this section we compare CL-KGA with CL-ASA and CL-CNG because obtaining the best results in a previous study [50].

Cross-Language Character N-Grams. The cross-language character n-gram (CL-CNG) model has shown to improve the performance of CL information retrieval for syntactically similar languages. This model typically uses character trigrams (CL-C3G) to compare documents in different languages [50].

Given a source document d written in a language L_1 and a suspicious document d' written in language L_2, the similarity $S(d, d')$ between the two documents is measured as follows:

$$S(d, d') = \frac{d \cdot d'}{|d| \cdot |d'|}, \tag{4}$$

where d and d' are the vector representation of documents d and d' into character n-gram space.

Cross-Language Alignment based Similarity Analysis. The cross-language alignment based similarity analysis (CL-ASA) model measures the similarity between two documents d and d', from two different languages L_1 and L_2 respectively, by aligning the documents at word level and determining the probability of d' being a translation of d. The similarity $S(d, d')$ between both documents is measured as in Eq. 5:

$$S(d, d') = l(d, d') * t(d|d'), \tag{5}$$

where $l(d, d')$ is the length factor defined in [56], which is used as normalization since two documents with the same content, in different languages do not have the same length. Moreover, $t(d|d')$ is the translation model defined in Eq. 6:

$$t(d|d') = \sum_{x \in d} \sum_{y \in d'} p(x, y), \tag{6}$$

where $p(x, y)$ is the probability of a word x from language L_1 being a translation of word y from L_2. These probabilities can be obtained using a bilingual statistical dictionary.

Cross-Language Knowledge Graph Analysis. The cross-language knowledge graphs analysis (CL-KGA) model uses knowledge graphs generated from a multilingual semantic network (MSN) in order to obtain a context model of text fragments in different languages. We employ BabelNet [38], although the graph-based model is generic and could be applied with other available MSNs such as EuroWordNet [68].

A knowledge graph is a weighted and labelled graph that expands and relates the original concepts of a set of words, providing a "context model". Using BabelNet to build the graphs we can have a multilingual dimension for each of the concepts. Therefore, we can compare directly pairs of graphs built from text fragments in different languages to detect CL plagiarism.

We can build a knowledge graph using a MSN such as BabelNet as follows: having a concept set C, we search BabelNet for paths connecting each pair $c, c^{\iota} \in C$, obtaining the set of paths P, where each $p \in P$ is a set of concepts and relations between concepts from C which include the conceptual expansion. The knowledge graph g is obtained after joining the paths from P including all its concepts and relations. Finally, to weight the concepts we use their degree of relatedness, i.e. the number of outgoing edges for each node. The relation weighting is performed also in function of the degree of relatedness of their source and target concepts.

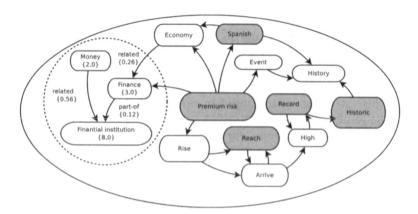

Fig. 5. Knowledge graph built from the sentence "Spanish premium risk reaches historic records", simplified without the multilingual dimension, and with labels and weights only inside the dashed circle.

Example 2. Having the English sentence of Fig. 5, we obtain its concepts $C = $ {Spanish, premium risk, reach, record, historic}. Using BabelNet to build a knowledge graph g from C, we obtain a concept set $C_g = C \cup C'$, where $C' = $ {economy, finance, history...} is the expanded concept set. In addition, we obtain a relation set $R \in$ {related-to, has-part, belong-to, is-a...} between concepts of C_g. We can observe the resulting graph g in Fig. 5.

To compare graphs we use a similarity function S that is an adapted version of flexible comparison of conceptual graphs similarity algorithm presented in [37].

$$S(g, g') = S_c(g, g') * (a + b * S_r(g, g')) \tag{7}$$

$$S_c(g, g') = \frac{\left(2 * \sum_{c \in g_{int}} w(c)\right)}{\left(\sum_{c \in g} w(c) + \sum_{c \in g'} w(c)\right)} \tag{8}$$

$$S_r(g, g') = \frac{\left(2 * \sum_{r \in N(c, g_{int})} w(r)\right)}{\left(\sum_{r \in N(c,g)} w(r) + \sum_{r \in N(c,g')} w(r)\right)} \tag{9}$$

where S_c is the score of the concepts, S_r is the score of the relations, a and b are smoothing variables to give the appropriate relevance to concepts and relations, c is a concept, r is a relation, g_{int} is the resulting graph of the intersection between g and g', and $N(c, g)$ is the set of all the relations connected to the concept c in a given graph g.

Evaluation Corpus and Measures. In our evaluation we use the German-English (DE-EN) and Spanish-English (ES-EN) CL plagiarism partitions of the PAN-PC'11 corpus [51]. We evaluate the performance of CL-KGA differentiating plagiarism cases between translated verbatim copies and paraphrase translations in which their structure was changed in order to hide the copying [51]. We compare the results obtained by CL-KGA with those provided by CL-ASA and CL-C3G (CL-CNG using 3-grams) for the same task. For CL-ASA model we use two statistical dictionaries: BabelNet's statistical dictionary (CL-ASA$_{BN}$ [15]) and a dictionary trained using the word-aligment model IBM M1 [42] on the JRC-Acquis [65] corpus.

With respect to the evaluation measures, apart from precision and recall at character level, the granularity measure was considered. In fact, due that neither precision nor recall account that plagiarism detectors sometimes report overlapping or multiple detections for a single plagiarism case, the granularity measure has been introduced in the PAN shared task:

$$granularity(S, R) = \frac{1}{|S_R|} \sum_{s \in S_R} |R_s|, \tag{10}$$

where $S_R \subseteq S$ are cases detected by detections in R, and $R_s \subseteq R$ are detections of s; i.e., $S_R = \{s \mid s \in S \text{ and } \exists r \in R : r \text{ detects } s\}$ and $R_s = \{r \mid r \in R \text{ and } r \text{ detects } s\}$.

Table 2. DE-EN cross-language plagiarism detection results for automatic and paraphrase translation cases, displayed in the decreasing order of the PlagDet score.

Model	German-english							
	Automatic translations				Paraphrase translations			
	PlagDet	Recall	Precision	Granularity	PlagDet	Recall	Precision	Granularity
CL-KGA	**0.5296**	**0.4671**	**0.6306**	**1.0188**	**0.1006**	**0.2101**	**0.0661**	**1.0**
CL-ASA$_{IBMM1}$	0.4230	0.3690	0.6019	1.1163	0.0462	0.0978	0.0303	1.0
CL-ASA$_{BN}$	0.3019	0.2363	0.5962	1.1753	0.0275	0.0796	0.0166	1.0
CL-C3G	0.0909	0.0564	0.3414	1.0913	0.0185	0.0389	0.0121	1.0

Finally, the three measures are combined into a single overall score to allow for a unique ranking among detection approaches [45]:

$$PlagDet(S, R) = \frac{F_1}{\log_2(1 + granularity(S, R))}, \tag{11}$$

Evaluation Results. As we can see in Table 2, for the DE-EN CL plagiarism detection, CL-C3G obtains the lowest results, being the baseline for this kind of experiments, due to the simplicity of the approach which uses n-grams. CL-ASA$_{BN}$ uses BabelNet's statistical dictionary, obtaining average results, despite many German words in the dictionary were not found. CL-ASA$_{IBMM1}$ outperforms the baseline $PlagDet$ by 365 % in automatic translations and 149 % in paraphrase translations. Finally, CL-KGA obtains the best values, increasing the baseline $PlagDet$ by 478 % in automatic translations and 443 % in paraphrase translations, along with better values for recall, precision and granularity.

Table 3. ES-EN cross-language plagiarism detection results for automatic and paraphrase translation cases, displayed in the decreasing order of the PlagDet score.

Model	Spanish-english							
	Automatic translations				Paraphrase translations			
	PlagDet	Recall	Precision	Granularity	PlagDet	Recall	Precision	Granularity
CL-KGA	**0.6087**	**0.5399**	**0.7036**	**1.0050**	**0.0993**	**0.1979**	**0.0662**	1.0
CL-ASA$_{BN}$	0.5793	0.5245	0.6631	1.0154	0.0738	0.1909	0.0457	1.0
CL-ASA$_{IBMM1}$	0.5339	0.4728	0.6911	1.0729	0.0612	0.1501	0.0384	1.0
CL-C3G	0.1756	0.1336	0.6158	1.3796	0.0289	0.0587	0.0192	1.0

As we can see in Table 3, for ES-EN CL plagiarism detection, the models performance was quite similar to the one obtained for DE-EN. CL-C3G is the baseline with the lowest values. CL-ASA$_{BN}$ increases the baseline $PlagDet$ by 230 % in automatic translations and 155 % in paraphrase translations. This time CL-ASA$_{BN}$ obtains better results than CL-ASA$_{IBMM1}$ showing that using BabelNet's statistical dictionary for ES-EN plagiarism detection allows to obtain a good performance. CL-KGA obtains the best values with all the measures, increasing the baseline $PlagDet$ by 246 % in automatic translations and 243 % in paraphrase translations. The $granularity$ for CL-KGA is the closest to 1.0, the best possible value. A more detailed analysis can be found in [16].

3 Related Work

3.1 Author Profiling

The study of how certain linguistic features vary according to the profile of their authors is a subject of interest for several different areas such as psychology, linguistics and, more recently, computational linguistics. In the first section we already mentioned how the teams that participated in the PAN shared task approached author profiling. In this section we describe some of the previous works.

Argamon et al. [1] analysed formal written texts extracted from the British National Corpus, combining function words with part-of-speech features and achieving approximately 80 % accuracy in gender prediction. Koppel et al. [27] studied the problem of automatically determining an author's gender in social media by proposing combinations of simple lexical and syntactic features, and achieving approximately 80 % accuracy. Schler et al. [61] studied the effect of age and gender in the writing style in blogs; they gathered over 71,000 blogs and obtained a set of stylistic features like non-dictionary words, parts-of-speech, function words and hyperlinks, combined with content features, such as word unigrams with the highest information gain. They modeled age in three classes — 10 s (13–17), 20 s (23–27) and 30 s (33–47)—obtaining an accuracy of about 80 % for gender identification and about 75 % for age identification. They showed that language features in blogs correlate with age, as reflected in, for example, the use of prepositions and determiners. Goswami et al. [19] added some new features as slang words and the average length of sentences, improving accuracy to 80.3 % in age group identification and to 89.2 % in gender detection. More recently, Nguyen et al. [40] studied the use of language and age among Dutch Twitter users. They modelled age as a continuous variable and used an approach based on logistic regression. They measured the effect of the gender in the performance of age identification, considering both variables as inter-dependent, and achieved correlations up to 0.74 and mean absolute errors between 4.1 and 6.8 years. Pennebaker et al. [44] connected language use with personality traits, studying how the variation of linguistic characteristics in a text can provide information regarding the gender and age of its author.

A shared task on computational personality recognition was recently organised at the WCPR workshop of ICWSM 2013[6] and at ACM Multimedia 2014.[7] Moreover, a shared task was also organised at the BEA-8 Workshop of NAACL-HLT 2013 on another aspect of author profiling: native language identification.[8] The number of shared tasks on different aspects of author profiling (gender and age identification, personality recognition, and native language identification) show the raising interest of the scientific community in this challenging problem.

3.2 Plagiarism Detection

In recent years, the evaluation of plagiarism detectors has been studied in the context of the PAN evaluation labs that have been organised annually since 2009.[9] During the first three labs, a total of 43 plagiarism detectors have been evaluated using this framework [47,49,51]. The two recent editions refocused on specific

---

[6] http://mypersonality.org/wiki/doku.php?id=wcpr13

[7] https://sites.google.com/site/wcprst/home/wcpr14

[8] https://sites.google.com/site/nlisharedtask2013/

[9] The corpora PAN-PC-2009/2010/2011 are available at http://www.webis.de/research/corpora

sub-problems of plagiarism detection: source retrieval and text alignment. Both source retrieval and text alignment have been identified as integral parts of plagiarism detection [63]. Instead of again applying a semiautomatic approach to corpus construction, a large corpus of manually generated plagiarism has been crowdsourced in order to increase the level of realism [18]. This corpus comprises 297 essays of about 5000 words length, written by professional writers. In this regard the writers were given a set of topics to choose from along with two more technical rules: (i) to use the ChatNoir search engine [53] to research their topic of choice, and (ii) to reuse text passages from retrieved web pages in order to compose their essay. The resulting essays represent the to-date largest corpus of realistic text reuse cases available, and they have been employed to evaluate another 33 plagiarism detectors in the past three labs [52,54,55].

Source Retrieval. In source retrieval, given a suspicious document and a web search engine, the task is to retrieve all source documents from which text has been reused whilst minimizing retrieval costs. To study this task, we employ a controlled, static web environment, which consists of a large web crawl and search engines indexing it. Using this setup, we built the previously described large corpus of manually generated text reuse in the form of essays, which serve as suspicious documents and which are fed into a plagiarism detector.

Table 4 shows the performances of the six plagiarism detectors that implemented source retrieval. Their cost-effectiveness is measured as average workload per suspicious document, and as average numbers of queries and downloads until the first true positive detection has been made. These statistics reveal if a source retrieval algorithm finds sources quickly, thus reducing its usage costs. Moreover, we measure precision and recall of downloaded documents with regard to the true source documents and compute F_1.

Table 4. Plagiarism detection: source retrieval results.

Team (alphabetical order)	Downloaded sources			Total Workload		Workload to 1st detection		No detect	Runtime
	F_1	precision	recall	Queries	Dwlds	Queries	Dwlds		
Elizalde	0.34	0.40	0.39	54.5	33.2	16.4	3.9	7	**04:02:00**
Kong	0.12	0.08	0.48	83.5	207.1	85.7	24.9	6	24:03:31
Prakash	0.39	0.38	**0.51**	60.0	38.8	8.1	3.8	7	19:47:45
Suchomel	0.11	0.08	0.40	**19.5**	237.3	**3.1**	38.6	2	45:42:06
Williams	**0.47**	**0.57**	0.48	117.1	**14.4**	18.8	**2.3**	4	39:44:11
Zubarev	0.45	0.54	0.45	37.0	18.6	5.4	**2.3**	3	40:42:18

None of the detectors dominates the others in terms of all of the employed measures, whereas three detectors share the top scores among them. The detector of Williams et al. [71] achieves the best trade-off between precision and recall in terms of F_1 as well as best precision, whereas the detector of Prakash and

Saha [57] achieves best recall. Suchomel and Brandejs [66]'s detector requires least query workload, least queries until first detection, and detects source documents for almost all of the test documents. The detector of Williams *et al.* [71], however, performs worst in terms of total querying workload, since it requires 117 queries on average. Posing a query to a search engine may entail significant costs, whereas downloading a document is considered much less costly. By comparison, the detector of Zubarev and Sochenkov [73] achieves a similarly good trade-off between precision and recall with much less querying costs and comparable downloading costs. This detector also competes in terms of workload until first true positive detection with less than 6 queries and about 2 downloads on average.

Text Alignment. In text alignment, given a pair of documents, the task is to identify all contiguous passages of reused text between them. Table 5 shows the overall performance of eleven plagiarism detectors that implemented text alignment. Performances are measured using precision and recall at character level as well as granularity (i.e., how often the same plagiarism case is detected) and PlagDet. The detectors are ranked by PlagDet. The best performing detector is that of Sanchez-Perez *et al.* [60], closely followed by the detectors of Oberreuter and Eiselt [41] and Palkovsii and Belov [43]. The detailed performances of each detector with regard to different kinds of obfuscation can be found in [55].

Table 5. Plagiarism detection: text alignment results.

Team	PlagDet	Recall	Precision	Granularity	Runtime
Sanchez-Perez	**0.87818**	**0.87904**	0.88168	1.00344	00:25:35
Oberreuter	0.86933	0.85779	0.88595	1.00369	00:05:31
Palkovskii	0.86806	0.82637	0.92227	1.00580	01:10:04
Glinos	0.85930	0.79331	**0.96253**	1.01695	00:23:13
Shrestha	0.84404	0.83782	0.85906	1.00701	69:51:15
R. Torrejón	0.82952	0.76903	0.90427	1.00278	**00:00:42**
Gross	0.82642	0.76622	0.93272	1.02514	00:03:00
Kong	0.82161	0.80746	0.84006	1.00309	00:05:26
Abnar	0.67220	0.61163	0.77330	1.02245	01:27:00
Alvi	0.65954	0.55068	0.93375	1.07111	00:04:57
Baseline	*0.42191*	*0.34223*	*0.92939*	*1.27473*	*00:30:30*
Gillam	0.28302	0.16840	0.88630	**1.00000**	00:00:55

4   Conclusions

In this paper we introduced the reader to author profiling and plagiarism detection as well as to the PAN shared tasks. The difficulties of both tasks have been highlighted together with the way the participating teams have approach them.

To improve the reproducibility of shared tasks, participants are asked at PAN to submit running softwares instead of their run output. To deal with the organisational overhead involved in handling software submissions, the TIRA web platform [17] helps to significantly reduce the workload for both participants and organizers, whereas the submitted softwares are kept in a running state. This year, 57 softwares have been submitted to our lab, and together with the 58 software submissions of last year, this forms the largest collection of softwares for our three tasks to date, all of which are readily available for further analysis.

In the future it would be interesting to approach author profiling in social media considering simultaneously several aspects such as gender, age and personality. With respect to plagiarism detection, recently it has been approached also in source code [12,13], and a PAN shared task on the detection of SOurce COde (SOCO) re-use has been organised at the Forum for Information Retrieval Evaluation.[10]

Acknowledgements. We would like to thank Yuri Chekhovich (Forecsys) and Mikhail Alexandrov (Russian Presidential Academy of national economy and public administration) for providing plagiarised cases in Russian for the APPD course at RuSSIR. We thank Irina Chugur (UNED) and Francisco Rangel (Autoritas Consulting) for helping with the author profiling corpus in Russian. The PAN shared tasks on author profiling and on plagiarism detection have been organised in the framework of the WIQ-EI IRSES project (Grant No. 269180) within the EC FP 7 Marie Curie People. The research work described in the paper was carried out in the framework of the DIANA-APPLICATIONS-Finding Hidden Knowledge in Texts: Applications (TIN2012-38603-C02-01) project, and the VLC/CAMPUS Microcluster on Multimodal Interaction in Intelligent Systems. Finally, we thank Hugo Jair Escalante (INAOE) for helping to improve this paper.

References

1. Argamon, S., Koppel, M., Fine, J., Shimoni, A.R.: Gender, genre, and writing style in formal written texts. TEXT **23**, 321–346 (2003)
2. Association of Teachers and Lecturers. School work plagued by plagiarism - ATL survey. Technical report, Association of Teachers and Lecturers, London, UK (2008). (Press release)
3. Barrón-Cedeño, A.: On the mono- and cross-language detection of text re-use and plagiarism. Ph.D. thesis, Universitat Politènica de València (2012)
4. Barrón-Cedeño, A., Rosso, P., Pinto, D., Juan, A.: On cross-lingual plagiarism analysis using a statistical model. In: Proceedings of the ECAI 2008 Workshop on Uncovering Plagiarism, Authorship and Social Software Misuse, PAN 2008 (2008)
5. Barrón-Cedeño, A., Gupta, P., Rosso, P.: Methods for cross-language plagiarism detection. Knowl. Based Syst. **50**, 11–17 (2013)
6. Barrón-Cedeño, A., Vila, M., Martí, M., Rosso, P.: Plagiarism meets paraphrasing: insights for the next generation in automatic plagiarism detection. Comput. Linguist. **39**(4), 917–947 (2013)

---

[10] http://www.dsic.upv.es/grupos/nle/soco/

7. Bogdanova, D., Rosso, P., Solorio, T.: Exploring high-level features for detecting cyberpedophilia. Comput. Speech Lang. **28**(1), 108–120 (2014)
8. Braschler, M., Harman, D.: Notebook papers of CLEF 2010 LABs and workshops. Padua, Italy (2010)
9. Cappellato, L., Ferro, N., Halvey, M., Kraaij, W.: CLEF 2014 labs and workshops, notebook papers. In: CEUR Workshop Proceedings (CEUR-WS.org), ISSN 1613–0073 (2014). http://ceur-ws.org/Vol-1180/
10. Comas, R., Sureda, J., Nava, C., Serrano, L.: Academic cyberplagiarism: a descriptive and comparative analysis of the prevalence amongst the undergraduate students at Tecmilenio University (Mexico) and Balearic Islands University (Spain). In: Proceedings of the International Conference on Education and New Learning Technologies (EDULEARN 2010), Barcelona (2010)
11. Flesch, R.: A new readability yardstick. J. Appl. Psychol. **32**(3), 221–233 (1948)
12. Flores, E., Barrón-Cedeño, A., Rosso, P., Moreno, L.: Desocore: detecting source code re-use across programming languages. In: Proceedings of 12th International Conference of the North American Chapter of the Association for Computational Linguistics: Human Language Technologies, NAACL-2012, pp. 1–4, Montreal, Canada (2012)
13. Flores, E., Barrón-Cedeño, A., Moreno, L., Rosso, P.: Uncovering source code re-use in large-scale programming environments. In: Computer Applications in Engineering and Education, Accepted (2014). doi:10.1002/cae.21608
14. Forner, P., Navigli, R., Tufis, D.: CLEF 2013 evaluation labs and workshop - working notes papers, 23–26 September. Valencia, Spain (2013)
15. Franco-Salvador, M., Gupta, P., Rosso, P.: Cross-Language plagiarism detection using a multilingual semantic network. In: Braslavski, P., Kuznetsov, S.O., Kamps, J., Rüger, S., Agichtein, E., Segalovich, I., Yilmaz, E., Serdyukov, P. (eds.) ECIR 2013. LNCS, vol. 7814, pp. 710–713. Springer, Heidelberg (2013)
16. Franco-Salvador, M., Gupta, P., Rosso, P.: Knowledge graphs as context models: improving the detection of cross-language plagiarism with paraphrasing. In: Ferro, N. (ed.) PROMISE Winter School 2013. LNCS, vol. 8173, pp. 227–236. Springer, Heidelberg (2014)
17. Gollub, T., Stein, B., Burrows, S.: Ousting Ivory tower research: towards a web framework for providing experiments as a service. In: Hersh, B., Callan, J., Maarek, Y., Sanderson, M., (eds.) 35th International ACM Conference on Research and Development in Information Retrieval (SIGIR 2012), pp. 1125–1126. ACM, August 2012. ISBN 978-1-4503-1472-5. doi:10.1145/2348283.2348501
18. Gollub, T., Hagen, M., Michel, M., Stein, B.: From keywords to keyqueries: content descriptors for the web. In: Gurrin, C., Jones, G., Kelly, D., Kruschwitz, U., de Rijke, M., Sakai, T., Sheridan, P., (eds.) 36th International ACM Conference on Research and Development in Information Retrieval (SIGIR 2013), pp. 981–984. ACM (2013)
19. Goswami, S., Sarkar, S., Rustagi, M.: Stylometric analysis of bloggers' age and gender. In: Adar, E., Hurst, M., Finin, T., Glance, N.S., Nicolov, N., Tseng, B.L., (eds.) ICWSM. The AAAI Press (2009)
20. Gressel, G., Hrudya, P., Surendran, K., Thara, S., Aravind, A., Prabaharan, P.: Ensemble Learning Approach for Author Profiling-Notebook for PAN at CLEF 2014. In: Cappellato, et al. [9]
21. Grozea, C., Popescu, M.: ENCOPLOT - performance in the Second International Plagiarism Detection Challenge lab report for PAN at CLEF 2010. In: Braschler and Harman [8]

22. Grozea, C., Gehl, C., Popescu, M.: ENCOPLOT: pairwise sequence matching in linear time applied to plagiarism detection. In: Stein et al., (ed.) Overview of the 1st International Competition on Plagiarism Detection, pp. 10–18 (2009)
23. Gunning, R.: The Technique of Clear Writing. McGraw-Hill Int. Book Co, New York (1952)
24. Gupta, P., Barrón-Cedeño, A., Rosso, P.: Cross-language high similarity search using a conceptual thesaurus. In: Catarci, T., Peñas, A., Santucci, G., Forner, P., Hiemstra, D. (eds.) CLEF 2012. LNCS, vol. 7488, pp. 67–75. Springer, Heidelberg (2012)
25. Honore, A.: Some simple measures of richness of vocabulary. Assoc. Lit. Linguist. Comput. Bull. **7**(2), 172–177 (1979)
26. IEEE. A Plagiarism FAQ. http://www.ieee.org/publications_standards/publical tions/rights/plagiarism_FAQ.htm (2008). Published: 2008; Last Accessed 25 November 2012
27. Koppel, M., Argamon, S., Shimoni, A.R.: Automatically categorizing written texts by author gender. Lit. Linguist. Comput. **17**(4), 401–412 (2002)
28. Liau, Y., Vrizlynn, L.: Submission to the author profiling competition at pan-2014. In: Proceedings Recent Advances in Natural Language Processing III (2014). http://www.webis.de/research/events/pan-14
29. Lopez-Monroy, A.P., Montes-Y-Gomez, M., Escalante, H.J., Villaseñor-Pineda, L., Villatoro-Tello, E.: INAOE's participation at PAN 2013: author profiling task-notebook for PAN at CLEF 2013. In: Forner, et al. [14]
30. Pastor López-Monroy, A., Montes y Gómez, M., Escalante, H.J., Villaseñor-Pineda, L.: Using Intra-profile information for author profiling-notebook for PAN at CLEF 2014. In: Cappellato, et al. [9]
31. Maharjan, S., Shrestha, P., Solorio, T.: A simple approach to author profiling in MapReduce-notebook for PAN at CLEF 2014. In: Cappellato, et al. [9]
32. Marquardt, J., Fanardi, G., Vasudevan, G., Moens, M.F., Davalos, S., Teredesai, A., De Cock, M.: Age and gender identification in social media-notebook for PAN at CLEF 2014. In: Cappellato, et al. [9]
33. Martin, B.: Plagiarism: policy against cheating or policy for learning? Nexus (Newsl. Aust. Sociol. Assoc.) **16**(2), 15–16 (2004)
34. Mcnamee, P., Mayfield, J.: Character n-gram tokenization for european language text retrieval. Inf. Retr. **7**(1), 73–97 (2004)
35. Meina, M., Brodzinska, K., Celmer, B., Czokow, M., Patera, M., Pezacki, J., Wilk, M.: Ensemble-based classification for author profiling using various features-notebook for PAN at CLEF 2013. In: Forner, et al. [14]
36. Eissen, S.M., Stein, B.: Intrinsic plagiarism detection. In: Tombros, A., Yavlinsky, A., Rüger, S.M., Tsikrika, T., Lalmas, M., MacFarlane, A. (eds.) ECIR 2006. LNCS, vol. 3936, pp. 565–569. Springer, Heidelberg (2006)
37. Montes y Gómez, M., Gelbukh, A.F., López-López, A., Baeza-Yates, R.A.: Flexible comparison of conceptual graphs. In: Proceedings DEXA, pp. 102–111 (2001)
38. Navigli, R., Ponzetto, S.P.: BabelNet: the automatic construction, evaluation and application of a wide-coverage multilingual semantic network. Artif. Intell. **193**, 217–250 (2012)
39. Nawab, R.M.A., Stevenson, M., Clough, P.: University of sheffield lab report for pan at clef 2010. In: Braschler and Harman [8]
40. Nguyen, D., Gravel, R., Trieschnigg, D., Meder, T.: "how old do you think i am?"; a study of language and age in twitter. In: Proceedings of the Seventh International AAAI Conference on Weblogs and Social Media (2013)

41. Oberreuter, G., Eiselt, A.: Submission to the 6th international competition on plagiarism detection, From Innovand.io, Chile (2014). http://www.webis.de/research/events/pan-14
42. Och, F.J., Ney, H.: A systematic comparison of various statistical alignment models. Comput. Linguist. **29**(1), 19–51 (2003)
43. Palkovskii, Y., Belov, A.: Developing high-resolution universal multi-type N-Gram plagiarism detector-notebook for PAN at CLEF 2014. In: Cappellato, et al. [9]
44. Pennebaker, J.W., Mehl, M.R., Niederhoffer, K.G.: Psychological aspects of natural language use: our words, our selves. Ann. Rev. Psychol. **54**(1), 547–577 (2003)
45. Potthast, M., Stein, B., Barrón-Cedeño, A., Rosso, P.: An evaluation framework for plagiarism detection. In: COLING 2010: Proceedings of the 23rd International Conference on Computational Linguistics, pp. 997–1005 (2010)
46. Potthast, M., Stein, B., Anderka, M.: A wikipedia-based multilingual retrieval model. In: Plachouras, V., Macdonald, C., Ounis, I., White, R.W., Ruthven, I. (eds.) ECIR 2008. LNCS, vol. 4956, pp. 522–530. Springer, Heidelberg (2008)
47. Potthast, M., Stein, B., Eiselt, A., Barrón-Cedeño, A., Rosso, P.:. Overview of the 1st international competition on plagiarism detection. In: Stein, B., Rosso, P., Stamatatos, E., Koppel, M., Agirre, E., (eds.) Proceedings of the SEPLN 2009 Workshop on Uncovering Plagiarism, Authorship, and Social Software Misuse (PAN 2009), pp. 1–9, 2009. CEUR-WS.org (September 2009). http://ceur-ws.org/Vol-502
48. Potthast, M., Barrón-Cedeño, A., Eiselt, A., Stein, B., Rosso, P.: Overview of the 2nd International Competition on Plagiarism Detection. In: Braschler and Harman [8]
49. Potthast, M., Barrón-Cedeño, A., Eiselt, A., Stein, B., Rosso, P.: Overview of the 2nd international competition on plagiarism detection. In: Braschler, M., Harman, D., Pianta, E., (eds.) Working Notes Papers of the CLEF 2010 Evaluation Labs (September 2010) 2010. http://www.clef-initiative.eu/publication/working-notes
50. Potthast, M., Barrón-Cedeño, A., Stein, B., Rosso, P.: Cross-language plagiarism detection. Lang. Resour. Eval. **45**(1), 45–62 (2011)
51. Potthast, M., Eiselt, A., Barrón-Cedeño, A., Stein, B., Rosso, P.: Overview of the 3rd international competition on plagiarism detection. In: Petras, V., Forner, P., Clough, P., (eds.) Working Notes Papers of the CLEF 2011 Evaluation Labs (September 2011) (2011). http://www.clef-initiative.eu/publication/working-notes
52. Potthast, M., Gollub, T., Hagen, M., Grabegger, J., Kiesel, J., Michel, M., Oberlander, A., Tippmann, M., Barrón-Cedeño, A., Gupta, P., Rosso, P., Stein, B.: Overview of the 4th international competition on plagiarism detection. In: Forner, P., Karlgren, J., Womser-Hacker, C., (eds.) Working Notes Papers of the CLEF 2012 Evaluation Labs (September 2012) (2012). http://www.clef-initiative.eu/publication/working-notes
53. Potthast, M., Hagen, M., Stein, B., Grabegger, J., Michel, M., Tippmann, M., Welsch, C.: Chatnoir: a search engine for the clueweb09 corpus. In: Hersh, B., Callan, J., Maarek, Y., Sanderson, M., (eds.) 35th International ACM Conference on Research and Development in Information Retrieval (SIGIR 2012), p. 1004 (2012)
54. Potthast, M., Gollub, T., Hagen, M., Tippmann, M., Kiesel, J., Rosso, P., Stamatatos, E., Stein, B.: Overview of the 5th international competition on plagiarism detection. In: Forner, et al. [14]   .
55. Potthast, M., Hagen, M., Beyer, A., Busse, M., Tippmann, M., Rosso, P., Stein, B.: Overview of the 6th International Competition on Plagiarism Detection. In: Cappellato, et al. [9]

56. Pouliquen, B., Steinberger, R., Ignat, C.: Automatic linking of similar texts across languages. In: Proceedings of Recent Advances in Natural Language Processing III, RANLP 2003, pp. 307–316 (2003)
57. Prakash, A., Saha, S.: Experiments on document chunking and query formation for plagiarism source retrieval-notebook for PAN at CLEF 2014. In: Cappellato, et al. [9]
58. Rangel, F., Rosso, P., Koppel, M., Stamatatos, E., Inches, G.: Overview of the author profiling task at PAN 2013-notebook for PAN at CLEF 2013. In: Forner, et al. [14]
59. Rangel, F., Rosso, P., Chugur, I., Potthast, M., Trenkman, M., Stein, B., Verhoeven, B., Daelemans, W.: Overview of the 2nd author profiling task at PAN 2014-notebook for PAN at CLEF 2014. In: Cappellato, et al. [9]
60. Sanchez-Perez, M., Sidorov, G., Gelbukh, A.: A winning approach to text alignment for text reuse detection at PAN 2014-notebook for PAN at CLEF 2014. In: Cappellato, et al. [9]
61. Schler, J., Koppel, M., Argamon, S., Pennebaker, J.W.: Effects of age and gender on blogging. In: AAAI Spring Symposium: Computational Approaches to Analyzing Weblogs, pp. 199–205. AAAI (2006)
62. Stamatatos, E.: Intrinsic plagiarism detection using character n-gram profiles. In: Stein, B., Rosso, P., Stamatatos, E., Koppel, M., Agirre, E., (eds.) Proceedings of the SEPLN09 Workshop on Uncovering Plagiarism, Authorship, and Social Software Misuse (PAN 2009), pp. 38–46, 2009. CEUR-WS.org, September 2009. http://ceur-ws.org/Vol-502
63. Stein, B., Meyer zu Eissen, S., Potthast, M.: Strategies for retrieving plagiarized documents. In: Clarke, C., Fuhr, N., Kando, N., Kraaij, W., de Vries, A., (eds.) 30th International ACM Conference on Research and Development in Information Retrieval (SIGIR 2007), pp. 825–826. ACM (2007)
64. Stein, B., Potthast, M., Rosso, P., Barrón-Cedeño, A., Stamatatos, E., Koppel, M.: Fourth international workshop on uncovering plagiarism, authorship, and social software misuse. ACM SIGIR Forum 45, 45–48 (2011)
65. Steinberger, R., Pouliquen, B., Widiger, A., Ignat, C., Erjavec, T., Tufis, D., Varga, D.: The jrc-acquis: a multilingual aligned parallel corpus with +20 languages. In: Proceedings of 5th International Conference on language resources and evaluation LREC 2006 (2006)
66. Suchomel, S., Brandejs, M.: Heterogeneous queries for synoptic and phrasal search-notebook for PAN at CLEF 2014. In: Cappellato, et al. [9]
67. Villena-Román, J., González-Cristóbal, J.C.: DAEDALUS at PAN 2014: Guessing Tweet Author's Gender and Age-Notebook for PAN at CLEF 2014. In: Cappellato, et al. [9]
68. Vossen, P.: Eurowordnet: a multilingual database of autonomous and language-specific wordnets connected via an inter-lingual index. Int. J. Lexicography 17, 161–173 (2004)
69. Wang, H., Lu, Y., Zhai, C.: Latent aspect rating analysis on review text data: a rating regression approach. In: Proceedings of the 16th ACM SIGKDD International Conference on Knowledge Discovery and Data Mining, pp. 783–792 (2010)
70. Weren, E.R.D., Moreira, V.P., de Oliveira, J.P.M.:. Exploring information retrieval features for author profiling-notebook for PAN at CLEF 2014. In: Cappellato, et al. [9]
71. Williams, K., Chen, H.H., Giles, C.: Supervised ranking for plagiarism source retrieval-notebook for PAN at CLEF 2014. In: Cappellato, et al. [9]

72. Yule, G.: The Statistical Study of Literary Vocabulary. Cambridge University press, Cambridge (1944)
73. Zubarev, D., Sochenkov, I.: Using sentence similarity measure for plagiarism source retrieval-notebook for PAN at CLEF 2014. In: Cappellato, L., et al. [9]

Young Scientists Conference Papers

Transformation of Categorical Features into Real Using Low-Rank Approximations

Alexander Fonarev[1,2]([X])

[1] Skolkovo Institute of Science and Technology, Skolkovo, Russia
newo@newo.su
[2] Yandex, Moscow, Russia

Abstract. Most of existing machine learning techniques can handle objects described by real but not categorical features. In this paper we introduce a simple unsupervised method for transforming categorical feature values into real ones. It is based on low-rank approximations of collaborative feature value frequencies. Once object descriptions are transformed, any common real-value machine learning technique can be applied for further data analysis. For example, it becomes possible to apply classic and powerful Random Forest predictor in supervised learning problems. Our experiments show that a combination of the proposed features transformation method with common real-value supervised algorithms leads to the results that are comparable to the state-of-the-art approaches like Factorization Machines.

Keywords: Categorical features · Low-rank approximations · Matrix factorization · Feature extraction · Factorization machines · Sparse data

1 Introduction

Let us have a dataset $\{X_i\}_{i=1}^n$, where every object $X_i = (X_i^1, ..., X_i^m) \in \mathcal{X}$. The element X_i^j is called the j-th feature of the object X_i [8]. The whole dataset can be represented as a matrix with size $n \times m$. Unsupervised problems involve finding regularities in the data. In supervised learning problems every element X_i of the training set has an object label $y_i \in \mathcal{Y}$, and the goal is to approximate $\mathcal{X} \to \mathcal{Y}$ function.

1.1 Goals of the Paper

Most of existing machine learning methods suppose that the features $\{X_i^j\}_{j=1}^m$ of the object X_i are in $\mathbb{R}$. But there are many problems, where the feature values come from a finite unordered set, but not from $\mathbb{R}$. These features are called categorical, nominal or factor. For example, the categorical feature *City* may have values from set $\{Moscow, New York, Paris, ...\}$ in some tasks. Such tasks become even more difficult with the increasing of the size q of this set, because of a huge data sparsity. There are very few methods that are directly suitable for

© Springer International Publishing Switzerland 2015
P. Braslavski et al. (eds.): RuSSIR 2014, CCIS 505, pp. 253–262, 2015.
DOI: 10.1007/978-3-319-25485-2_7

categorical data analysis, e.g. naive Bayes based methods for supervised problems. It means that many widely used and very powerful real-value techniques, e.g. Random Forests [3], can't be efficiently applied in these tasks. So we aim at finding a method of efficient transformation of categorical features into real ones for further using real-value techniques.

1.2 Categorical Feature Applications

There are a lot of tasks with categorical features. We point out some of them in this Section.

Collaborative filtering is one of the most popular examples [18]. Each object X_i in the training set is a corresponding user rating description. It has categorical features such as *User, Item, Context*, etc. Object labels y_i are rating scores. The task is to predict rating scores for unseen objects.

Furthermore, many existing information retrieval techniques can be improved by using the categorical features. For example, the learning to rank task can be generalized to the personalized learning to rank just by adding a the single categorical feature *User* to (*Query, Document*) pairs.

Categorical features are also applied in the natural language processing [10]. E.g. the categorical feature *Word* appears in such tasks. Highly demanded language modeling problem can be reduced to the task of probability $p(w_{t-n+1}, ..., w_t)$ approximation, where w_i is the word in i-th position and n is the size of the language model. Then the task can be viewed as the task with n categorical features [2].

2   Existing Approaches

We are going to propose a method that transforms a categorical data matrix $X \in Cat^{\,n \times m}$ into a real matrix $Z \in \mathbb{R}^{n \times m'}$, where Cat is a set of categorical values and m' is a number of features after the transformation. The real-value matrix Z can be further used with classical machine learning methods.

In this work we numerate categorical feature values. There is an example for feature *City*:

$$Moscow \;\rightarrow\; 1,$$
$$New\ York \;\rightarrow\; 2,$$
$$Paris \;\rightarrow\; 3,$$
$$London \;\rightarrow\; 4,$$
$$... \;\rightarrow\; ... \;.$$

So we use $X \in \mathbb{N}^{n \times m}$ instead of $X \in Cat^{\,n \times m}$.

Certainly, the categorical feature value set is unordered, so this simplest transformation of categorical features into real ones can not be effectively used with common real-value machine learning techniques directly, because a non-existing order of feature values is significantly considered this way. E.g. there is no sense in the inequality *New York < London*, but most of real-value machine learning algorithms will take into a account this wrong knowledge.

2.1   Decision Tree Based Methods

Actually Decision Trees [4], Random Forests [3] and other Decision Tree based methods can handle categorical features. They work in the following way. Training process tries to find the optimal split of the dataset into two parts in each node. The best split is one that increases purity of the subsets. But it is very computational expensive to find the optimal split in case of categorical features with large number of unique values. Actually there are $O(2^{q_j})$ different splits for the feature X^j with q_j distinct values. So even if $q_j = 100$, it becomes impossible to find the optimal split because of a computational complexity.

2.2   One-Hot Encoding, Dummy Encoding, etc

The very popular method of transformation of categorical matrix X into real-value matrix Z is the so called one-hot encoding [13]. Let some feature X^j have q_j unique values $\{a_1, ..., a_{q_j}\}$. The X^j feature is expanded into q_j new binary features Z^{j,a_k}, where $k \in \{1, ..., q_j\}$, in this way:

$$Z_i^{j,a_k} = 1[X_i^j = a_k], \quad X \in \mathbb{N}^{n \times m}, \quad Z \in \mathbb{R}^{n \times m'},$$

where $m' = \sum_{j=1}^m q_j$. Here $1[A]$ is an indicator of A, i.e.

$$1[A] = \begin{cases} 1, & \text{if } A \text{ is true,} \\ 0, & \text{else.} \end{cases}$$

We get new transformed matrix Z by applying this procedure for every feature X^j.

Despite one-hot transformation is natural and easy to use, it has many drawbacks. The major one is a high increasing of object space dimensionality m'. Because of most of binary feature values are zeros, it's necessary to use sparse data representation, especially if q is large. Only a few methods can efficiently work with such representations, e.g. sparse linear methods. Thus, using one-hot encoding leads to big limitations in choosing further tools of data analysis.

There many other categorical encoding approaches that are very similar to the one-hot encoding, e.g. dummy encoding, effects encoding and other [9]. They are basically designed for further using with linear methods. All these approaches have the same drawbacks as the previously discussed one-hot encoding.

2.3   Factorization Machines

Factorization Machine (FM) [14] is the state-of-the-art technique for a modeling recommender data. FMs use a polynomial approximation methods and the one-hot encoding. The advantages of FMs are in their scalability and in the fact, that they can mimic the most successful approaches for the task of collaborative filtering, including SVD++ [11], PITF [17] and FPMC [16]. Note that FM is a complete supervised prediction model, that can handle categorial features, but it is not a categorical feature transformation method.

3   Proposed Methods

3.1   Transformation Using Direct Feature Value Frequencies

Suppose we have the training data matrix $X \in \mathbb{N}^{n \times m}$ and every feature X^j is categorical. Direct using of categorical values is not effective as mentioned in the Sect. 2. But we can use feature values co-occurrences, they involve an important information about the dataset.

A transformed real-value matrix Z can be obtained from categorical matrix X by using collaborative feature value frequencies in the following way. Every transformed feature Z^{j_1,j_2} is computed for the every pair of features X^{j_1} and X^{j_2} as their co-occurrence value frequency:

$$Z_i^{j_1,j_2} = \frac{1}{n} \sum_{k=1}^{n} \mathbf{1}[X_i^{j_1} = X_k^{j_1}] \cdot \mathbf{1}[X_i^{j_2} = X_k^{j_2}], \quad \forall j_1, j_2, \quad Z \in \mathbb{R}^{n \times m'},$$

where $m' = \frac{m(m-1)}{2}$. This matrix Z provides the statistical information about interactions [15] between pairs of the features.

Note that this transformation is unsupervised, because it doesn't need the actual object labels y_i. This transformation considers a generative structure of the object space and can be used for a wide range of machine learning tasks.

3.2   Transformation Using Low-Rank Frequency Approximations

We propose new transformation approaches that are based on the previously discussed transformation, but do not use collaborative feature value frequencies directly. Instead of them, low-rank approximations [6] can be used. Let the pair of features X^{j_1} and X^{j_2} has unique values $\{a_k\}_{k=1}^{q_{j_1}}$ and $\{b_l\}_{l=1}^{q_{j_2}}$ respectively. Then we have a matrix P of a collaborative feature value frequencies:

$$P_{k,l} = \frac{1}{n} \sum_{i=1}^{n} \mathbf{1}[X_i^{j_1} = a_k] \cdot \mathbf{1}[X_i^{j_2} = b_l], \quad X \in \mathbb{N}^{n \times m}, \quad P \in \mathbb{R}^{q_{j_1} \times q_{j_2}}.$$

The model is regularized in the following way. Matrix P can be approximated as a product of two matrices with low rank r. If j-th feature has a large number q_j of distinct values than $r \ll q_j$ usually. Let us have a rank r approximation of P:

$$P \approx GH, \quad G \in \mathbb{R}^{q_{j_1} \times r}, \quad H \in \mathbb{R}^{r \times q_{j_2}}.$$

The scheme of the factorization process is illustrated in the middle part of Fig. 1. This matrix factorization provides the reduction of the transformation model parameters number from the quadratic $O(q_{j_1} \cdot q_{j_2})$ to the linear $O(r \cdot (q_{j_1} + q_{j_2}))$, so it seems to be helpful in avoiding an extra overfitting.

Each row $G_{a,:}$ and column $H_{:,b}$ corresponds to the values a and b of the features X^{j_1} and X^{j_2} respectively. So the scalar product of vectors $G_{a,:} \in \mathbb{R}^{1 \times r}$

and $H_{:,b} \in \mathbb{R}^{r \times 1}$ can be used as an approximated frequency of pair (a, b) in a features encoding:

$$Z_i^{j_1, j_2} = \left(G_{X_i^{j_1}, :}\right) \cdot \left(H_{:, X_i^{j_2}}\right), \quad \forall j_1, j_2, \quad Z \in \mathbb{R}^{n \times m'},$$

where $m' = \frac{m(m-1)}{2}$. Applying this transformation to every pair of the features gives the transformed matrix Z. The scheme of the full transformation process is illustrated in Fig. 1. Because of that many values in P are zeros, it's helpful to use the sparse low-rank factorization algorithms [1].

3.3 Transformation Using Low-Rank Latent Vectors

The vectors $G_{a,:}$ and $H_{:,b}$ are called latent vectors [11], corresponding to the values a and b. They provide a hidden low-dimensional representation of the feature values. They may be used for a direct feature transformation without matrix multiplication. In this case the transformed feature consists of the following latent vectors concatenation:

$$Z_i^{j_1, j_2} = \text{concat}\left(\left(G_{X_i^{j_1}, :}\right), \left(H_{:, X_i^{j_2}}\right)^T\right), \quad \forall j_1, j_2, \quad Z \in \mathbb{R}^{n \times m'},$$

where $m' = 2r \cdot \frac{m(m-1)}{2} = rm(m-1)$. Thus, every pair of features is transformed into new $2r$ real features. This approach is unsupervised as well. The scheme of this process for all features is illustrated in Fig. 2.

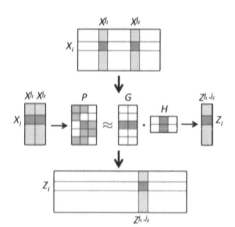

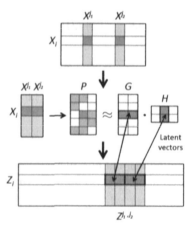

Fig. 1. Scheme of using low-rank feature value frequency approximations for transformation

Fig. 2. Scheme of using low-rank feature value latent vectors for transformation

3.4   Matrix Similarity Measures

The matrix approximation task

$$P \approx GH = Q,$$

supposes the following optimization problem

$$D(P, Q) \rightarrow \min_{G,H},$$

where $D(P, Q)$ is a matrix similarity measure. The choice of a particular measure strongly influences the result of matrix low-rank approximation process. We examine several similarity measures in this work.

The most simple way is to use the Frobenius norm of matrix:

$$\parallel P - Q \parallel_F^2 = \sum_{i,j}(P_{i,j} - Q_{i,j})^2.$$

However, as we are approximating the probabilistic distribution by the given frequencies, it is much more natural to use the generalized Kullback-Leibler divergency or I-divergency [5]:

$$D_{\mathrm{KL}}(P \parallel Q) = \sum_{i,j}(P_{i,j} \cdot \log \frac{P_{i,j}}{Q_{i,j}} - P_{i,j} + Q_{i,j}).$$

These measures are the particular cases of so called β-divergency:

$$D_{\mathrm{Beta}}^{\beta}(P \parallel Q) = \sum_{i,j}(P_{i,j} \cdot \frac{P_{i,j}^{\beta-1} - Q_{i,j}^{\beta-1}}{\beta - 1} - \frac{P_{i,j}^{\beta} + Q_{i,j}^{\beta}}{\beta}).$$

Actually, these are the cases when $\beta \rightarrow 1$:

$$D_{\mathrm{Beta}}^{\beta}(P \parallel Q) \rightarrow D_{\mathrm{KL}}(P \parallel Q),$$

and when $\beta = 2$:

$$D_{\mathrm{Beta}}^{\beta}(P \parallel Q) = \parallel P - Q \parallel_F^2.$$

Computational and algorithmic details about sparse matrix approximations can be found in [1]. The algorithm for β-divergency optimization is described in [7].

4   Experimental Setup

We examine our categorical feature transformation approach in supervised learning tasks to check its effectiveness.

4.1   Datasets

We used two supervised datasets in our experiments. The first dataset was published at the international data mining competition *Amazon.com – Employee Access Challenge*, which was held in 2013. We call this dataset shortly *Amazon* in our work. It provides a binary classification task. Training set has 32769 objects and $\approx 94\%$ of them belong to class 1, other belong to class 0. Every object corresponds to the employee access request to some resource. Target label shows if the corresponding request was approved by a supervisor. Task is to provide model that will automatically predict approvement.

Every object has a categorical feature description. The features are *Employee ID, Resource ID, Department ID*, etc. There are 9 features in total. A number of unique values for each feature is shown in Table 1.

Table 1. Number of unique values for every feature in *Amazon* dataset

Number of the feature	1	2	3	4	5	6	7	8	9
Number of unique values	7518	4243	128	177	449	343	2358	67	343

The second dataset is *Movie Lens 100K*, we call it shortly *Movie Lens*. The dataset provides 100000 user film ratings, each object is a single rating. It is described by the following categorical features: *User ID, Item ID, User social information, Genre information*. The original dataset includes many binary features corresponding to genre indicators. Binary features are a very simple case of categorical features, so we merge these features into the one categorical feature. Every unique value of this new feature corresponds to every unique genre combination. Object labels are the binary indicators those show if rating equals to 5. There are 100000 objects in the training set, and $\approx 64\%$ of the objects belong to the class 1. A number of unique values for the every feature is shown in Table 2.

Table 2. Number of unique values for every feature in *Movie Lens* dataset

Number of the feature	1	2	3	4	5	6
Number of unique values	943	1682	2	21	795	216

4.2   Prediction Quality Estimation

Let us have a true label vector $y \in \{0,1\}^n$ and a vector of predictions $\tilde{y} \in \mathbb{R}^n$ where $\tilde{y}_i \in [0,1]$ (e.g. it can be an estimated probability of belonging to the class 1). The quality of this prediction is computed as an area under the ROC-curve (AUC-ROC):

$$AUC(y, \tilde{y}) = \frac{\sum_{i=1}^{n} \sum_{j=1}^{n} \mathbf{1}[y_i < y_j] \cdot \mathbf{1}[\tilde{y}_i < \tilde{y}_j]}{\left(\sum_{i=1}^{n} \mathbf{1}[y_i = 0]\right) \cdot \left(\sum_{i=1}^{n} \mathbf{1}[y_i = 1]\right)} \in [0,1].$$

We use cross-validation technique for quality estimation. We set 7-fold cross-validation for the *Amazon* dataset and 5-fold for the *Movie Lens* dataset. These numbers of folds show a rather low variance of the quality estimation. We compare the cross validated quality results of different algorithms using Mann-Whitney U statistical test [12] with p-value = 0.05.

4.3 Algorithms

We have implemented three proposed in Sect. 3 categorical feature transformation methods using [1,7] and combined them with the popular supervised machine learning techniques: Random Forest and Support Vector Machines. Some of existing algorithms were used in the experiments as well: Factorization Machines, sparse logistic regression, Naive Bayes with additively smoothed probability estimations and Random Forest learned with numerated feature values.

5   Results of the Experiments

Experiments show that using low-rank latent vectors transformation Sect. 3.3 is the most effective approach among the proposed ones. The performance of 10-rank transformations in the combination with different supervised classifiers are presented in Table 3. Also you can see there the results for the already existing approaches. All results are statistically significantly different, what is proved by Mann-Whitney U test.

Note that the simple numeration of categorical values provides very poor results. Also we can see that our transformation approach in combination with RF outperforms FMs on the *Amazon* dataset and inferiors on *Movie Lens*. FMs are designed as a collaborative filtering method, so they show the better result on the domain dataset *Movie Lens*. So our transformation approach allows us to get comparable to the state-of-the-art performance with the quite common machine learning techniques (RF).

Figure 3 shows the dependency of the prediction quality on the rank r of matrix approximation on the *Amazon* dataset. According to the results, it is not necessary to use rank much more than 10. The situation at *Movie Lens* dataset is quite similar.

An another question is a choice of a matrix similarity measure for getting more accurate results. Figure 4 shows the dependence of the prediction AUC on the parameter β in β-divergency. We see that the optimization of Kullback-Leibler divergency ($\beta \rightarrow 1$) is actually more effective than the Frobenius norm optimization ($\beta = 2$). But the optimization of Kullback-Leibler divergency works much slower (more than 10^4 times slower) than the Frobenius norm optimization, because of very simple MSE derivative calculating. We don't even have computed results for *Amazon* dataset in case of the KL-divergency. So we can not talk about its efficiency in spite of a potentially good prediction performance. All previous experiments use the Frobenius norm as a matrix similarity measure.

Table 3. Comparison of proposed and existing approaches. All differences in the quality are statistically significant.

Method	Amazon	Movie Lens
Direct frequency transformation + RF	0.8503	0.7597
Low-rank approximated transformation + RF	0.8472	0.7539
Low-rank latent transformation + RF	**0.8817**	0.7702
Low-rank latent transformation + SVM	0.8442	0.7174
Simple numeration + RF	0.5703	0.5311
Sparse logistic regression	0.8691	0.7958
Smoothed naive Bayes	0.8776	0.7744
Factorization machines	0.8765	**0.8116**

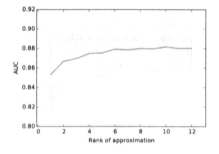 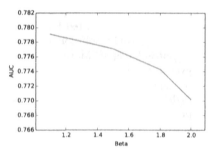

Fig. 3. Dependence AUC on parameter r with 2σ errorbar on *Amazon* dataset

Fig. 4. Dependence AUC on parameter β with 2σ errorbar on *Movie Lens* dataset

6   Conclusion and Future Work

We have introduced the domain independent method of transformation of categorical features into real ones that uses the latent vectors of low-rank multipliers. The data transformed this way can be efficiently analyzed by common machine learning techniques. We have shown that the proposed transformation method in the combination with widely used Random Forest binary classifier leads the results comparable to the state-of-art approaches like Factorization Machines. This fact proves the effectiveness of this transformation method. Furthermore, our unsupervised approach can also be directly used for the regression, multiclass classification, cluster analysis and other machine learning problems.

   In future work we are going to explore a generalization of the proposed method for a handling missing feature values and an incorporating real feature values. Real values can only be transformed into binary (thus into categorical) features, obtained by comparing feature values with thresholds, in the currently proposed approach. Furthermore, choosing of the matrix similarity measure is very important as we have shown in Fig. 4. So we will also examine which particular measure makes the best choice in our future work.

References

1. Bader, B.W., Kolda, T.G.: Efficient MATLAB computations with sparse and factored tensors. SIAM J. Sci. Comput. **30**(1), 205–231 (2007)
2. Bengio, Y., Schwenk, H., Senécal, J.S., Morin, F., Gauvain, J.L.: Neural probabilistic language models. In: Holmes, D.E., Jain, L.C. (eds.) Innovations in Machine Learning. SFSC, vol. 194, pp. 137–186. Springer, Heidelberg (2006)
3. Breiman, L.: Mach. Learn. Random forests **45**(1), 5–32 (2001)
4. Breiman, L., Friedman, J., Stone, C.J., Olshen, R.A.: Classification and Regression Trees. CRC Press, Boca Raton (1984)
5. Cichocki, A., Zdunek, R., Phan, A.H., Amari, S.I.: Nonnegative matrix and tensor factorizations: applications to exploratory multi-way data analysis and blind source separation. Wiley, United Kingdom (2009)
6. D'yakonov, A.G.: Solution methods for classification problems with categorical attributes. Comput. Math. Model. **46**, 1–21 (2015)
7. Févotte, C., Idier, J.: Algorithms for nonnegative matrix factorization with the β-divergence. Neural Comput. **23**(9), 2421–2456 (2011)
8. Friedman, J., Hastie, T., Tibshirani, R.: The Elements of Statistical Learning. Springer, Heidelberg (2001)
9. Hardy, M.A.: Regression with dummy variables. No. 93, Sage (1993)
10. Jurafsky, D., James, H.: Speech and language processing an introduction to natural language processing, computational linguistics, and speech (2000)
11. Koren, Y.: Factorization meets the neighborhood: a multifaceted collaborative filtering model. In: Proceedings of the 14th ACM SIGKDD international conference on Knowledge discovery and data mining, pp. 426–434. ACM (2008)
12. Mann, H.B., Whitney, D.R.: On a test of whether one of two random variables is stochastically larger than the other. Ann. Math. Stat. **18**, 50–60 (1947)
13. Omlin, C.W., Giles, C.L.: Stable encoding of large finite-state automata in recurrent neural networks with sigmoid discriminants. Neural Comput. **8**(4), 675–696 (1996)
14. Rendle, S.: Factorization machines. In: 2010 IEEE 10th International Conference on Data Mining (ICDM), pp. 995–1000. IEEE (2010)
15. Rendle, S.: Factorization machines with libfm. ACM Trans. Intell. Syst. Technol. (TIST) **3**(3), 57 (2012)
16. Rendle, S., Freudenthaler, C., Schmidt-Thieme, L.: Factorizing personalized markov chains for next-basket recommendation. In: Proceedings of the 19th International Conference on World Wide Web, pp. 811–820. ACM (2010)
17. Rendle, S., Schmidt-Thieme, L.: Pairwise interaction tensor factorization for personalized tag recommendation. In: Proceedings of the Third ACM International Conference on Web Search and Data Mining, pp. 81–90. ACM (2010)
18. Ricci, F., Rokach, L., Shapira, B.: Introduction to recommender systems handbook. In: Ricci, F., Rokach, L., Shapira, B., Kantor, P.B. (eds.) Recommender Systems Handbook, pp. 1–35. Springer, Heidelberg (2011)

A Comparative Evaluation of Statistical Part-of-Speech Taggers for Russian

Rinat Gareev[✉] and Vladimir Ivanov

Kazan Federal University, Kazan, Russia
gareev-rm@yandex.ru, nomemm@gmail.com

Abstract. Part-of-speech (POS) tagging is an essential step in many text processing applications. Quite a few works focus on solving this task for Russian; their results are not directly comparable due to the lack of shared datasets and tools. We propose a POS tagging evaluation framework for Russian that comprises existing third-party resources available for researchers. We applied the framework to compare several implementations of statistical classifiers: HunPos, Stanford POS tagger, OpenNLP implementation of MaxEnt Markov Model, and our own re-implementation of Tiered Conditional Random Fields. The best tagger that was trained on a corpus with less than one million words achieved an accuracy above 93 % .We expect that the evaluation framework will facilitate future studies and improvements on POS tagging for Russian.

1 Introduction

Part-of-speech (POS) tagging is one of the earliest and important steps in a text processing pipeline for many applications. In this paper, we focus on Russian POS taggers that solve the task in the widely accepted definition: to label each word with its morphosyntactic category. Russian is an inflective language, and it has a rich morphology, which leads to an extensive tagset with hundreds of tags. Every tag encodes a part of speech and a possible assignment of values to different grammatical categories: number, case, gender, tense, and others.

While a number of publications is devoted to computational aspects of Russian morphology, such as tagset construction, lemmatization, prediction of grammatical category for rare (nondictionary) words, rule-based tag disambiguation, and others, much fewer works report directly comparable results for POS tagging. Most researchers use training and testing corpora of different versions, sizes, and tagsets due to lack of a unified framework for training and evaluation of POS taggers for Russian.

In 2013, the Russian National Corpus (RNC) community has made available a part of the RNC with disambiguated morphological annotations;[1] this boosted a comparative evaluation activity. The OpenCorpora project provides a comprehensive and emerging morphological dictionary, which is freely available.[2]

---

[1] http://ruscorpora.ru/en/corpora-usage.html.
[2] http://opencorpora.org/dict.php.

© Springer International Publishing Switzerland 2015
P. Braslavski et al. (eds.): RuSSIR 2014, CCIS 505, pp. 263–275, 2015.
DOI: 10.1007/978-3-319-25485-2_8

We adopt both the RNC dataset and the OpenCorpora dictionary as a basis for our contribution.

The key contributions of this paper are (i) an evaluation framework for Russian POS tagging and (ii) experimental results of several taggers according to this framework. The evaluation framework consists of available third-party resources as well as our alignment of them. This framework facilitates a comparative study of different taggers and their properties. Focusing on accuracy, we examined the performance of the six statistical taggers, including those that are successfully applied for other languages: HunPos [6], OpenNLP MaxEnt,[3] Stanford POS tagger [19], and Tiered Conditional Random Fields [14]. Some of these taggers have not been evaluated for Russian. We apply all of them in the unified experimental setup of the proposed framework: using the same corpus split for training, optimization, and evaluation, as well as the same dictionary. The developed software components are open-source.

The rest of the paper is structured as follows. The next section provides an overview of related work. Section 3 describes resources that constitute an evaluation framework. Section 4 presents taggers under evaluation, parameters of their models, and results of their adjustment on a development set. Section 5 provides final evaluation results and their analysis.

2 Related Work

In [9], Ljashevskaja et al. reported results of the RU-EVAL-2010 forum. The forum was devoted to the evaluation of morphological parsers for Russian. Several tracks were held including one for POS tagging, but the tagset was limited to lexical categories.[4] A full tagset was considered in the other track; however, complete disambiguation was not required there. The evaluation of systems was performed on a golden standard that contains about 2,000 words, but no training corpus was provided to participants.

In [17], Sokirko and Toldova described an HMM-tagger 'Trigram', which, in turn, is heavily based on a statistical component described in [5]. The tagger uses a trigram transition probability model, a lexical probability model (where a word is conditioned on a tag bigram), and different smoothing methods for both models. Authors trained models on the RNC with 3.3 million words.[5] This tagger achieved 94.46 % accuracy for a full tagset (829 tags).

In [16], Sharoff et al. introduced a tagset for Russian as a part of MULTEXT-East specification. The result tagset contains about 600 tags. Furthermore, authors evaluated three statistical taggers on the RNC with five million words: TnT, TreeTagger, and SVMTool. The best result (95.28 %) was obtained by the TnT tagger using 90 % of the corpus for training and the rest for testing.

In [1], Antonova and Soloviev compared three POS taggers: a CRF-based one, Stanford POS tagger, and TreeTagger. However, the tagset was limited to

---

[3] http://opennlp.apache.org/.

[4] i.e., parts of speech: noun, verb, adjective, etc.

[5] The size is approximate; it was estimated from a number of sentences.

lexical categories, and experiments were carried out on an in-house corpus. In [8], Lakomkin et al. compared accuracies of Hidden Markov Models and Maximum Entropy Markov Models. They used the RNC consisting of six million words. While their experimental results are worth mentioning, the limited information about the implementation and the configuration of the taggers was given. The contribution of the morphological analyzer to the provided results is also unclear.

3   Evaluation Framework

An evaluation framework for statistical POS taggers defines a corpus, its split, a tagset, and a dictionary. We propose one that uses easily obtainable resources.

We used the available part of the RNC with disambiguated morphological tags for experiments. It contains one million words. We used 80 % of documents for training, 10 % for parameter tuning (the development set), and 10 % for final testing. Properties of these sets are given in Table 1. Here, word is a token that holds a tag in the corpus. The tagset design of this corpus is described in [10].

Table 1. Statistics of the corpus splits

Set	Documents	Sentences	Words	Unseen[a]
Training	426	76805	815654	0
Development	53	8020	92040	12252
Testing	53	8415	93369	9506

[a] A count of words that do not occur in the training set

As a dictionary, we used the one from OpenCorpora project [2]. It contains about 390,000 lexemes and over five million word forms. Its tagset differs from the one of the RNC, though both of them are based on Zaliznjak's model of grammatical categories [20]. Because the taggers under evaluation are able to use a dictionary, we implemented the corpus preprocessing procedure that replaces each RNC tag with its OpenCorpora alternative. This mapping procedure mainly does operations of three kinds. First, it relabels values of grammatical categories. For the sake of brevity, we will use 'grammeme' instead of 'grammatical category value'. For example, 'sg' in the RNC corresponds to 'sing' (singular) in OpenCorpora. Second, the mapping removes certain grammemes in particular cases when an RNC tag contains more grammatical categories. For example, RNC tags of different verb forms (finite, infinitive and participles) contain voice grammemes, while the dictionary defines them for participles only. Third, if a word has a tag that is not in the dictionary tagset, but there is a single dictionary tag for this word that complements the corpus tag, then the procedure will apply the dictionary tag. So this procedure always preserves the interpretations of original corpus tags.

The next preprocessing step retains only grammatical categories that are a matter of evaluation. We considered only the following categories: part of speech, number, gender, case, person, aspect, tense and voice. There are also two special tags created by RNC conventions: 'NON-LEX' (for non-Russian words) and 'Init' (for name initials). We treated these two as separate parts of speech.

Technically, each tag is a concatenation of grammeme labels in a fixed order (e.g., 'NOUN&femn&plur&nomn').[6] The alignment we carried out resulted in 503 such tags in the corpus. There are also the 17 separate tags for punctuation and special tokens that are assigned using the unambiguous mapping for characters. For example, all opening parentheses and brackets (square, curly and others) get the tag '(', all closing parentheses and brackets get the tag ')' and so on. While all punctuation tokens were excluded from the evaluation, the taggers that we present below treat them as observations in an input sequence.

One should perform the following steps to use this framework:

- obtain the corpus and the dictionary from their websites;
- apply preprocessing procedures that are described above; their implementation and detailed documentation are given at a webpage;[7]
- and apply the same split of the corpus to train, tune and test a tagger; lists of corpus document names in each set are also available at the webpage.

4   Taggers

We compared the performance of two well-known baselines and four open-source taggers, focusing on their accuracy.

4.1   Frequency-Based Baseline

This baseline is one of the simplest forms of tagging. During the training phase, the baseline counts tag distribution for each word and its suffix. During the tagging phase, the baseline labels each word with its most frequent tag. Unless the word is trained, the tagger will assign the most frequent tag for the suffix of this word. The algorithm processes numbers in a slightly different way: they are assigned the most frequent tag for any number. Suffix length is a configuration parameter that we tuned on the development set. In the range 2–6, the baseline achieved the best accuracy (81.68 %) for suffix length set to 4. In this setting, about 10 % of errors was caused by the lack of training data.

4.2   Dictionary-Based Baseline

This baseline tagger is included to reveal the coverage of a data set by the rather exhaustive morphological dictionary. The baseline assigns a tag to each word in input by looking it up in a dictionary. If the dictionary produces several tags, the baseline will apply the most frequent one for this word in a training set. This baseline processes numbers in the same way as the previous tagger. It achieved

---

[6] The labelling of gramemmes can be found at http://opencorpora.org/dict.php.

[7] http://github.com/CLLKazan/UIMA-Ext.

an accuracy of 77.06 % on the development set where 4.4 % of words are not in the dictionary and 5.4 % of words are ambiguous and unseen in the training set. Note, that there are corpus tag assignments that do not comply with the dictionary: 1.5 % of training set words have such tags.

4.3   HunPos

HunPos [6] is an open-source reimplementation of TnT [3] that allows the additional adjustment of configuration parameters. As previously mentioned, the original TnT implementation performed the best in [16]. By default, HunPos trains a second-order HMM where transition and lexical probabilities depend on previous tag pairs. It also incorporates tag guessing for unknown words by memorizing the tag distribution for each suffix among rare words.

We varied the following parameters: an order of transition probabilities (from 1 to 3), an order of lexical probabilities (1 or 2), and a frequency threshold (5, 10, 20) to treat a word as rare. The best accuracy of 88.65 % was achieved when second-order transition and first-order emission probabilities are estimated with a rare word threshold of 5. HunPos is also able to use a morphological dictionary in the tagging phase to predict tags for unseen words. Enabling this option increased the accuracy up to 89.81 % on the development set.

4.4   Stanford POS Tagger

The current version of Stanford POS tagger is based on cyclic dependency network [19]. We tested several predefined feature architectures that incorporate features about tags on the left side of a classification target.[8] So these tagger configurations correspond to Maximum Entropy Markov Models (MEMM). Brief descriptions of these feature sets and their performance are given in Table 2, where the following labels for feature types are used:

- words(b, e): individual features for each word at positions from b to e, where position 0 stands for the current word;
- tag(i): an individial feature for each tag at position i;
- biwords(b, e): individial features for each word bigram in the word sequence from b to e;
- wordTag(w, t): a feature combining the word at w and the tag at t.

In every configuration, the following features for rare words were used: one feature for each suffix of lengths 1–3 of the current word, one feature for each suffix of lengths 1–3 of the previous word, the Unicode shape of the current word. We defined each tag for pronouns, conjunctions, prepositions, and particles as a closed class tag so they are not considered as a candidate for unseen words. Training optimization parameters were left with their default values.

---

[8]   There are also several predefined bidirectional architectures but we experienced technical issues with them.

Table 2. Performance of the different Stanford POS tagger feature architectures on the development set

Macro	Features	Accuracy
left3words	words(−1,1), tag(−2)+tag(−1), tag(−1)	89.31 %
left5words	words(−2,2), tag(−2)+tag(−1), tag(−1)	**89.51 %**
generic	words(−1,1), tag(−2)+tag(−1), tag(−1), biwords(−1,0), wordTag(0,−1)	89.46 %

4.5 OpenNLP MaxEnt

The Apache OpenNLP library provides an implementation of MEMM for a variety of NLP tasks. While the Stanford POS tagger configurations that are presented above are based on MEMM as well, there are two main reasons to include OpenNLP MaxEnt into this study. First, extensibility of this tool allows incorporating features based on a morphological dictionary. Second, its implementation of beam search [15] enables a sequence validation.

We used the following feature set for experiments:

1. surface form features of a current token: its string, its string in the lower case, its suffixes (up to length 3), capitalization type (ALL_UPPERCASE, all_lowercase, Initial_uppercase or MiXed), and number type (all digits, year digits, alphanumeric, roman numeral, or other);
2. [D] tags for a current word from dictionary (one feature for each tag);
3. [D] indicator of an out-of-dictionary word;
4. [D] agreement in case, number, and gender (or different combinations of these categories, one feature for each agreement type) of a current word dictionary interpretation with the previous tag;
5. context token features: for each token $t_{i-l} \ldots t_{i-1}$, $t_{i+1} \ldots t_{i+r}$ the features from item (1) (without capitalization type) are generated; l and r are configuration parameters that restricts size of context;
6. previous tags: one feature for each tag at position $-1 \cdots - p$, where p is a configuration parameter.

During the tagging phase, the beam search algorithm starts from the first token of an input sequence. At each token i, it finds B most probable tag assignments for the subsequence $1..i$ given B most probable assignments for the previous subsequence $1..(i-1)$. A sequence validator allows to exclude certain tag assignments on a subsequence from this search. We used the latter capability to implement the sequence validator based on lookup of the morphological dictionary. The sequence validator that is used in the experiments rejects the following tag assignments:

- the candidate tag for a current word is not among its dictionary tags;
- the dictionary does not contain a current word and the candidate tag belongs to a closed lexical class.

There are few exceptions. The validator always allows 'Init'[9] and 'Prnt' tags as well as nominalization of adjectives.

We tuned the following parameters of this tagger on the development set: the number of previous tags used as a feature ($p = 1 \ldots 3$) and sizes of the left ($l = 1 \ldots 3$) and the right ($r = 1 \ldots 2$) context. Beam size B was set to 10. Only the features that occur more than 5 times in the training set were used to train models. The best results are shown in Table 3.

Table 3. Performance of the different OpenNLP MaxEnt tagger configurations on the development set

Tagger	The best configuration	Accuracy
Base	$p = 2, l = 2, r = 1$	87.82%
Base + [D] features	$p = 2, l = 2, r = 1$	90.87%
Base + [D] features + the sequence validator	$p = 3, l = 2, r = 1$	91.51%

4.6 Tiered CRF

Tagset size for inflecting languages makes direct application of a linear-chain conditional random fields (CRF) not practical because computational complexity of its inference and learning routines depends quadratically on a number of output classes [18]. Radziszewski [14] alleviated this obstacle for Polish through the application of successive CRF models where each model outputs only grammatical categories of a particular tier. So tiers represent a partition of all grammatical categories encoded in the tagset. For the experiments, we fixed the following tier composition: (1) part of speech, (2) number, (3) gender, (4) case, (5) person, (6) aspect, (7) tense, and (8) voice.

We implemented the tagger algorithm of [14] and its training procedure with minor changes. The details are depicted in Algorithms 1 and 2. So there are two main differences between our implementation and the original work [14]. First, our implementation will not reject the label of a word from any CRF if it contradicts the dictionary entry for this word. Second, our implementation does not generate a list of most frequent tags for nondictionary words.

For each tier, we used the same set of base features as for MaxEnt tagger that are described in items 1 and 5 of Sect. 4.5. In addition, we used the following features:

- grammemes (one feature for each value) assigned on previous tiers;
- additional context features: for each token $t_{i-l} \ldots t_{i-1}, t_{i+1} \ldots t_{i+r}$ extract grammemes assigned on previous tiers (one feature for each value and token position);

---

[9] The special POS tag for name initials in the RNC.

Algorithm 1. Training of TCRF tagger

$prevTiers = \{\}$
for $tier \in tiers$ **do**
    $corpus_{tier} \leftarrow corpus$ with all tags trimmed to $prevTiers \cup tier$
    for $sentence \in corpus_{tier}$ **do**
      for $token \in sentence$ **do**
        $f\_values \leftarrow \{f(token, sentence) : f \in features(tier)\}$
        if $token\_tag$ contains $tier$ values **then**
          $label \leftarrow$ value of $tier$ in $token\_tag$ {the class label for current tier CRF}
        else
          $label \leftarrow$ NULL {the special label of class}
        store training example (f_values, $label$) for $tier$
    add $tier$ to $prevTiers$
for $tier \in tiers$ **do**
    train CRF with training data for $tier$

Algorithm 2. TCRF tagging of a single sentence

$result\_tags = []$
for $token \in sentence$ **do**
    append empty tag to $result\_tags$
for $tier \in tiers$ **do**
    $sent\_repr \leftarrow []$ {sentence representation}
    for $token \in sentence$ **do**
      $f\_values \leftarrow \{f(token, sentence) : f \in features(tier)\}$
      append f_values to $sent\_repr$
    $label\_list \leftarrow$ classify $sent\_repr$ with CRF trained for $tier$
    for $token \in sentence$ **do**
      $token\_label \leftarrow label\_list[token]$
      if $token\_label \neq$ NULL **then**
        add grammeme $token\_label$ to $result\_tags[token]$

- [D] grammemes of the current tier (one feature for each) that are defined in the dictionary entries for the current word; only the dictionary entries that are compatible with output grammemes of the previous tiers are used to calculate these features;
- [D] indicator of a nondictionary word.

We tuned l ($= 1 \ldots 3$) and r ($= 1 \ldots 2$) parameters and the regularization coefficient on the development set. The same feature count cut-off (> 5) was used as for experiments with the previous two taggers. The best results are shown in Table 4.

4.7   Implementation Details

Our tagger implementations, dictionary, and corpus preprocessing procedures and data structures are based on Apache UIMA. Third-party taggers were also

Table 4. Performance of the TCRF taggers on the development set

Tagger	The best configuration	Accuracy
Base features	$l = 2, r = 1$	87.20 %
Base + [D] features	$l = 3, r = 2$	92.46 %

wrapped into UIMA annotators. DKPro-Lab framework [4] was used to configure a work-flow of experiments and sweeping of tagger parameters. Implementation of Tiered CRF tagger uses crfsuite [13]. Feature generation for OpenNLP MaxEnt and TCRF taggers is implemented with ClearTK-ML [12].

5   Evaluation Results

As a final test, the taggers were applied to the test set, with the optimal parameter settings on development data. The results are shown in Table 5. The given accuracy is a percentage of words that are assigned a correct tag by a tagger. Punctuation and other special tokens are excluded from the evaluation. We checked statistical significance of observed differences in tagger performances by using the approximate randomization test [11].

As expected, a morphological dictionary improves the accuracy of taggers on unseen words with smaller improvement on seen words. The best accuracy is achieved by the TCRF tagger. We suppose that there are two main reasons for this. First, CRF allows to build more sophisticated probabilistic models and has several advantages over HMM and MEMM. A more detailed explanation with experiments on English POS tagging is given in the work [7]. Second, the TCRF tagger resolves morphological ambiguity among hundreds of tags gradually: the first CRF resolves general lexical categories (parts of speech) of input words, and the consecutive CRF models have narrower search space and use output of previous classifiers as additional features.

The Stanford POS tagger performed the best among all tagger configurations that does not use an external dictionary. Given two MEMM implementations here, Stanford POS tagger and OpenNLP MaxEnt, the best configuration of the former uses a bit different feature set, different algorithm of parameter estimation (L-BFGS vs Generalized Iterative Scaling), and different strategy of feature value generation: the Stanford tagger generates additional features for rare and unseen words. Evaluating the contribution of each of these distinctions to the observed improvement of accuracy requires additional experiments. The high accuracy of the Stanford tagger comes at a price: tagging the testing set takes about 90 min with it, while the other evaluated taggers takes 1–5 min for the same task.

Note that the implementation of TCRF tagger has the potential to perform better. Currently, it does not filter tag sequences in Viterbi search of the most

Table 5. Tagging accuracy on the test set. Values that are marked by * represent a statistically significant ($p < 0.01$) improvement over an accuracy in the previous table entry.

Tagger	Overall	Seen	Unseen
D-Baseline	80.51 %	84.61 %	44.30 %
F-Baseline	83.44 %*	86.22 %	58.94 %
Tiered CRF	88.68 %*	90.90 %	69.10 %
MaxEnt	89.06 %*	91.65 %	70.01 %
HunPos	89.59 %*	91.58 %	72.06 %
HunPos+Dict	90.61 %*	91.62 %	81.73 %
Stanford (without Dict)	**90.74 %**	92.91 %	71.55 %
MaxEnt+Dict+Seq validator	92.12 %*	93.00 %	84.35 %
Tiered CRF+Dict	**93.34 %***	**94.30 %**	**84.87 %**

probable one as in the beam search of the OpenNLP MaxEnt tagger. It also does not have any special treatment of rare words in a training set like in the Stanford tagger.

The top entries from the confusion matrix of the most accurate tagger are shown in Table 6. Plenty of them represents a well-known disambiguation problem for nominative, accusative, and genitive cases of nouns and adjectives. Other entries of this table are related to pronominal adverbs and composite conjunctions. Table 7 presents the tagger accuracies for each part of speech and their distribution in the test and train sets. About 55 % of errors are produced on nouns and full form adjectives (ADJF). The quite low accuracy is achieved for full form participles (PRTF). We suppose that this is a consequence of training data sparsity for this part of speech. There are 146 tags for PRTF, almost third of the tagset, while only 1.1 % of words have these tags in the training data.

Table 6. The top 10 entries from the confusion matrix of the TCRF output sorted by frequency. The third column contains a percentage among all incorrectly tagged words.

Gold	TCRF		Example words
NOUN&masc&sing&accs	NOUN&masc&sing&nomn	2.75%	рынок, результат, проект
NOUN&neut&sing&accs	NOUN&neut&sing&nomn	1.99%	место, кино, занятие
ADVB	CONJ	1.71%	как, когда, пока
CONJ	PRCL	1.59%	ни, же, ли
NOUN&masc&sing&nomn	NOUN&masc&sing&accs	1.58%	крем, бас, час
ADJF&plur&accs	ADJF&plur&nomn	1.34%	новые, дачные, сельские
NOUN&masc&plur&accs	NOUN&masc&plur&nomn	1.30%	этапы, участки, стаканы
NOUN&femn&plur&nomn	NOUN&femn&sing&gent	1.19%	системы, коллекции
CONJ	ADVB	1.13%	как, когда, пока
NOUN&neut&sing&nomn	NOUN&neut&sing&accs	1.09%	кино, поле, начало

Table 7. The relative accuracy of the TCRF tagger for each POS on the test set. The second and third columns show a distribution of parts of speech in the test and training sets.

POS	Acc%	Test%	Train%	POS	Acc%	Test%	Train%
NOUN	90.18	29.62	29.21	PRTF	76.98	1.27	1.10
PREP	99.88	10.43	10.21	ADJS	83.66	0.83	0.69
VERB	98.35	9.94	10.75	PRED	83.70	0.79	0.76
ADJF	91.02	8.25	8.16	COMP	87.94	0.49	0.38
CONJ	96.78	8.18	7.59	Prnt	97.51	0.47	0.56
NPRO	92.67	6.81	7.56	PRTS	96.94	0.46	0.47
ADVB	94.43	5.96	6.05	ADJF&Anum	78.82	0.43	0.47
ADJF&Apro	89.98	4.66	4.79	Non-Lex	89.58	0.43	0.81
PRCL	94.26	4.57	4.53	GRND	98.02	0.38	0.45
NUMR	94.38	2.91	2.49	RNC_INIT	97.06	0.33	0.39
INFN	98.63	2.73	2.48	INTJ	82.98	0.05	0.11

6 Conclusion

We proposed the POS tagging evaluation framework for Russian. The framework consists of the corpus and its split, the tagset, the dictionary, and auxiliary tools. All these resources are available for researchers. The resources and software components that have been developed for this study are open-source and available at http://github.com/CLLKazan/UIMA-Ext.

We used the proposed framework to compare accuracy of several implementations of statistical classifiers: HunPos, Stanford POS tagger, OpenNLP implementation of MaxEnt Markov Model, and Tiered Conditional Random Fields. To our knowledge, only few of them have reports about their application for Russian POS tagging. We confirmed that the integration of the external morphological dictionary significantly improves accuracy on words that are absent in the training set. Our comparison revealed that the TCRF tagger performed the best while it still has the potential for improvements. For example, in experiments without a dictionary, the Stanford tagger demonstrated the best result, so distinguishing properties of this tagger motivate recommendations for further elaboration of TCRF tagger.

Though, this comparison does not cover all available implementations, our study can be exploited in the following cases. End users are able to use provided experimental results to select the most appropriate tool. NLP researchers are recommended to use the provided framework to evaluate other existing and future developments for Russian POS tagging, that will facilitate reproducible, comparative and incremental research.

Acknowledgments. This work was financially supported by the Russian Science Foundation (grant 15-11-10019).

References

1. Antonova, A.Y., Soloviev, A.N.: Conditional random field models for the processing of Russian. In: Computational Linguistics and Intellectual Technologies: Papers From the Annual Conference "Dialogue" (Bekasovo, 29 May – 2 June 2013), vol. 1, pp. 27–44. RGGU, Moscow (2013) (in Russian)
2. Bocharov, V., Bichineva, S., Granovsky, D., Ostapuk, N., Stepanova, M.: Quality assurance tools in the OpenCorpora project. In: Computational Linguistics and Intellectual Technologies: Papers From the Annual Conference "Dialogue" (Bekasovo, 25–29 May 2011), pp. 101–109. RGGU, Moscow, Russia (2011)
3. Brants, T.: TnT: a statistical part-of-speech tagger. In: Proceedings of the Sixth Conference on Applied Natural Language Processing, ANLC 2000, pp. 224–231. Association for Computational Linguistics, Stroudsburg, PA, USA (2000)
4. de Castilho, R.E., Gurevych, I.: A lightweight framework for reproducible parameter sweeping in information retrieval. In: Agosti, M., Ferro, N., Thanos, C. (eds.) Proceedings of the 2011 Workshop on Data Infrastructures for Supporting Information Retrieval Evaluation, DESIRE 2011, pp. 7–10. ACM, New York (2011)
5. Hajič, J., Krbec, P., Květoň, P., Oliva, K., Petkevič, V.: Serial combination of rules and statistics: a case study in Czech tagging. In: Proceedings of the 39th Annual Meeting on Association for Computational Linguistics, ACL 2001, pp. 268–275. Association for Computational Linguistics, Stroudsburg, PA, USA (2001)
6. Halácsy, P., Kornai, A., Oravecz, C.: HunPos: an open source trigram tagger. In: Proceedings of the 45th Annual Meeting of the ACL on Interactive Poster and Demonstration Sessions, ACL 2007, pp. 209–212. Association for Computational Linguistics, Stroudsburg, PA, USA (2007)
7. Lafferty, J.D., McCallum, A., Pereira, F.C.N.: Conditional random fields: probabilistic models for segmenting and labeling sequence data. In: Proceedings of the Eighteenth International Conference on Machine Learning, ICML 2001, pp. 282–289. Morgan Kaufmann Publishers Inc., San Francisco, CA, USA (2001)
8. Lakomkin, E.D., Ryzhova, D.A., Puzyrevskij, I.: Analiz statisticheskix algoritmov snyatiya morfologicheskoj omonimii v russkom yazyke. In: Доклады всероссийской научной конференции АИСТ' 2013. Moscow (2013) (in Russian)
9. Ljashevskaja, O.N., Astaf'eva, I., Bonch-Osmolovskaja, A., Garejshina, A., Grishina, J., D'jachkov, V., Ionov, M., Koroleva, A., Kudrinskij, M., Litjagina, A., Luchina, E., Sidorova, E., Toldova, S., Savchuk, S., Koval, S.: NLP evaluation: Russian morphological parsers. In: Computational Linguistics and Intellectual Technologies: Papers From the Annual Conference "Dialogue" (Bekasovo, 26–30 May 2010), pp. 318-326 (2010) (in Russian)
10. Ljashevskaja, O.N., Plungjan, V.A., Sichinava, D.V.: O morfologicheskom standarte Nacional'nogo korpusa russkogo jazyka. In: Национальный корпус русского языка: 2003–2005. Результаты и перспективы, pp. 111–135. Indrik, Moscow, Russia (2005) (in Russian)
11. Noreen, E.: Computer-Intensive Methods for Testing Hypotheses: An Introduction. A Wiley-Interscience publication, Wiley (1989)

12. Ogren, P.V., Wetzler, P.G., Bethard, S.J.: ClearTK: a framework for statistical natural language processing. In: Unstructured Information Management Architecture Workshop at the Conference of the German Society for Computational Linguistics and Language Technology (2009)
13. Okazaki, N.: CRFsuite: a fast implementation of conditional random fields (CRFs) (2007). http://www.chokkan.org/software/crfsuite/
14. Radziszewski, A.: A tiered CRF tagger for polish. In: Bembenik, R., Skonieczny, L., Rybiński, H., Kryszkiewicz, M., Niezgódka, M. (eds.) Intell. Tools for Building a Scientific Information. SCI, vol. 467, pp. 215–230. Springer, Heidelberg (2013)
15. Ratnaparkhi, A.: A maximum entropy model for part-of-speech tagging. In: Brill, E., Church, K. (eds.) Proceedings of the Empirical Methods in Natural Language Processing, pp. 133–142 (1996)
16. Sharoff, S., Kopotev, M., Erjavec, T., Feldman, A., Divjak, D.: Designing and evaluating a Russian tagset. In: Chair, N.C.C., Choukri, K., Maegaard, B., Mariani, J., Odijk, J., Piperidis, S., Tapias, D. (eds.) Proceedings of the Sixth International Conference on Language Resources and Evaluation (LREC 2008). European Language Resources Association (ELRA), Marrakech, Morocco (2008)
17. Sokirko, A., Toldova, S.: Sravnenie effektivnosti dvuh metodik snyatiya lexicheskoy i morfologicheskoy neodnoznachnosti dlya russkogo yazyka. Technical report (2005). http://www.aot.ru/docs/RusCorporaHMM.htm, in Russian
18. Sutton, C., McCallum, A.: An introduction to conditional random fields. Found. Trends Mach. Learn. 4(4), 267–373 (2012)
19. Toutanova, K., Klein, D., Manning, C.D., Singer, Y.: Feature-rich part-of-speech tagging with a cyclic dependency network. In: Proceedings of the 2003 Conference of the North American Chapter of the Association for Computational Linguistics on Human Language Technology, NAACL 2003, vol.1, pp. 173–180. Association for Computational Linguistics, Stroudsburg, PA, USA (2003)
20. Zaliznjak, A.A.: Grammaticheskij slovar' russkogo jazyka. Slovoizmenenie. Грамматический словарь русского языка. Словоизменение Russkij jazyk, Moscow, 3 edn. (1987) (in Russian)

Recommendation of Ideas and Antagonists for Crowdsourcing Platform Witology

Dmitry I. Ignatov[1]([✉]), Maria Mikhailova[1],
Alexandra Yu. Zakirova[1,3], and Alexander Malioukov[2]

[1] National Research University Higher School of Economics, Moscow, Russia
dignatov@hse.ru
http://www.hse.ru/en
[2] Witology, Moscow, Russia
http://www.witology.com/en
[3] Yandex, Moscow, Russia
http://company.yandex.com

Abstract. This paper introduces several recommender methods for crowdsourcing platforms. These methods are based on modern data analysis approaches for object-attribute data, such as Formal Concept Analysis and biclustering. The use of the proposed techniques is illustrated by the results of recommendation of ideas and antagonists for crowdsourcing platform Witology. In particular we show how the quality of antagonists recommender can be improved by usage of biclusters as focal areas for distance and similarity calculation.

Keywords: Crowdsourcing · Recommender systems · Biclustering · Formal concept analysis · Witology · Social innovation platforms

1 Introduction

The success of modern collaborative technologies is marked by the appearance of many so called Social Innovation Platforms for distributed brainstorming or carrying out so called "public examination". There are many such crowdsourcing companies in the USA (Mindjet[1], BrightIdea[2], InnoCentive[3] etc.) and Europe (Imaginatik[4]) [1]. There is also a Kaggle platform[5] which is very beneficial for data practitioners and companies that want to select the best solutions for their data mining problems and crowdsourcing Internet marketplace Amazon Mechanical Turk[6] for Human Intelligence Tasks solution. In 2011 Russian companies

---

[1] http://www.mindjet.com/spigitengage/.
[2] http://www.brightidea.com/.
[3] http://www.innocentive.com/.
[4] http://www.imaginatik.com/.
[5] http://www.kaggle.com.
[6] http://www.mturk.com/.

© Springer International Publishing Switzerland 2015
P. Braslavski et al. (eds.): RuSSIR 2014, CCIS 505, pp. 276–296, 2015.
DOI: 10.1007/978-3-319-25485-2_9

decided to launch business in crowdsourcing area as well. Witology[7] and Wikivote[8] are the most representative examples of such Russian companies, however the latter is rather sophisticated voting sytem with crowdsourced components; system 4i of TEKORA company is also an example of idea and innovation management system[9] on Russian market. As for Witology company, over 30 large-scale projects have already been finished successfully among those all-Russian ongoing projects like Sberbank-21[10] and National Entrepreneurial Initiative[11] as well as city-level projects like project of Moscow Department of Transportation "Our Routes" or project "My Office of Government Services" for Moscow Government[12].

As a rule, while participating in a project, users of such crowdsourcing platforms discuss and solve one common problem, propose their ideas and evaluate ideas of each other as experts [2]. As a result of this process the platform owners can get a reliable ranking of ideas and users who generated them. From data model point of view, socio-semantic networks constitute the core of such crowdsourcing systems [3–6], where user-to-user, user-to-item relations as well as user-to-item relations are presented. In our previous work we proposed promising approaches to analyze data underlying these networks [7]. Paper [7] is devoted to the new methodological base for the analysis of data generated by collaborative systems, which uses modern data mining and artificial intelligence models and methods. Special means are needed in order to develop a deeper understanding of users' behavior, adequate ranking criteria and perform complex dynamic and statistic analyses. Traditional methods of clustering, community detection and text mining should be adapted to this end. Moreover, these methods require ingenuity and efforts for their effective and efficient use that will allow to obtain non-trivial results. In our previous work we briefly described models of data used in crowdsourcing projects in terms of Formal Concept Analysis (FCA) [8]. Furthermore, we presented a prototype of the data analysis system CrowDM (Crowd Data Mining) for collaborative platform, its architecture and methods underlying the key steps of data analysis.

Recommendations for crowdsourcing is a relatively new phenomenon. The recent literature mainly discuss or propose methods for task recommendation in systems similar to Amazon Mechanical Turk [9–11]. The comprehensive survey [12] investigates recent papers on the topic taking into account such aspects as context of recommendations, source of knowledge and evaluation process. An interesting prototype for choosing a proper crowdsourcing system for micro-task solution is present in [13].

The main part of current paper is devoted to recommender system and algorithms for crowdsourcing platforms. The setting, in which these platforms work,

---

[7] http://witology.com/.

[8] http://www.wikivote.ru/.

[9] http://www.tekora.ru/Products/44i/.

[10] http://sberbank21.ru/.

[11] http://www.asi.ru/en/.

[12] http://crowd.mos.ru/.

is quite different from the one of online-shops or specialized music/films recommender websites. Crowdsourcing projects consist of several stages and results of each stage greatly depend on the results of a previous one and users can evaluate the proposed ideas and each other at different stages of the project. This is the reason why existing recommender models should be considerably adapted. We have proposed new methods for making recommendations based on well-known mathematical approaches and data analysis techniques. Namely, we propose methods of idea recommendation (for voting stage), like-minded persons recommendation (e.g. for collaboration in teams) and antagonists recommendation (for counter-idea generation stage). The last recommendation type is firstly proposed and it is very important for Witology platform because it features a stage of counter-idea generation where each idea need an active and sound criticism. We have developed and tested several original methods for antagonists recommendation based on similarity measures and object-attribute biclustering. Thus, this paper substantially extends our short industry paper presented at ICCS 2014 [14], where we only shortly outlined the proposed techniques and presented results on antagonists' recommendation. In the current paper we bridge the gap and present the algorithms with their pseudo-codes, full results of experiments including 3D plots for optimal parameters tuning. However, since we presented data analysis system CrowDM along with FCA-based models for crowdsourcing data analysis earlier, we omit these matters here and refer the reader to [7].

One more remark goes to the conventional collaborative filtering approaches. Since the very beginning they operate on only matrices of ratings or user-item binary matrices but all our tasks require to find similarity (distance) in two related matrices, both binary and numeric ones. Taking this into account conventional user-based and item-based approaches [15] cannot be directly applied. Moreover, there is one more reason why industry standard SVD-based techniques [16] cannot be directly applied: our data are rather small and sparse to factorise them to find similarity in a new feature space. So, our approaches can be classified according to the taxonomy of recommender systems [17] as hybrid ones that combine features of neighbourhood-based and content-based methods.

The paper consists of 6 sections. In Sect. 2 introduces the platform and Witology project. Section 3 explains the reasons why the company needs recommender systems. Section 4 introduces basic notions of Formal Concept Analysis, describes recommender models and algorithms based on those notions. Evaluation of algorithms on real Witology data is discussed in Sect. 5. Section 6 concludes the paper.

2   Crowdsourcing Platform and Project Witology

One Russian proverb says "Two heads are better than one" and when dealing with crowdsourcing projects we can count on several thousands of heads. The term "crowdsourcing" is a portmanteau of "crowd" and "outsourcing", coined by Jeff Howe in 2006 [2]. There is no general definition of crowdsourcing, however it has a set of specific features. Thus, the company defines crowdsourcing as a technology allowing to harness the intellectual potential and expertise of a large amount of people for solving various tasks in an online environment.

We shortly describe the methodology of the Witology crowdsourcing company.

The clients of Witology company are banks (Sberbank), large firms (Moscow United Energy Company, RosAtom), Government institutions (Moscow Government, Zelenograd prefecture), etc. The platfrom interface is simple and friendly: there are sections with the description of a project and tasks of particular stage, personal account of a user, rating tables etc. The main interface unit is a "project tree", a hierarchical structure which is growing during the project. The root of the tree is the project theme, below the root the problems are listed; each problem is connected with the related ideas how to solve the problem and so on.

There are predefined rules to maintain a project on Witology platform. The client sets up the project objectives. The project participants, crowdsourcers, undergo preliminary surveys, which include specific questions on the domain. Thus, taking into account requirements of a client and the project goal, crowdsourcers are not necessary are experts in the domain and have the experience in crowdsourcing projects. The project team is formed by a client and Witology at the preliminary project stage, which usually takes about two weeks along with problematization stage.

There are several roles of the project team members: project manager, facilitator, manager of communities, helpdesk staff, methodologist, content analytic, and network analytic. Facilitators (from the Latin word facilis easy, comfortable) are the key figures in the management of crowdsourcing projects; they are persons who ensures the whole running of a project on the crowdsourcing platform, without influencing the content and help participants communicate in an effective way so they achieve project goals.

The main stages of the project are "Solution's generation", "Selection of similar solutions", "Generation of counter-solutions", "Total voting", "Solution's improvements", "Solution's stock", "Final improvements" and finally "Solution's review". In total they may take from 1 up to 3 months. After "Selection of similar solutions" by participants, analytical operations including comparison, clustering, and filtering of the ideas are performed. As a result of this stage, solutions are selected. The stage "Total voting" resulted in the selection of less number of solutions. After the stage "Solution's stock" again less number of solutions is left. From about 10–15 remaining solutions after "Solution's review" about 3–5 solutions are nominated as the best ones. The first stage "Solution's generation" is performed individually by each user. A key difference between traditional brainstorming and the "Solution's generation" stage is that nobody can see or listen to the ideas of other participants. The main similarity is the absence of criticism which was moved to later stages. In the "Selection of similar solutions" phase participants are selecting similar ideas (solutions) and their aggregated opinions are transformed into clusters of similar ideas. Counter-solutions generation includes criticism (pros and cons) and evaluation of the proposed ideas by means of communication between an author and experts. During this stage the author of the idea can invite other experts to his team taking into account their contribution to discussion and criticism. Total voting is performed by evaluation of each proposed idea by all users in terms of their attitude and quality levels of the solution (marks are integers between −3 and 3). Two stages, i.e. "Solution's improvements" and "Final improvements", involve active collaboration of experts and authors who improve their solutions together.

The system maintains 7 composite ratings of participants that published on the platform in real time and over 200 ratings for each user based on their activity; these ratings include "Popularity", "Social capital", "Performance", "Gamer", "Actor", "Judge", "Commentator", "Importance", "Influence", and "Reputation". For texts the company uses the following ratings: "Significance", "Influence", "Popularity", "Quality", "Attitude" and also "Reputation".

Solution's stock is one of the most interesting game stages of the project when all the participants with positive reputation rate accumulated at the previous stages receive money in internal currency "wito" and can take part in stock trade. The solutions with the highest price become winners. Finally, during the review of the solutions, experts with high reputation make their final evaluations based on several criteria in −3 to 3 scale: "Solution efficiency", "Solution originality", "Solution performance", and "Return on investment".

3   Recommender System for Witology: Rationale

All the crowdsourcing projects that we analysed were very well thought out both in terms of stages and semantic content of the project. It is known that every stage is connected to a special problem which users should solve. The completion and quality of this solution influence a lot on users ratings. Some of these ratings coincide with the motivation system for users, others show the leadership and the winners of the project can be defined with the help of them. The authors of the work [18] have checked out the Pareto-law which says that 20 % of the users generate 80 % of the content (in terms of Witology system). The results have shown that the large part of the registered users are absolutely inactive and do not participate in generation of the content. Partly it happens because of the time- and effort-consuming stages. At some moment of the project most of the users understand that their every-day life and work are more important and stop taking part in the project. This leads to the situation when the project becomes a club of small amount of very active people but not the project that should have benefited from the "synergy effect" and collaborative work. This is why decreasing laboriousness of some stages is a good solution. Recommender system may help to achieve this goal.

3.1   Idea Recommendation

Let us consider the total voting stage. All the users evaluate the ideas of each other (Sect. 2). Sometimes the total number of ideas exceeds hundreds. Moreover, each idea should be evaluated according to two parameters: attitude (shows how much the user agrees with the idea) and quality (scores the idea description and validity). While evaluating the first ideas of the pool, users are really motivated and well-concentrated. At a later stage they are divided into 3 "groups": (1) continue evaluating with high quality; (2) continue doing this but only just to end the task as quickly as possible, (3) stop evaluating at all.

This work suggests to allow users evaluating not all the ideas but only the most interesting for them. Thus, we have modelled a recommender system that defines the most interesting ideas for a particular user. This algorithm is described in Sect. 4.2.

3.2   User Recommendation

In addition to the idea recommendation we can recommend users to each other. The collaborative work of the "closest" users leads to more reasonable results than the work of a bitty crowd. The stage of team work is one of the stages of active user interaction. At this stage users are grouped in teams over the ideas in order to describe them better. As we have mentioned above, there is a big amount of ideas and users may have problems choosing the most interesting ones to work at. We assume that the presence of similar users in one command can help in productive discussion. Thus, the recommender system can show the other users suitable to collaborate with or the existing commands where the closest users are. All this will both help in choosing the most interesting ideas and make the team work more effective.

One more important stage of the crowdsourcing project is the stage of idea criticism (counter-idea generation). As we have mentioned, at this stage users should criticize the ideas of each other in order to make the resulting ones. However, we may face the same problem: there are too many ideas, criticizing all of them is very time- and effort-consuming. Sometimes people choose to criticize some ideas only because they have had a previous discussion with their authors. This situation can lead to misunderstandings and non-constructive criticism. We suggest showing only a subset of all the ideas. To choose the most interesting ideas we need an algorithm that can identify so-called antagonists: the furthest users from the target one. Then we can use this information to suggest to criticize the ideas of these antagonists. The description of the algorithm is given in Sect. 4.3.

We do not address the problem of cold-start in our algorithms and recommend something only to those users who have a behavior history. However, we suggest the methodological solution of the cold-start problem for each case.

3.3   Evaluation of the Recommender System

The recommendation quality is measured by precision@N and recall:

$$precision = \frac{|relevant\ recommended\ items|}{|recommended\ items|},$$

$$recall = \frac{|relevant\ recommended\ items|}{|relevant\ items|}.$$

Recall for recommendations deserves a further discussion. In contrast to precision where we have a number of recommendations in the denominator, for recall we have a total number of relevant items that have almost no sense because the aim of all the above-mentioned algorithms is to decrease efforts of project participants. This means that we are planning to recommend to users doing something (group in teams, evaluate and criticize ideas) and recommend him/her only relevant tasks. So the recall will end up being low. Anyway we will observe the recall dynamics depending on different parameters of the algorithm, this should allow us to assess the adequacy of the algorithm.

It is important to say that the interpretation of precision and recall is not the same as usually in recommender systems: some data is hidden (test set) and then one compares the results of the algorithm with real values that were hidden [19]. Firstly, all our algorithms are trained using one type of data (e.g. marks of user comments) and recommend another type of data (e.g. ideas). Secondly, the main aim of the recommender system in a crowdsourcing project is to help users and to increase the effectiveness of their work. Probably, now users group in teams not the best way and recommendations may improve that. In this case, the low precision will not necessary mean the bad result 100 %. It seems more adequate to use online evaluation schemes [20]. The detailed results of precision and recall estimation are presented in Sect. 5.

4   Recommender Model and Algorithms

At the initial stage of the project on mining of Witology crowdsourcing data, among 7 prospective directions we proposed recommendation models based on multimodal clustering (biclustering, triclustering, and spectral clustering [21–24]) and FCA (concept lattices, implications, association rules) along with its extensions for multimodal data, triadic, for instance [25]. It is not the first usage of these approaches for recommender systems ([26–34]).

4.1   Formal Concept Analysis and Object-Attribute-biclustering

Methods described in this paper mainly rely on the multimodal clustering block at the analysis scheme in [7]. The main actors of crowdsourcing projects (and corresponding collaborative platforms) are platform users (project participants) and we consider them as *objects* for analysis. Moreover, each object can possess a certain set of *attributes*. User's attributes can be: topics that the user discussed, ideas that he generated or voted for, or even other users. The main instrument for analysis of such object-attribute data is Formal Concept Analysis [8]. Let us give a formal definition. A *formal context* in FCA is a triple $\mathbb{K} = (G, M, I)$, where G is a *set of objects*, M is a *set of attributes*, and the relation $I \subseteq G \times M$ shows which object possesses which attribute. For any $A \subseteq G$ and $B \subseteq M$ one can define *Galois operators*:

$$A' = \{m \in M \mid gIm \text{ for all } g \in A\}, \tag{1}$$
$$B' = \{g \in G \mid gIm \text{ for all } m \in B\}.$$

The operator $''$ (applying the operator $'$ twice) is a *closure operator*: it is idempotent ($A'''' = A''$), monotonous ($A \subseteq B$ implies $A'' \subseteq B''$) and extensive ($A \subseteq A''$). A set of objects $A \subseteq G$ such that $A'' = A$ is called a closed set. The same properties hold for closed attribute sets, i.e. subsets of the set M. A couple (A, B) such that $A \subset G$, $B \subset M$, $A' = B$ and $B' = A$, is called a *formal concept* of a context $\mathbb{K}$. The sets A and B are closed and are called *extent* and *intent* of a formal concept (A, B) respectively. For a set of objects A, a set of their common attributes A' describes a similarity of objects of the set A, and a closed set A'' is a group of similar objects with the set of common attributes A'. The relation "to be more general concept" is defined as follows: $(A, B) \geq (C, D)$ iff $A \subseteq C$. By $\mathfrak{B}(G, M, I)$ we denote a set of all concepts of a formal context $\mathbb{K} = (G, M, I)$. The concepts of a formal context $\mathbb{K} = (G, M, I)$ ordered by extensions inclusion form a lattice, which is called a *concept lattice*. For its visualization a *line diagram* (Hasse diagram) can be used, i.e. the cover graph of the relation "to be a more general concept".

To represent datasets with numerical (e.g., age, word frequency, number of comments, marks) and categorical (e.g., gender, job) attributes there are many-valued contexts. A *many-valued context* $\mathbb{K} = (G, M, W, I)$ consists of sets G, M and W and a ternary relation $I \subseteq G \times M \times W$ for which it holds that

$$(g, m, w) \in I \text{ and } (g, m, v) \in I \Rightarrow w = v.$$

The elements of G are still called objects, those of M (many-valued) attributes and the elements of W attribute values. Sometimes we write $m(g) = w$ to show that the object g has the value w of the attribute m.

We can transform the many-valued context into a one-valued one by means of *conceptual scaling* [8].

In the worst case (Boolean lattice), the number of concepts is equal to $2^{\{min |G|, |M|\}}$, thus, for large contexts, FCA can be used only if the data is sparse. Moreover, one can use different ways of reducing the number of formal concepts (choosing concepts by stability [35] index or extent size). An alternative approach is a relaxation of the definition of formal concept as a maximal rectangle in an object-attribute matrix which elements belong to the incidence relation. The notion of object-attribute bicluster (OA-bicluster) [22] is one of such relaxations. If $(g, m) \in I$, then (m', g') is called *object-attribute bicluster* with the density $\rho(m', g') = |I \cap (m' \times g')| / (|m'| \cdot |g'|)$.

The main features of OA-biclusters are listed below:

1. For any bicluster $(A, B) \subseteq 2^G \times 2^M$ it is true that $0 \leq \rho(A, B) \leq 1$.
2. OA-bicluster (m', g') is a formal concept iff $\rho = 1$.
3. If (m', g') is a bicluster, then $(g'', g') \leq (m', m'')$.

Let $(A, B) \subseteq 2^G \times 2^M$ be a bicluster and ρ_{min} be a non-negative real number such that $0 \leq \rho_{min} \leq 1$, then (A, B) is called *dense*, if it fits the constraint $\rho(A, B) \geq \rho_{min}$. The above mentioned properties show that OA-biclusters differ from formal concepts since unit density is not required. Graphically it means that not all the cells of a bicluster must be filled by a cross (see Fig. 1).

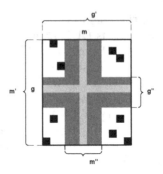

Fig. 1. OA-bicluster.

Besides concept lattice construction and its visualization by means of Hasse diagrams, one can use implications and association rules for detecting dependencies between sets of attributes in data. Then, using the obtained rules, it is easy to form recommendations (for example, offering users the most interesting discussions for them).

4.2 Idea Recommendation – Before the Total Voting Stage

The task of this stage is to evaluate all the ideas (using marks in the range of $-3 \cdots 3$) by two parameters: attitude and quality (Subsect. 4.1) Attitude shows user's opinion about the idea and indicates whether the user agrees with the proposed idea or not. Whereas the quality evaluation shows whether a user considers the proposed idea sound and well-grounded.

A problem that arises at this stage is that not all users will evaluate all the ideas; it often happens so that users start to evaluate the proposed ideas but drop the task before finishing all evaluations. There are only a few users who evaluate all the ideas till the end.

We suggest a solution that can help to deal with the problem mentioned above: not to show all the ideas, but only the top ranked (most interesting) ones. The idea is considered to be an interesting one for a user if he/she has evaluated other posts and comments of the ideas' author before.

Necessary data for the recommendation consists of a multi-valued context $\mathbb{K}_{UU}^S$ (user–user–sum relation) with the absolute values of evaluations sum before the total voting stage, a context $\mathbb{K}_{UI}$ (idea–author relation), which indicates who is the author of each idea.

The algorithm uses the multi-valued context $\mathbb{K}_{UU}^S$ which aggregates the history of users evaluation of each others posts and comments before the total voting stage. It is important to mention that unlike the ideas evaluation (total voting) which is a special stage at a given time, posts and comments evaluation is possible during the whole time of the project. The hypothesis is that if the user 1 was interested in posts and comments of the user 2, user 1 will be interested in ideas of user 2 as well. The objects of context $\mathbb{K}_{UU}^S$ are the users-evaluators, the columns – the evaluated users (authors) and $u(v)$ is the sum (absolute value)

Algorithm 1. Idea recommendation algorithm

Input: multi-valued context $\mathbb{K}^S_{UU} = (U, U, \mathbb{N}, S)$ (relation user–user–sum), context $\mathbb{K}_{UI} = (U, I, P)$ (relation author-idea), target user $u0 \in U$, number of nearest users n, number of returned ideas k

Output: list of top k ideas $topI$

1: $indUsers = $ sort U by $u(u_0)$ for all $u \in U \setminus u_0$ desc
2: $Unn = $ top$(n, indUsers)$
3: $Inn = \bigcup_{u \in Unn} \{i | i \in u'\}$
4: $indIdeas = $ sort Inn by $u(u_0)$ for all $(u, i) \in P \cap Unn \times Inn$ desc
5: $topI = $ top$(k, indIdea)$
6: **return** topI

of all the evaluations which user u has given to all the posts and comments of the user v before the stage of total voting. The absolute value of the evaluation sum is an aggregate measure of the users' interest in each other: we can see the number of evaluations and their values. The high value of $u(v)$ means that the user u has given many evaluations of one sign to the user v thus (s)he will be able to evaluate his idea adequately.

Let us look at the suggested algorithm step-by-step. Lines 1–2 define a neighborhood Unn of n the most interesting users for a target user u_0. Line 3 calculates a set of potentially interesting ideas Inn: the ideas of the users from Unn. After that the ideas from Inn are sorted (line 4) according to the sum of marks of their authors for the target user (from the context $\mathbb{K}^S_{UU}$). Then k the most interesting ideas are selected (line 5).

If the target user has not evaluated (or has not evaluated many) comments and posts of the others, the system may offer him/her to evaluate the ideas that have not yet been evaluated at all or were evaluated by just small amount of users.

4.3 Antagonists Recommendation for the Idea Criticism – Before Counter-Idea Generation Stage

The task of this stage is to criticize the ideas according to two parameters: what the idea is lacking and how it can be improved. A problem that we can encounter: the user may decide to criticize only the ideas of their rivals and this can lead to project conflicts. The way we suggest to address this problem: not to show all the ideas for criticism but only the ideas of the authors with the most opposed views (so-called antagonists). The user 1 is considered to be the antagonist for the user 2 if at the total voting stage they evaluated the most ideas of others with opposite signs. The input data for this stage will be a context $\mathbb{K}^E_{UI}$ (user–idea–evaluation) of the total voting evaluations.

Biclustering-Based Antagonists Recommendation. The algorithm consists of two stages: data preprocessing (in order to find OA-biclusters) and antagonists recommendation itself. Lines 1–5 describe the first stage. As far as the algorithm obtains the context $\mathbb{K}_{UI} = (U, I, J)$ from the context $\mathbb{K}^E_{UI} = (U, I, \mathbb{Z}, E)$ indicating whether user evaluated idea or not (i.e. $(u, i) \in J$ iff $(u, i, z) \in E$), it can find all the OA-biclusters $\mathbf{B}$. After that for any bicluster we can count so called "bicluster mass": its density multiplied by its size. According to the density definition (Subsect. 4.1), a bicluster mass is just a number of non-empty cells. Then we should take top-m biclusters sorted by their mass for the future steps. The bicluster mass is a good measure in this case to sort by because it can help to discover biclusters with big density and/or big size. After the top-m massive biclusters are defined, the second stage of algorithm starts. Similarly to [27] the algorithm finds k nearest biclusters (lines 6–9) to the target user. The similarity between a bicluster $b = (X, Y)$ and a user u is counted by the formula:

$$sim(u, b) = \frac{|u' \cap Y|}{|Y|}, \text{ where}$$

Y is the set of bicluster items and u' is the set of items evaluated by user u.

After calculation of k nearest biclusters (B_k) the next step, antagonists finding, starts (lines 10–18). Inside each bicluster b from B_k we count the distance between each user of the bicluster and a target user. This time we do it not with the one-valued context $\mathbb{K}_{UI}$ but with the multi-valued context $\mathbb{K}^E_{UI}$. The distance is counted using the Pearson correlation (or cosine similarity) just by the ideas in bicluster b. Then we multiply this distance by the similarity of a target user and a bicluster b and get a matrix R (users-biclusters). An element $R(u_i, b)$ means the rating of users u_i for the target user in bicluster b. Finally, taking the mean value of this ratings by all the biclusters, we get the furthest users (by distance) from the target user and can take top-n of them to recommend.

Besides the Pearson correlation we can use Hamming distance inside the bicluster $b = (X, Y)$:

$$d(u, v|b) = \sum_{i \in Y} [sign(i(u)) \neq sign(i(v))].$$

In this case we give 0 for evaluations of one sign and 1 for evaluations of opposite signs.

Correlation-Based Antagonists Recommendation. A simpler way of antagonists' recommendation is based on simple Pearson correlation (without biclusters). We just take the top-n users with the smallest (negative) correlation and recommend them. In this case we count the correlation by all the ideas (not only those which are in bicluster).

Hamming Distance-Based Antagonists Recommendation. One more way which we have tested is the simple Hamming distance (without biclusters data preprocessing). We should just count the Hamming distance for target user with all other users and take top-n users with the largest distance as antagonists.

Algorithm 2. Biclustering-based algorithm for recommendation of antagonists

---

Input: multi-valued context $\mathbb{K}_{UI}^{E} = (U, I, \mathbb{Z}, E)$ (relation user-idea-mark) and its associated one-valued context $\mathbb{K}_{UI} = (U, I, J)$ (relation user-idea), number of the most
massive biclusters m, number of the most similar biclusters to a target user k, size
of the recommended list of users n

Output: topN

1: $\mathbf{B} = \text{OA-Bicluster}(\mathbb{K}_{UI})$
2: **for all** $b \in \mathbf{B}$ **do**
3:    $mass(b) = dens(b) \cdot size(b)$
4: **end for**
5: $B_m = $ sort $\mathbf{B}$ by $mass$ desc
6: $B_m = top(m, B_m)$
7: **for all** $b \in B_m$ **do**
8:    $sim(b) = sim(b, u_0)$
9: **end for**
10: $B_k = $sort B_m by sim desc
11: $B_k = top(k, B_k)$
12: **for all** $(X, Y) \in B_k$ **do**
13:    **for all** $u \in X$ & $u \neq u0$ **do**
14:        $R(u, b) = dist(u_0, u|b) \cdot sim(u_0, b)$
15:    **end for**
16: **end for**
17: **for all** $u \in U$ **do**
18:    $aggr(u) = mean(R(u, :))$
19: **end for**
20: $U_{sort} = $ sort U by $aggr$ desc
21: $topN = top(n, U_{sort})$
22: **return** $topN$

If at the total voting stage the target user has not evaluated (or evaluated too few) ideas of others, the system may recommend him/her to criticize the ideas that have not been evaluated till that moment.

4.4   Like-Minded People Recommendation for Teams Joining

The goal of the stage is to group in teams over the ideas to have an active discussion.

There is a problem: Usually users group in teams because of the interest to a particular idea and do not care about effectiveness of work with others who are in the team.

Problem solution: to show the target user the list of the closest users to group in teams with. The collaboration of likeminded people in terms of teams will do command stage more effective. The user 1 is considered to be the like-minded for the user 2 if at the total voting stage they evaluated the ideas of others the same way. Necessary data: multi-valued context $\mathbb{K}_{UI}^{E}$ (user–idea evaluation relation) of total voting evaluations.

The algorithm to find like-minded people is absolutely the same as the antagonists recommendation but instead of counting users distance we count users similarity (the biggest Pearson correlation instead of the smallest). Even though we describe the methodology of like-minded persons recommendation in this work, we provide only results for the two aforementioned methods.

5   Data and Experiments

For experiments we have used data of crowdsourcing project "Sberbank: No. Queues!" managed by Witology. The project took place from 03.09.2012 to 23.10.2012. The main topic was the problem of big queues at Sberbank offices. During the project crowdsourcers discussed the reasons why queues are appearing and the ways to get rid of them. The number of registered users was 5,946 (25 of them are Witology staff). The experiment have shown that there were only about 200 more or less active users. They have generated 880 queues reasons and 1,137 ideas how to solve this problem. The number of user posts (including comments) and evaluation pairs (attitude, quality) amounted to 20,326 and 130,000 respectively.

There were following stages at the project: (1) Problem determination; (2) Total voting for problems; (3) Idea generation; (4) Selection of similar ideas; (5) Idea verification; (6) Team work at ideas; (7) Counter-idea generation; (8) Total voting for ideas; (9) Command work at top-50 ideas; (10) Top-50 ideas review.

The recommendation algorithms were performed using Matlab. An additional supporting program, that finds oa-biclusters was implemented in Python. As a result of data preprocessing we have found 19,405 biclusters (without density threshold). After that we have used 20 largest biclusters ($m = 20$).

5.1   Idea Recommendation

For the idea recommendation algorithm we have used a matrix of the user-user type (rows are for the evaluators and columns are for the evaluated users), where each cell shows absolute value of all the evaluations of the respective evaluator-evaluated user pair before the idea voting stage. Generally there were 117 evaluators and 139 evaluated users. The assumption is that if the user 1 was interested in some posts and comments of user 2 before the total voting stage for ideas, his ideas will be interesting for the user 1 as well (at the total voting stage). In order to evaluate the algorithm we have used the one-valued context user-idea which shows whether a particular user supports a particular idea (from the total voting stage for ideas). We have looked at classic measures: precision and recall. Two parameters were being changed: $topN$, the number of the closest neighbors, and $topI$, the number of recommended ideas.

We tuned these parameters one by one and fix others guided by 3D plots on Fig. 2.

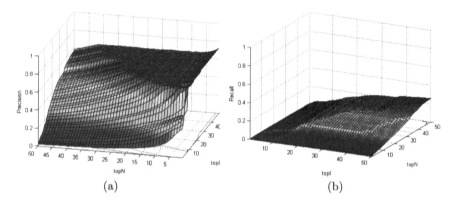

Fig. 2. The precision (a) and recall (b) dependence of neighborhood size and the number of recommended ideas

The Precision and Recall Dependence of Neighborhood Size. The precision and recall are shown in Fig. 3, where the number of recommendations is fixed: $topI = 20$.

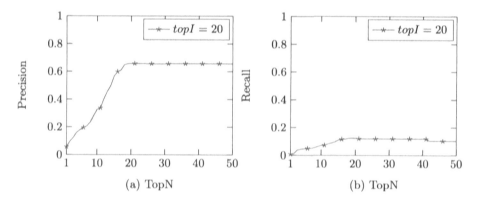

Fig. 3. The precision (a) and recall (b) dependence of neighborhood size for idea recommendation

So, the recall is really not high as we have expected. The precision stops growing at about 0.65 value. There are two possible explanations. Firstly, our hypothesis may be not that strong and the are users that evaluate ideas regardless their previous evaluations. Secondly, the total voting stage is really very hardworking and time consuming. Moreover, this stage is obligatory and each user should evaluate all the ideas. Often users do it not very thoroughly because of a big work load. Let us look at the distribution of evaluations numbers in Fig. 4 (at the total voting stage):

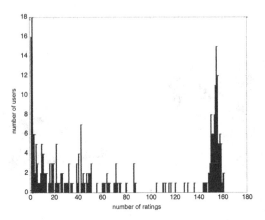

Fig. 4. The distribution of evaluations of ideas number at total voting stage.

In fact, a great number of users either evaluates almost all the ideas (172 in our case) or just less than 10 ideas. The moment when the precision and recall stops changing (about 20 users in neighborhood) may mean that the algorithm reach the necessary number of user neighborhood to give relevant recommendations.

The Precision and Recall Dependence on Recommendations Number. The precision and recall are shown in Fig. 5, where the number of recommendations is fixed: $topN = 20$.

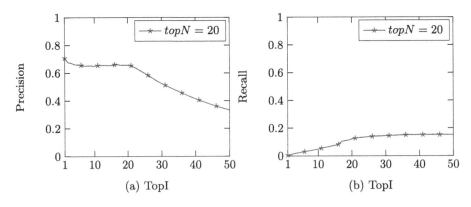

Fig. 5. The precision (a) and recall (b) dependence of number of idea recommendations.

Firstly, let us point out a high level of precision. It decreases while the recommendation number grows. It is explained quite well: the more ideas we recommend the less is the part of guessed correctly and the more is the probability to make correct recommendations for particular user (recall increasing). As it was foreseen, the recall is quite low. But the tendency of its growth let us assume that the algorithm works correctly.

5.2 Antagonists Recommendation for the Idea Criticism

For antagonists recommendation a matrix of idea evaluations at total voting stage was used (user-idea). In order to evaluate recommendations quality we have generated a one-valued context $\mathbb{K}_{UU} = (U, U, C)$ (user-user relation) where $(u, v) \in C$ means that user u has participated in the idea criticism of user v. Finally we got a context 62×73 because some users did not participate at this stage at all. Quality measures are the same: precision and recall. Algorithm parameters are the following:

- $minE$, the minimum required number of evaluated ideas for giving recommendation to a person;
- $topk$, the number of the closest biclusters in the target user's neighborhood;
- $topN$, the number of recommended items.

The average number of ideas that user criticized is 20, that is why when $topN$ is fixed it equals to 20.

Precision and Recall Dependence of the Closest Biclusters Number. Fixed algorithm parameters $minE = 5$ and $topN = 20$ give us the following results Fig. 6 (a).

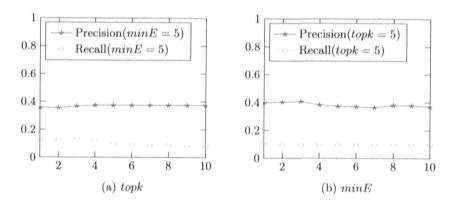

(a) $topk$ (b) $minE$

Fig. 6. Precision@20 and recall dependence (a) on the closest biclusters number and (b) the minimal required number of evaluations for antagonists' recommender

The precision is acceptable. The decreasing tendency is observed: that may happen because of some noise presence in biclusters' neighborhood. We do not show the results of like-minded people recommendations in this paper, but we should note that precision here is higher. For the new methodology of recommendations (antagonist), it is a good indicator.

Precision and Recall Dependence on Minimal Required Number of Evaluations. Fixed algorithm parameters: $topk = 5$, $topN = 20$ (Fig. 6 (b)).

The same result is for like-minded people recommendations: this parameter is not meaningful.

Precision and Recall Dependence of Recommendations Number. In case of fixed algorithm parameters $topk = 5$ and $minE = 5$, the precision is almost the same but recall increases in quite a predictable way (Bicluster curve in Fig. 7).

Comparison with Other Distance Measures. The most interesting dynamic result is observed depending on the recommendations number. Changing this parameter, we will compare the quality of this antagonists recommendation algorithm (based in biclusters) with others: based on only Pearson correlation (CosineBased) and based on only Hamming distance (SimpleHamming). Moreover, the similarity (distance) in biclusters-based algorithm is calculated using two ways: with a help of Pearson correlation (Bicluster, as it was described before) and with a help of Hamming distance (BiclusterHamming).

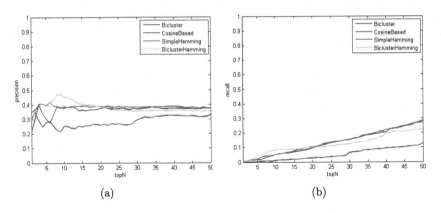

(a)   (b)

Fig. 7. Precision (a) and recall (b) for different antagonists' recommendation algorithms

Figure 7 shows that when the recommendations number is low the best result has BiclusterHamming and when the recommendations number is high two algorithm are good: Bicluster and SimpleHamming. The initial aim was to shorten user effort and time, so we are interested in low recommendations number. In this case BiclusterHamming is the real winner: the precision is almost 0.5. Now let us compare algorithms quality and their work time.

Despite the rather high precision, BiclusterHamming has the worst time results. However, when the recommendations number is low (our case) it may

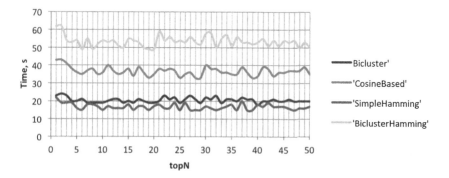

Fig. 8. Time efficiency of antagonists' recommenders

be not so critical. Let us also point out that in comparison with information retrieval algorithms where CosineBased is one of the best, here it is both slower (Fig. 8) and worse in terms of quality than SimpleHamming (Fig. 7).

6 Conclusion

This work describes the recommender system based on Formal Concept Analysis and OA-biclustering. The recommendations of ideas and users-antagonists were proposed. The last type of recommendations is a methodologically novel one. As for antagonists recommendations, we have tested several algorithms: with data pre-processing (OA-biclustering) and without it. The results show that usage of biclusters improves recommender system quality. Moreover, two ways of antagonists determination were tested: using negative Pearson correlation and Hamming distance. When the recommendations number is low, the best is BiclusterHamming (based on biclustering and using Hamming distance). As for the idea recommendations, the precision of this algorithm is much higher than the same one used for antagonists recommendations. The algorithm's originality is that users' similarity is calculated based on other types of objects (posts and comments) than ideas' evaluation. Experiments and observations at a real crowdsourcing projects let us conclude that the recommender system implementation should be started from idea recommendation at a total voting stage. Firstly, the precision is higher, secondly this stage is very time consuming and sometimes boring, therefore decreasing a number of ideas for evaluation should make crowdsourcing projects more attractive for participants. A cold start problem is not addressed in this work – in case of previous users inactivity she gets no recommendations. A solution for the cold start problem will be addressed in the future work, here we only outlined one possible solution.

Acknowledgment. The study was conducted in the framework of the Basic Research Program at the National Research University Higher School of Economics (HSE) in 2013 and 2014 and in the Laboratory of Intelligent Systems and Structural Analysis at HSE as a part of the project "Mathematical Models, Algorithms, and Software Tools for Intelligent Analysis of Structural and Textual Data". First author was also supported by Russian Foundation for Basic Research (grant #13-07-00504).

References

1. Rozwell, C., Harris, K., Mesaglio, M.: Who's who in innovation management technology. Gartner (2010)
2. Howe, J.: The rise of crowdsourcing. Wired Mag. **14**(6), 1–4 (2006)
3. Roth, C.: Generalized preferential attachment: Towards realistic socio-semantic network models. In: ISWC 4th International Semantic Web Conference, Workshop on Semantic Network Analysis, CEUR-WS Series (ISSN 1613–0073), vol. 171, pp. 29–42, Galway, Ireland (2005)
4. Cointet, J.P., Roth, C.: Socio-semantic dynamics in a blog network. In: CSE, vol. 4, pp. 114–121. IEEE Computer Society (2009)
5. Roth, C., Cointet, J.P.: Social and semantic coevolution in knowledge networks. Soc. Netw. **32**, 16–29 (2010)
6. Yavorsky, R.: Research challenges of dynamic socio-semantic networks. In: Ignatov, D., Poelmans, J., Kuznetsov, S., (eds.) CEUR Workshop proceedings, CDUD 2011 - Concept Discovery in Unstructured Data, vol. 757, pp. 119–122 (2011)
7. Ignatov, D.I., Kaminskaya, A.Y., Bezzubtseva, A.A., Konstantinov, A.V., Poelmans, J.: FCA-based models and a prototype data analysis system for crowdsourcing platforms. In: Pfeiffer, H.D., Ignatov, D.I., Poelmans, J., Gadiraju, N. (eds.) ICCS 2013. LNCS, vol. 7735, pp. 173–192. Springer, Heidelberg (2013)
8. Ganter, B., Wille, R.: Formal Concept Analysis: Mathematical Foundations, 1st edn. Springer-Verlag New York Inc., Secaucus (1999)
9. Ambati, V., Vogel, S., Carbonell, J.G.: Towards task recommendation in micro-task markets. In: (2011) Human Computation, Papers from the 2011 AAAI Workshop, vol. WS-11-11. AAAI, San Francisco, August 8, 2011
10. Yuen, M.C., King, I., Leung, K.S.: Taskrec: A task recommendation framework in crowdsourcing systems. Neural Process. Lett. **41**(2), 1–16 (2014)
11. Lin, C.H., Kamar, E., Horvitz, E.: Signals in the silence: Models of implicit feedback in a recommendation system for crowdsourcing. In: Brodley, C.E., Stone, P. (eds.) Proceedings of the Twenty-Eighth AAAI Conference on Artificial Intelligence, pp. 908–915. AAAI Press, Canada (2014)
12. Geiger, D., Schader, M.: Personalized task recommendation in crowdsourcing information systems - current state of the art. Decis. Support Syst. **65**, 3–16 (2014)
13. Chander, D., Bhattacharya, S., Celis, L.E., Dasgupta, K., Karanam, S., Rajan, V., Gupta, A.: Crowdutility: a recommendation system for crowdsourcing platforms. In: Bigham, J.P., Parkes, D.C., (eds.) Proceedings of the Seconf AAAI Conference on Human Computation and Crowdsourcing, HCOMP 2014, November 2–4, 2014. AAAI, Pittsburgh (2014)
14. Ignatov, D.I., Kaminskaya, A.Y., Konstantinova, N., Malioukov, A., Poelmans, J.: FCA-based recommender models and data analysis for crowdsourcing platform witology. In: Hernandez, N., Jäschke, R., Croitoru, M. (eds.) ICCS 2014. LNCS, vol. 8577, pp. 287–292. Springer, Heidelberg (2014)
15. Desrosiers, C., Karypis, G.: A comprehensive survey of neighborhood-based recommendation methods. In: Ricci, F., Rokach, L., Shapira, B., Kantor, P.B. (eds.) Recommender Systems Handbook, pp. 107–144. Springer, Heidelberg [36] (2011)

16. Koren, Y., Bell, R.M.: Advances in collaborative filtering. In: Ricci, F., Rokach, L., Shapira, B., Kantor, P.B. (eds.) Recommender Systems Handbook, pp. 145–186. Springer, Heidelberg [36] (2011)
17. Ricci, F., Rokach, L., Shapira, B.: Introduction to recommender systems handbook. In: Ricci, F., Rokach, L., Shapira, B., Kantor, P.B. (eds.) Recommender Systems Handbook, pp. 1–35. Springer, Heidelberg [36] (2011)
18. Bezzubtseva, A., Ignatov, D.I.: A new typology of collaboration platform users. In: Tagiew, R., Ignatov, D.I., Neznanov, A.A., Poelmans, J. (eds.) CEUR Workshop proceedings, EEML 2012 - Experimental Economics and Machine Learning, vol. 757, pp. 9–19 (2012)
19. Ignatov, D.I., Poelmans, J., Dedene, G., Viaene, S.: A new cross-validation technique to evaluate quality of recommender systems. In: Mitra, S., Mazumdar, D., Kundu, M.K., Pal, S.K. (eds.) PerMIn 2012. LNCS, vol. 7143, pp. 195–202. Springer, Heidelberg (2012)
20. Radlinski, F., Hofmann, K.: Practical online retrieval evaluation. In: Serdyukov, P., Braslavski, P., Kuznetsov, S., Kamps, J., Ruger, S., Agichtein, E., Segalovich, I., Yilmaz, E. (eds.) Advances in Information Retrieval. LNCS, vol. 7814, pp. 878–881. Springer, Heidelberg (2013)
21. Barkow, S., Bleuler, S., Prelic, A., Zimmermann, P., Zitzler, E.: BicAT: a biclustering analysis toolbox. Bioinformatics $22(10)$, 1282–1283 (2006)
22. Ignatov, D.I., Kaminskaya, A.Y., Kuznetsov, S., Magizov, R.A.: Method of biclusterzation based on object and attribute closures. In: Proceedings of 8-th international Conference on Intellectualization of Information Processing (IIP 2011), pp. 140–143. MAKS Press, Cyprus, Paphos, 17–24 October 2010 (in Russian)
23. Ignatov, D.I., Kuznetsov, S.O., Magizov, R.A., Zhukov, L.E.: From triconcepts to triclusters. In: Hepting, D.H., Ślęzak, D., Mirkin, B.G., Kuznetsov, S.O. (eds.) RSFDGrC 2011. LNCS, vol. 6743, pp. 257–264. Springer, Heidelberg (2011)
24. Ignatov, D.I., Kuznetsov, S.O., Poelmans, J., Zhukov, L.E.: Can triconcepts become triclusters? Int. J. Gen. Syst. $42(6)$, 572–593 (2013)
25. Jäschke, R., Hotho, A., Schmitz, C., Ganter, B., Stumme, G.: TRIAS-an algorithm for mining iceberg tri-lattices. In: Proceedings of the Sixth International Conference on Data Mining. ICDM 2006, pp.907–911. IEEE Computer Society, Washington (2006)
26. du Boucher-Ryan, P., Bridge, D.G.: Collaborative recommending using formal concept analysis. Knowl. Based Syst. $19(5)$, 309–315 (2006)
27. Symeonidis, P., Nanopoulos, A., Papadopoulos, A.N., Manolopoulos, Y.: Nearest-biclusters collaborative filtering based on constant and coherent values. Inf. Retr. $11(1)$, 51–75 (2008)
28. Ignatov, D.I., Kuznetsov, S.O.: Concept-based recommendations for internet advertisement. In: Belohlavek, R., Kuznetsov, S.O (eds.) Proceedings of CLA 2008, CEUR WS., Palacky University, vol. 433, pp. 157–166, Olomouc (2008)
29. Ignatov, D., Poelmans, J., Zaharchuk, V.: Recommender system based on algorithm of bicluster analysis RecBi. In: Ignatov, D., Poelmans, J., Kuznetsov, S. (eds.) CEUR Workshop proceedings, CDUD 2011 - Concept Discovery in Unstructured Data, vol. 757, pp. 122–126 (2011)
30. Ignatov, D.I., Kuznetsov, S.O., Poelmans, J.: Concept-based biclustering for internet advertisement. In: (2012) Vreeken, J., Ling, C., Zaki, M.J., Siebes, A., Yu, J.X., Goethals, B., Webb, G.I., Wu, X. (eds.) 12th IEEE International Conference on Data Mining Workshops, ICDM Workshops, pp. 123–130. IEEE Computer Society, Brussels, 10 December 2012

31. Nanopoulos, A., Rafailidis, D., Symeonidis, P., Manolopoulos, Y.: MusicBox: personalized music recommendation based on cubic analysis of social tags. IEEE Trans. Audio Speech Lang. Process. **18**(2), 407–412 (2010)
32. Jelassi, M.N., Ben Yahia, S., Mephu Nguifo, E.: A personalized recommender system based on users' information in folksonomies. In: Proceedings of the 22nd International Conference on World Wide Web Companion. WWW 2013 Companion, Republic and Canton of Geneva, Switzerland, International World Wide Web Conferences Steering Committee, pp. 1215–1224 (2013)
33. Ignatov, D.I., Nenova, E., Konstantinova, N., Konstantinov, A.V.: Boolean matrix factorisation for collaborative filtering: an FCA-based approach. In: Agre, G., Hitzler, P., Krisnadhi, A.A., Kuznetsov, S.O. (eds.) AIMSA 2014. LNCS, vol. 8722, pp. 47–58. Springer, Heidelberg (2014)
34. Alqadah, F., Reddy, C., Hu, J., Alqadah, H.: Biclustering neighborhood-based collaborative filtering method for top-n recommender systems. Knowl. Inf. Syst. **44**(2), 1–17 (2014)
35. Kuznetsov, S.O.: On stability of a formal concept. Ann. Math. Artif. Intell. **49** (1–4), 101–115 (2007)
36. Ricci, F., Rokach, L., Shapira, B., Kantor, P.B.: Recommender Systems Handbook. Springer, Hedielberg (2011)

Modelling Movement of Stock Market Indexes with Data from Emoticons of Twitter Users

Alexander Porshnev$^{(\boxtimes)}$, Ilya Redkin, and Nikolay Karpov

National Research University Higher School of Economics,
Nizhny Novgorod, Russia
{aporshnev,nkarpov}@hse.ru, ilya-redkin@yandex.ru

Abstract. The issue of using Twitter data to increase the prediction rate of stock price movements draws attention of many researchers. In this paper we examine the possibility of analyzing Twitter users' emoticons to improve accuracy of predictions for DJIA and S&P500 stock market indices. We analyzed 1.6 billion tweets downloaded from February 13, 2013 to May 19, 2014. As a forecasting technique, we tested the Support Vector Machine (SVM), Neural Networks and Random Forest, which are commonly used for prediction tasks in finance analytics. The results of applying machine learning techniques to stock market price prediction are discussed.

Keywords: Prediction · Emoticons · DJIA · S&P500 · Twitter · Mood · Support vectors machine · Neural networks · Random forest · Behavioral finance

1 Introduction

Moods and emotions influence our decisions; in psychological experiments Johnson and Tversky report that psychological states invoked by reading stories can affect the evaluation of risk level [1]. While in a good mood, an individual tends to make decisions expecting positive consequences and, vice versa, bad moods lead to pessimistic choices [2–4]. Moods and emotions determine the choice of basic heuristics, which can be done unconsciously. For example, individuals in a good mood are more likely to eliminate alternatives that fail to meet a criterion for an important dimension, which leads to increased efficiency [5].

It should be mentioned that other people's states play a big role not only in shaping individual moods, but also influence decision making. Nofsinger suggests the idea that the general level of optimism/pessimism shared in society is connected with economic activity [6]. Whereas Nofsinger supposes that the stock market itself can be a direct measure of social mood [6], in our research we decided to focus on an additional measure of shared emotions in Twitter.

Following Nofsinger, we will regard the economy as a complex system of human interactions, in which moods and irrationalities can play a significant role. This point can be supported by observing the informational cascades phenomenon in the stock market [7–9].

P. Braslavski et al. (eds.): RuSSIR 2014, CCIS 505, pp. 297–306, 2015.
DOI: 10.1007/978-3-319-25485-2_10

Twitter sentiment analysis gained in popularity in the financial domain thanks to the works by Bollen and his colleagues [10]. However, the possibility of predicting the stock market by means of analysis based on the wisdom of the crowds still triggers questions.

There are three main approaches to the use of Twitter data for financial forecasting. The first one is based on news analysis. For example, Reuters data shows that even fake news from a reliable source (Twitter account of Associated Press) can change the market, which means that information published in Twitter was used in a real trading strategy [11, 12]. The second approach is to analyze positive or negative sentiments about a company or a company's stock prices [13]. The third approach focuses on measuring the public mood and following the logic of behavioral finance used to improve stock market price forecasts. We know that the first and second approaches can be used in a trading strategy, but as far as the third approach is concerned, the situation is still unclear. There are several works on this topic and results vary from 87.4 % of accuracy in works by Bollen and his colleagues [10] to 51.8 % in those by Ding et al. [14]. In our research using the same methodology we found S&P500 data from Twitter to be capable of significantly improving forecast accuracy to 68.63 % [15]. Thus we chose to follow the third approach by testing the hypothesis within a bigger time frame and tried to change the methodology and concentrate on emoticons rather than on words.

In our research we regarded the amounts of Twitter emoticons as possible measures of social mood, and tested the hypothesis that it would be possible to use the analyses of moods expressed in tweets to increase prediction accuracy for stock market indicators.

The article has the following structure. The introduction is followed by Sect. 2 that describes the main design decisions and overviews the prediction system methodology. Section 3 contains a description of the dataset used in our research. Section 4 provides analysis of DJIA prediction and S&P500 indexes using additional information from Tweets. Section 5 compares the findings of our method and the approaches applied in the previous research. Section 6 concludes the work defining open research issues for further investigation.

2 Methodology

While analyzing online social networks using emoticons, Boia, Faltings and others found emoticons to closely coincide with the sentiment of the entire message. Tweets and their evaluation show that emoticons have a very good classification power [16] and that accuracy of emoticon-based sentiment classification is higher than 90 % (for tweets with emoticons) [16]. Impressed by the results obtained by Boia and coauthors, we decided to analyze amounts of different emoticons as a measurement of public mood.

In our research, we calculated the amount of different emoticons by days, the most frequent ones being ":(" and ":)" (see Table 4 in Appendix 1 for a complete list of emoticons analyzed in our study). Rare emoticons, for example, ":-c", with an average occurrence of less than 10 per day, were excluded from the analysis.

We created two datasets – Basic and Emoticons. The Basic Dataset contains stock prices data for three previous days. The Emoticons Dataset contains a normalized frequency of emoticons in tweets on each day in addition to the Basic Dataset.

The standard supervised machine learning techniques were used to classify days by shifts in stock market indices: Support Vector Machine, Neural Networks and Random Forest. These techniques were chosen as the most common ones with the best performance in the field. We trained a model with one part of data and tested the created model for prediction with another part of data.

We used RapidMiner (http://rapidminer.com/) for data handling, which is one of the key data mining tools according to www.KDnuggets.com poll in 2013.

In RapidMiner, the Support Vector Machine algorithm uses the Java implementation of the support vector machine mySVM by Stefan Rueping [17]. The SVM implementation in RapidMiner supports the following types of kernels: dot, radial, polynomial, neural, anova, epachnenikov, gaussian combination and multiquadric. We tested dot, radial, polynomial and neural kernels to establish the baseline.

We used the Neural Net operator to create the Neural Networks Model in Rapid-Miner. The Neural Net operator implemented a feed-forward neural network trained by a back propagation algorithm.

The Random Forests technique is implemented in the Random Forest operator of RapidMiner, but our preliminary tests showed that W-Random Forest from Weka (from the Weka extension of RapidMiner) provided better performance. Thus, we used this implementation of the Random Forest technique.

To establish the baseline for prediction rates, we trained models on historical index price data (information of three previous days). First, we used all data and made 1,000 cycles of validation using 90 randomly chosen days for training and one random day for testing (1,000 predictions in total). This validation allowed us to test the hypothesis about the predictability level of a stock market and to choose the technique that had demonstrated the best performance. Next, we ran the optimization parameters operator to establish the baseline performance.

Second, we used the best modeling technique, trained it on our sequence of 89 trading days and tested it on the next (90th) day following the chosen period. Our dataset allowed us to carry out 189 experiments, which means that we validated our models on 189 days. The same type of parameters optimization was employed to establish comparativeness with the baseline performance. We used this validation as we intended this study to be one of the steps to devising a trading strategy. We wanted to model the actual situation in the stock market, where we made a prediction for tomorrow's stock price movement, based on 89 observed days.

According to our hypothesis that emoticons can provide additional information, we expect the techniques trained on the Emoticons Dataset to exhibit better accuracy.

3   Data Description

By making use of Twitter API we managed to download more than 1.6 billion unfiltered tweets over the period from 13/02/2013 to 19/05/2014 (we downloaded an average of 3,483,642 tweets per day) and that is approximately 1 % of the total amount, according to API limitations. All the English tweets (where the user's "lang" parameter value equals "en") were sorted by day and analyzed automatically according to the counts of the emoticons (the complete set of emoticons and their frequency are presented in Table 4, Appendix 1).

We chose two stock market indicators whose prediction accuracy could be improved. The first one is the Dow Jones Industrial Average (DJIA), one of the oldest US stock market indices. The second one is Standard & Poor's 500 (S&P500), a stock market index based on the market capitalizations of 500 large companies having common stock listed on the New York Stock Exchange[1] or in the National Association of Securities Dealers Automated Quotations System[2].

For the DJIA and S&P500 stock market prices data we used the Yahoo Finance website[3], which provides opening and closing historical prices as well as the volume for any given trading day.

To apply the machine learning techniques, we divided the days into two equal groups by the difference between closing and opening prices. If on a day the opening price minus the closing price exceeded 50 % of all the differences for the period from 13/02/2013 to 19/05/2014, then "shift" was equal to 1, and when it was lower than 50 % it was 0. The Basic Dataset consisted of 16 columns (variables: shift (information about index shift 1 or 0), $Open_{-1\ day}$, $Close_{-1\ day}$, $Min_{-1\ day}$, $Max_{-1\ day}$, $Volume_{-1\ day}$, $Open_{-2\ day}$, $Close_{-2\ day}$, $Min_{-2\ day}$, $Max_{-2\ day}$, $Volume_{-2\ day}$; $Open_{-3\ day}$, $Close_{-3\ day}$, $Min_{-3\ day}$, $Max_{-3\ day}$, $Volume_{-3\ day}$) and was employed to establish the baseline accuracy.

The Emoticons Dataset was created by adding columns about normalized frequencies of emoticons for the previous day (one day – 12 columns). To calculate normalized frequencies for each day, we divided the number of tweets with selected emoticons by the total number of tweets downloaded on this day.

The whole period from 13/02/2013 to 19/05/2014 contained 277 working days with available stock market information. This period was used for the first validation.

For the second validation the whole dataset was divided into sets with data of 90 days. The period of 90 days was chosen to enable the use of 89 days for training and 1 day for prediction. Our period of time allowed us to perform at least 189 prediction experiments for each time lag (from one to seven days), depending on the availability of stock market data.

The most frequent emoticons in 1.6 billion tweets were ":)" and ":(" – the same as in the study of the Twitter emoticon dictionary [18].

4 Analysis

4.1 DJIA Stock Market Prediction

First, to find the baseline accuracy we trained Neural Networks, Support Vector Machine and Random Forest on the Basic DJIA data with one-day time lags (Table 1).

---

[1] http://www.nyse.com.

[2] http://www.nasdaq.com.

[3] http://www.finance.yahoo.com.

Table 1. "Shift" value prediction for DJIA. Accuracy of the support vector machines, random forest and neural networks trained on the basic dataset

Model	Accuracy	Kappa	RMSE	Calculation time
SVM (dot)	48.80 %	−0.038	0.54	6 s
SVM (radial)	48.60 %	−0.038	0.522	6 s
SVM (polynomial)	48.80 %	−0.038	0.543	6 s
SVM (neural)	52.20 %	0.043	0.61	7 s
W-Random Forest	51.30 %	0.017	0.54	32 s
Neural Net	47.80 %	−0.049	0.553	6 min

The best accuracy was demonstrated by the Support Vector Machine technique with neural kernel (52.20 %). We ran parameter optimization for this technique, which helped us increase prediction accuracy to 53.20 % (kappa = 0.018, RMSE = 0.500).

That level of performance became the baseline for our analysis. As the Support Vector Machine technique provided better performance, we used it in further analyses.

It is worth mentioning that Random Forest demonstrated compatible performance and the calculation time was reasonable (in comparison with SVM). In our next study we plan to focus more on applying Random Forest, as it allows the multiclass classification.

Next we extended prediction datasets with Twitter information and the train selected machine learning model – Support Vector Machine. Prediction accuracy for the SVM model trained on the Emoticons Dataset with different time lags is presented in Table 2.

Table 2. "Shift" value prediction for DJIA. Accuracy of the support vector machine trained on the emoticons dataset with different time lags

Lag in days	Accuracy	Kappa	RMSE
1	52.38 %	0.049	0.642
2	**57.59 %**	**0.154**	**0.511**
3	47.62 %	−0.052	0.506
4	52.33 %	0.046	0.524
5	41.75 %	−0.165	0.532
6	48.68 %	−0.024	0.533
7	**41.80 %**	**−0.164**	**0.529**

Although the model showed better performance, the additional Tweeter information failed to significantly increase accuracy (Chi-square = 1.085, p = 0.297).

4.2 S&P500 Index Prediction

To establish the baseline accuracy, we ran the Support Vector Machine (neural kernel) on historical data with parameter optimization. The results showed that the Support Vector Machine provided a baseline accuracy of 50.70 %.

Addition of Twitter information significantly improved our prediction accuracy (Chi-square = 5.189, p < 0.05). The best performance was achieved using the Emoticons Dataset with a two-day lag Table 3.

Table 3. "Shift" value prediction for S&P500. Accuracy of the support vector machine trained on the emoticons dataset with different time lags

Lag in days	Accuracy	Kappa	RMSE
1	52.91 %	0.044	0.517
2	**59.69 %**	0.192	0.504
3	47.62 %	−0.054	0.527
4	49.22 %	−0.023	0.529
5	50.00 %	−0.002	0.518
6	59.26 %	0.186	0.507
7	52.38 %	0.051	0.051

5 Discussion

In our previous research we found that Twitter sentiment analysis data could significantly improve forecasting for the S&P500 index, and the new results with emoticons support our findings. Addition of Twitter information allowed us to increase accuracy from 50.70 % (baseline) to 59.69 % (SVM trained Emoticons data). It should be mentioned that a more complex approach we took in our previous research allowed forecast improvement of up to 68.63 %.

Compared to other studies which used Twitter data analysis, we obtained lower accuracy. Bollen and his colleagues obtained an 86.7 % accuracy for determining stock market movement [10]. Aanalyzing prediction of stock prices movements for the Apple company, Vu, Shu, Thuy and Nigel demonstrated 82.93 % [19]. It should be mentioned, however, that these results were obtained on a relatively small number of testing days (21 days in the study by Bollen et al., and 41 days in that of Vu et al.) In our study the test sample was 189 days.

Comparison with other results in the stock market prediction field showed that what we demonstrated was almost the average performance. For example, usage analysis of Financial news gained 57 % of directional accuracy [20]. Mahajan et al. taking the same approach obtained an accuracy of 60 % [21]. While analyzing ad hoc announcements, Groth and Muntermann reached an accuracy of 75 % [22].

Also worthy of mention is that we did not simulate any trading strategy based on our results, but expected that it would ultimately deliver more than 2.06 % (demonstrated by simulation in a study by Schumaker and Chen who obtained an accuracy of 57.1 % [20]).

Therefore, the emoticons approach may be used alone if only we need small improvement, but it is not suitable for more complex calculations.

SVM techniques exhibit not only the best performance for our classification task, but also enable the best calculation speed.

It should be also mentioned that analysis of correlations between normalized frequencies of different emoticons showed that they were highly related. For example, the correlation between the most frequent emoticons ":)" and ":(" is 0.965 (Fig. 1). Such a high correlation between the appearance of sad and happy emoticons remains yet to be accounted for. We can only suppose that it may be connected with emotionality, and a rise in public emotionality will lead to increased emotions, whether good or bad.

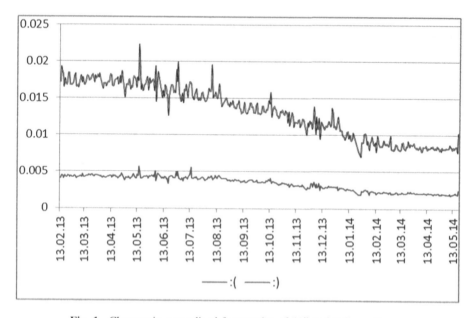

Fig. 1. Changes in normalized frequencies of ":)" and ":(" emoticons.

Interestingly, the best performance was demonstrated by SVM trained on the Emoticons Dataset with a two-day lag for both DJIA and S&P500 indices. In the research undertaken by Bollen, Mao, and Zeng the best results were achieved on a database with a four-day lag [10]. In our previous research, the lag for the best accuracy rate was 5 days. Such instability of results can be explained by changes occurring on a stock market and model mishitting could be a sign of information leakage, emerging policies or other events.

6   Conclusions

In our research we planned to test the hypothesis that even a simple sentiment analysis of Twitter data, such as that of emoticons frequency, can provide additional information capable of enhancing prediction accuracy for DJIA and S&P500. Analysis of 1.6 billion tweets downloaded over a period from 13/02/2013 to 19/05/2014 allowed us to significantly improve forecasts using the SVM technique. The obtained results suggest that our hypothesis can be confirmed. However, we found no significant improvement in accuracy, so our further research will combine both lexical and emoticons techniques in order to receive more information from Twitter.

High correlations between sad and happy emoticons also need special attention, as does the instability of time lag in which public emotions are expressed in Twitter influence on the stock market prices. Our future research will also deal with shaping and testing a trading strategy based on Twitter data analysis.

Appendix 1

See Table 4.

Table 4. List of emoticons analyzed in the study

Emoticon	M	SD	Included in analysis	M (normalized)	Total amount
:)	10,176.87	36,170.15	+	0.00977	15,625,504
:(	3,643.26	12,539.19	+	0.00338	5,416,930
:))	1,415.28	4,748.58	+	0.00128	2,051,390
:-)	1,066.40	3,380.31	+	0.00091	1,460,296
:'(	720.09	2,197.49	+	0.00059	949,318
:)))	449.71	1560.92	+	0.00042	674,320
:((	382.91	1,417.75	+	0.00038	612,471
:(((	187.83	701.45	+	0.00018	303,027
:-(	142.69	580.67	+	0.00015	250,852
o_o	135.94	360.83	+	9.8383E-05	155,881
:-))	26.24	120.45	+	3.2152E-05	52,036
:-|	18.19	37.01	+	1.0073E-05	15,992
:-o	9.27	24.73		6.6889E-06	10,686
:-&	5.44	9.03		2.4352E-06	3,904
:'-(	3.65	8.86		2.3701E-06	3,828
x-(	3.28	5.28		1.4477E-06	2,285
:-@	2.74	4.63		1.2474E-06	2,002
:-!	2.65	4.23		1.1599E-06	1,829
:o	2.58	2.53		6.7428E-07	1,094
:(::	1.11	1.06		2.802E-07	461
:-t	0.50	0.21		5.7265E-08	94
:-1	0.49	0.16		4.2563E-08	72
:-c	0.30	0.07		1.9534E-08	34
:-o	0.27	0.06		1.6285E-08	28
:-(:::	0.04	0.002		6.4163E-10	1
>:o	0	0		0	0

References

1. Johnson, E.J., Tversky, A.: Affect, generalization, and the perception of risk. J. Pers. Soc. Psychol. **45**, 20 (1983)
2. Isen, A.M., Patrick, R.: The effect of positive feelings on risk taking: When the chips are down. Organ. Behav. Hum. Perform. **31**, 194–202 (1983)
3. Mayer, J.D., Gaschke, Y.N., Braverman, D.L., Evans, T.W.: Mood-congruent judgment is a general effect. J. Pers. Soc. Psychol. **63**, 119 (1992)
4. Schwarz, N., Clore, G.L.: Mood, misattribution, and judgments of well-being: informative and directive functions of affective states. J. Pers. Soc. Psychol. **45**, 513 (1983)
5. Isen, A.M., Means, B.: The influence of positive affect on decision-making strategy. Soc. Cogn. **2**, 18–31 (1983)
6. Nofsinger, J.R.: Social mood and financial economics. J. Behav. Fin. **6**, 144–160 (2005)
7. Bikhchandani, S., Hirshleifer, D., Welch, I.: A theory of fads, fashion, custom, and cultural change as informational cascades. J. Polit. Econ. **100**, 992–1026 (1992)
8. Salganik, M.J., Dodds, P.S., Watts, D.J.: Experimental study of inequality and unpredictability in an artificial cultural market. Science **311**, 854–856 (2006)
9. Bikhchandani, S., Hirshleifer, D., Welch, I.: Learning from the behavior of others: conformity, fads, and informational cascades. J. Econ. Perspect. **12**, 151–170 (1998)
10. Bollen, J., Mao, H., Zeng, X.: Twitter mood predicts the stock market. J. Comput. Sci. **2**, 1–8 (2011)
11. Selyukh, A.: Hackers send fake market-moving AP tweet on White House explosions | Reuters. http://www.reuters.com/article/2013/04/23/net-us-usa-whitehouse-ap-idUSBRE93 M12Y20130423
12. Market reaction to Tuesday's erroneous tweet. http://pdf.reuters.com/pdfnews/pdfnews.asp? i=43059c3bf0e37541&u=2013_04_23_07_12_0ae1bd28b07544d5a23c965af0b0ac10_ PRIMARY.jpg
13. Sprenger, T.O., Tumasjan, A., Sandner, P.G., Welpe, I.M.: Tweets and trades: the information content of stock microblogs. Eur. Fin. Manag. **20**, 926–957 (2013)
14. Ding, T., Fang, V., Zuo, D.: Stock market prediction based on time series data and market sentiment (2013). http://murphy.wot.eecs.northwestern.edu/~pzu918/EECS349/final_ dZuo_tDing_vFang.pdf
15. Porshnev, A., Redkin, I., Shevchenko, A.: Improving Prediction of Stock Market Indices by Analyzing the Psychological States of Twitter Users. Social Science Research Network, Rochester (2013)
16. Boia, M., Faltings, B., Musat, C.-C., Pu, P.: A :) Is Worth a Thousand Words: How People Attach Sentiment to Emoticons and Words in Tweets, pp. 345–350. IEEE. http://doi.org/10. 1109/SocialCom.2013.54
17. Rüping, S.: SVM kernels for time series analysis. Technical Report, SFB 475: Komplexitätsreduktion in Multivariaten Datenstrukturen, Universität Dortmund (2001)
18. Schnoebelen, T.: Do you smile with your nose? Stylistic variation in Twitter emoticons. University of Pennsylvania Working Papers in Linguistics, vol. 18, p. 14 (2012)
19. Vu, T.-T., Chang, S., Ha, Q. T., Collier, N.: An experiment in integrating sentiment features for tech stock prediction in Twitter. In: Proceedings of the Workshop on Information Extraction and Entity Analytics on Social Media Data (pp. 23–38). Mumbai, India: The COLING 2012 Organizing Committee (2012). http://www.aclweb.org/anthology/W12-5503

20. Schumaker, R.P., Chen, H.: Textual analysis of stock market prediction using breaking financial news: the AZFin text system. ACM Trans. Inf. Syst. **27**, 1–19 (2009)

21. Mahajan, A., Dey, L., Haque, S.M.: Mining financial news for major events and their impacts on the market. In: IEEE/WIC/ACM International Conference on Web Intelligence and Intelligent Agent Technology, 2008. WI-IAT 2008, pp. 423–426. IEEE (2008)

22. Groth, S.S., Muntermann, J.: An intraday market risk management approach based on textual analysis. Decis. Support Syst. **50**, 680–691 (2011)

ImSe: Exploratory Time-Efficient Image Retrieval System

Ksenia Konyushkova[1,2](✉) and Dorota Głowacka[1,2]

[1] Helsinki Institute for Information Technology, Espoo, Finland
konyushkova@gmail.com
[2] Department of Computer Science, University of Helsinki, Helsinki, Finland

Abstract. We consider the problem of Content-Based Image Retrieval (CBIR) with interactive user feedback when the user is unable to query the system with natural language text. We employ content-based techniques with Relevance Feedback mechanism to capture the precise need of the user and interactively refine the query. We apply the Exploration/Exploitation trade-off with Hierarchical Gaussian Process Bandits and pseudo feedback in order to tackle the problem of optimization in face of uncertainty and to improve the quality of multiple images selection. We tackle the scalability issue with Self-Organizing Map as a preprocessing techniques. A prototype system called *ImSe* was developed and tested in experiments with real users in different types of search tasks. The experiments show favorable results and indicate the benefits of proposed aprroach.

Keywords: Relevance feedback · Exploration/Exploitation · Content-based image retrieval · Gaussian process bandits · Self-organizing maps

1 Introduction

With the growth of the Internet and the associated amount of available images, there has been a growing interest in new tools for managing, analyzing and retrieving them [3,15,18]. Image retrieval techniques are required in many domains, such as medicine, photography, advertising and design. With the number of available images growing very fast, image retrieval cannot be simply reduced to text search utilizing only metadata such as keywords, tags or natural language text [10]. Veltkamp and Tanase (2002) [18] list over 50 different Content-Based Image Retrieval (CBIR) systems that rely on visual features automatically extracted from images, such as color, shape and texture. The first CBIR experiments date back to 1992 [7] when color and shape features were used for automatic image retrieval. Since then, many CBIR image retrieval systems have been developed [3,10,18]. Kosch and Maier (2008) [10] analyze recently developed image retrieval systems presented by Veltkamp and Tanase (2002) [18] and show that many of them rely on some form of Relevance Feedback mechanism. Relevance feedback [19] was first used in document retrieval but soon it was applied in multimedia retrieval including image

© Springer International Publishing Switzerland 2015
P. Braslavski et al. (eds.): RuSSIR 2014, CCIS 505, pp. 307–319, 2015.
DOI: 10.1007/978-3-319-25485-2_11

retrieval. Early expertiments show that the process of image retrieval can be greatly improved by actively involving the user into the search loop and utilizing his knowledge in the iterative search process for better target modeling and personalized search (e.g. *PicHunter* [2] or *PicSOM* [11]). Relevance Feedback mechanism can be viewed as an active learning framework, where the algorithm chooses the appropriate image subset for annotation [19]. It can be also used as input for Exploration/Exploitation algorithms which attempt to predict "the most informative" and "the most positive image" at the same time [19]. For example, *PinView* [1] is based on contextual multi-arm bandit algorithm LinRel [13]. Application of Exploration/Exploitation to retrieval problems seems to be promising but it has not been throughly studied in real-life systems and different types of retrieval. Datta et al. (2005) [3] indicate that most of CBIR research so far has concentrated on feature extraction, Relevance Feedback mechanisms and system design, however, the new trend will be to focus on more application-oriented aspects, such as interface design, visualization, scalability and system evaluation. We cannot but agree that multimedia retrieval systems require real-time response and therefor image retrieval systems must be time-efficient and easily scalable to large datasets. To overcome scalability limitations one can use primarily retrieved images or precomputed clusters to choose the most representative samples in an active learning scenario [3].

In this paper, we introduce *ImSe*, an interactive image retrieval system that tackles the problem of scalable time-efficient exploratory search. We propose a novel approach based on a hierarchical Gaussian Process (GP) bandits [14] combined with Self-Organizing Maps (SOM) [8] to speed up the computation. The proposed algorithm is called GP-SOM and forms an integral part of the *ImSe* system. We demonstrate the benefits of employing Exploration in an image retrieval system and show that the proposed algorithm outperforms similar appraoches.

The paper is organized as follows. In the next section, we give a detailed description of the proposed system and the GP-SOM algorithm. In the following section, we evaluate the system in real-life experiments and in simulations. We ran a small user study in order to evaluate the advantages of employing an Exploration/Exploitation strategy in a CBIR system (Sect. 3). Next, we ran a number of simulations (Sect. 4) to compare GP-SOM algorithm against 2 other bandit algorithms and we prove the computational efficiency of the proposed framework. Finally, we discuss the results and future plans in Sect. 5.

2   System Design

In this section, we describe the design of the *ImSe* system – multimedia retrieval system for images in large databases. The system operates through a sequence of rounds, where n images are displayed at each iteration and the user indicates which images are the closest to the ideal target image and which are the most different from the ideal target. We will focus on the key features of the *GP-SOM* algorithm (the integral part of the proposed system): the Relevance Feedback mechanism with diversification in results, time-efficiency of the on-line computational step and scalability to large databases. At every iteration, the algorithm balances between exploiting available knowledge to make the best current prediction and exploring other possibilities to decrease uncertainty given only a limited feedback from the user. Classical Exploration/Exploitation trade-off algorithms are designed to predict one object per iteration. In hierarchical bandit settings, we face the question of selecting multiple objects per round. We tackle this problem by utilizing the Gaussian Process belief on the potential feedback as a temporal feedback.

The main flow of the system is presented in Fig. 1. First, we extract features $H = h_1, h_2, ..., h_N$ from a database of N images $P = p_1, p_2, ..., p_N$ and compute the distances between them for digital processing of visual information. Distances are presented in a matrix D_P of dimensions $N \times N$. Next, we build a SOM $M = \{V, A_V\}$ based on the extracted features, which results in an image hierarchy with m model vectors $V = \{v_1, ..., v_m\}$. A_V shows image assignments to model vectors and $A_V = \{a_1, ..., a_m\}$, where $a_i = \{p_j : p_j \in cluster(v_i)\}$. Next, we calculate a kernel $K_P \subset \mathbb{R}^{N \times N}$ for images and model vectors. The preprocessing techniques from Sect. 2.1 are crucial for the system to be able to perform retrieval in on-line manner. In Sect. 2.2, we describe how the pre-build SOM is used as levels of hierarchy in GP bandits. In order to ensure diversity of the images presented at each iteration, we apply pseudo relevance feedback when sampling the images.

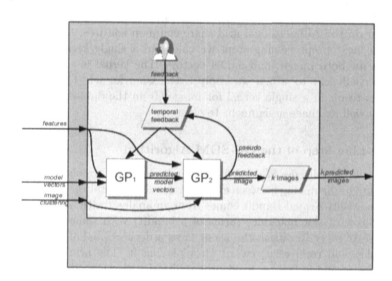

Fig. 1. Dataflow overview of ImSe system

2.1   Preprocessing

The aim of the pre-processing step is to keep the time needed for the on-line calculations almost independent of the linear increase in the size of the database and it is done off-line. The preprocessing starts with feature extraction and similarity matching. We rely on MPEG-7 standards [12] for feature extraction. A probability density metrics Hellinger distance [4] was selected for the similarity matching. Hellinger distance is a measure of similarity of probability distributions of histograms. If H^i and H^j are histograms of images i and j, N_b is a number of histogram bins and H_k^i is the value of the kth bin in the histogram of image i, then the Hellinger coefficient $d(H^i, H^j)$ for discrete distribution in bins can be calculated as

$$d(H^i, H^j) = \sqrt{1 - \frac{1}{\sqrt{H^i \cdot H^j \cdot (N_b)^2}} \cdot \sum_{k=1}^{N_b} \sqrt{H_k^i \cdot H_k^j}} \qquad (1)$$

Next, we create a discretization of the input space topology through a SOM [8]. SOM is an unsupervised method for dimensionality reduction by constructing an artificial neural network of instances that reflects their topological order. The character of features used in images makes it natural to employ topological order of features. SOM provides the so-called model vectors that are treated in our algorithm as clustering of the input space. In our case, the number of clusters m is determined to be $\sqrt{N}$, where N is the number of images in the database. Our calculation of SOM is based on EM algorithms and Gaussian kernel was chosen to be a neighborhood function. An example of SOM for 1000 images and 36 model vectors is presented in Fig. 2. We can clearly see that images close to each other on the 2-dimensional grid share common features.

As the next preprocessing step, we calculate a single kernel for Gaussian Processes for both images and model vectors. The kernel is also based on the Hellinger coefficient to which we apply a Gaussian kernel. The preprocessing operations result in a single kernel for images from the dataset and the model vectors as well as image assignment to clusters.

2.2   On-Line Step of the GP-SOM Algorithm

In this section, we discuss Multi-Armed Bandit problem formulation as a variant of algorithm for balancing between exploration and exploitation [17]. An intuitive idea of the Multi-Armed Bandit comes from an analogy with a gambler (agent) playing a row of slot machines (arms of a bandit) with some unknown prizes (rewards). At every iteration, the agent faces a decision-making problem which arm to play and receives a reward after playing it. The arm rewards can be modeled as a sample from a fixed probability distribution. The agent aims to maximize the reward but as he does not know how the rewards are generated, he has to not only exploit the most promising arms but also explore different potential actions.

Instead of straightforward optimization in the image space, we utilize hierarchical multi-armed bandits [14]. Model vectors that are produced as a result of clustering become the first level of instances for hierarchical multi-armed bandits and images associated with the model vectors become the second level. We apply a 2-layer bandit settings: first we select a model vector and then an image from a pool of images associated with a chosen model vector (Fig. 1).

Fig. 2. An example of a Self-Organizing Map for an image dataset.

There might be two causes for an the image to be selected – it might have high predicted user interest or we might want to get feedback on it to reduce our uncertainty about it. When we need to obtain multiple images in a single iteration, we want to ensure that they are not only different but also diverse. However, Exploration/Exploitation algorithms are designed to select one image per iteration. In our system, we present multiple images at each iteration. Thus, when sampling the ith image, we assume that we received pseudo feedback in the form of GP belief for the previously sampled images $1, 2, ..., i - 1$ that is added to the training set. The sampling procedure is described in Algorithm 1.

3   Real-Life Experiments

We designed a set of experiments (1) to demonstrate the benefits of employing Exploration/Exploitation strategy in CBIR and (2) to assess what type of search GP-SOM is most suited for. We benchmark GP-SOM against Exploitation only and Random policies. Exploitation-based retrieval is an example of the most common techniques in CBIR based on similarity matching. It predicts images that share features with positive user selections and attempts to avoid features from negative selections. If $F = \{f^1, ..., f^r\}$ is the feedback on a set of images $S = \{p^1, ..., p^r\}$ and $D = \{d^{1,k}, ..., d^{r,k}\}$ is the vector of distances from image k to all presented images S, then each image k gets the score $f^1 \cdot d^{1,k} + ... + f^r \cdot d^{r,k}$

Algorithm 1. On-line step of *GP-SOM*

Require: SOM $M = \{V', A_V\}$, where $V' = \{v'_1, ..., v'_m\}$ is a set of kernelized model
vectors, $v'_i \subset \mathbb{R}^{N \times 1}$ and $A_V = \{a_1, ..., a_m\}$ is a set of image assignments to model
vectors, $a_i = \{p_j : p_j \in cluster(v_i)\}$, where p_j is an image;
set of shown images p^i_{t-1} up to iteration $t - 1$: $S = \{p^1_{t-1}, ..., p^r_{t-1}\}$;
feedback F on S: $F = \{f^1_{t-1}, ..., f^r_{t-1}\}$, where $-1 \leq f^i_{t-1} \leq 1$.
set of selected images: $R \leftarrow \{\}$
  repeat
    Arms are model vectors $V' = \{v'_1, ..., v'_m\}$, choose an arm v^*: $v^* \leftarrow \max\{$GP-
    UCB$(V', S, F)\}$
    Arms are images associate with v^* - $A_{V^*} = \{p_j : p_j \in cluster(v^*)\}$, choose an
    image p^* and get its predicted relevance p^*_f: $p^*_f, p^* \leftarrow \max\{$GP-UCB$(A_{V^*}, S, F)\}$
    Update the set of shown images S with the selected image p^*: $S \leftarrow S \cup p^*$
    Update the feedback F with pseudo feedback p^*_f: $F \leftarrow F \cup p^*_f$
    Add the selected image p^* to the result set R: $R \leftarrow R \cup p^*$
  until n images are sampled: $|R| = n$
  return set of n selected images R

---

and Exploitation policy selects n images with the highest score. The Random
strategy helps us to assess the empirical difficultly of the task.

We test the performance of GP-SOM on three types of search [2]. In the
target search the user is looking for a particular image in the database, for
instance, a very specific image of himself/herself with their favorite cat or some
well known historical picture of a famous cathedral in the city. When performing
the *category* search the user will be satisfied with any of the images from the
desired category, for example, picture of a young girl or a building in baroque
style. The last scenario is search by *association* also known as *open* search. The
user is browsing a big collection of images having only a vague concept in mind,
such as an illustration for the concept of "youth" or for an essay about the
history of some city a century ago.

In the experiments reported below, we used the color and texture features
as they have a high discriminating power [6] and it is the base of many image
retrieval systems [18]. The real life experiments were based on the first measure
as it was demonstrated to be the most intuitive for the users in pilot studies. We
used the MIRFLICKR-25000 dataset [5] with 2 sets of visual descriptors: color
feature is presented by color histograms and texture by homogeneous texture
descriptor and edge histogram descriptor [12].

We decided to use the common HCI methodology of user studies as the evaluation procedure. 18 participants performed 9 tasks each from 3 tasks categories
designed by us. *ImSe* has a very simple user interface designed to be used without
any initial training (Fig. 3). When a mouse hovers around an image, two control panels appear. To navigate the system, the slidebar at the bottom of each
image is used. The feedback ranges from -1 to 1, where -1 indicates that a given
image is very different from the ideal target and 1 indicates that a given image
is very similar to the ideal target. The database we are using is not annotated

appropriately for every type of retrieval and thus we have to rely on the user's labels. The users were asked to mark all the appropriate images with labels characterizing 3 degrees of their similarity to the target – *Satisfactory*, *Good* and *Excellent* and we rely on these assessments as if it is a ground truth for the search task. Through this slidebar, the users provide their relevance feedback on the displayed images and radio buttons are used for the system evaluation and were added only for testing.

Fig. 3. ImSe system interface

The number of iterations to complete the task is often considered to be a performance measure in evaluating Exploration/Exploitation strategies. Table 1 summarizes the average number of iterations that it took the users to complete each type of task (the first number in each cell) as well as the average time spent on each iteration (the second number) for different types of settings. In all types of tasks, the average number of iterations for GP-SOM and Exploitation algorithms was approximately the same, but clearly smaller than for the Random algorithm. In general, the *target* search took the longest reflecting the fact that looking for a specific target takes more iterations than a less specific type of search. The results on the time spent on each iteration show that the users need to spend more time at every iteration of GP-SOM in order to observe how the results have changed, compared to Exploitation, where there is little variation in the type of images presented at subsequent iterations.

Taking into consideration the images bookmarked in different categories, we define the following measure as a quality of search: $(3m_e + 2m_g + m_s)/n_i$, where m_e, m_g and m_s is the number of *Excellent*, *Good* and *Satisfactory* images correspondingly and n_i is the number of iterations. Table 2 compares the performance for the three algorithms in terms of the above measure for different categories of search. We report the average cumulative score per iteration. GP-SOM outperforms Exploitative and Random algorithms for all types of searches. The best result by GP-SOM was obtained in *category* search. The pure exploitation algorithm performs worse than GP-SOM in all tasks indicating that GP-SOM helps the user to find more suitable images. In the proposed retrieval scenario

Table 1. Number of iterations (first value) and time in seconds spent at each iteration (last value) for GP-SOM, Exploitation and Random algorithms for different types of tasks.

Algorithm	*target* search	*category* search	*open* search
Random	20.8; 0.43	14.4; 0.43	19.1; 0.37
Exploitation	16.1; 0.37	12.2; 0.28	13.8; 0.25
GP-SOM	15.4; 0.43	12.6; 0.42	13.8; 0.31

Exploitation strategy often ends up being stuck in the local minimum of the search space, so, this strategy depends on the fist image selection and is prone to over-exaggerating the limited user feedback. The trade-off between exploration and exploitation ensures that the system does not only show the images that it believes best reflect the user interest, but also displays diverse images on which it has not received much feedback.

Table 3 shows detailed information on how many images of each type were bookmarked by users on average per one iteration. Users of GP-SOM were more successful than those of other algorithms in finding *Excellent* and *Good* images. Below, we discuss the results for all types of search separately.

Table 2. Comparison of the performance of GP-SOM, Exploitation and Random algorithms in different types of tasks.

Search type	Algorithm	Weighted average of bookmarked images per iteration
Target	Random	0.95
	Exploitation	1.56
	GP-SOM	1.64
Category	Random	1.51
	Exploitation	2.06
	GP-SOM	2.85
Open	Random	1.34
	Exploitation	2.06
	GP-SOM	2.58

Target Search. In *target* search, images marked as *Excellent* were either the exact target image or close substitutes. In Fig. 4, we can see that the number of "excellent" images found by the users grows with the number of iterations in GP-SOM, Exploitation and Random experiments. We clearly see that the strategy based on Exploitation has converged quite soon and after 12 iterations most of the

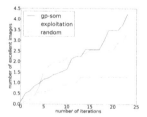 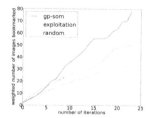 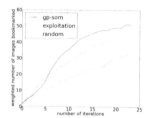

Fig. 4. Cumulative number of "excellent" bookmarked images in target search

Fig. 5. Weighted cumulative number of bookmarked images in category search

Fig. 6. Weighted cumulative number of bookmarked images in open ended search

users could not find many *Excellent* images. GP-SOM, where exploration is also present, continues to find suitable images throughout the search session.

Category Search. We obtained category annotations for some of the image classes from MIRFLICKR dataset and we used this information for additional evaluation of the performance of the *category* search. We calculated the precision for all three algorithms and obtained on average 12 % precision for Random browsing, 23 % for Exploitation search and 26 % for the GP-SOM based system. The graph of cumulative quality score for *category* search is shown in Fig. 5. It took only 5 iterations for GP-SOM to outperform the Exploitation setting (that represents a type of the popular similarity based retrieval). Table 3 shows that GP-SOM obtains the highest score for *Excellent* images but it is slightly worse at finding *Good* images, while the amount of *Satisfactory* images retrieved is approximately the same in all settings.

Open-ended Search. When analyzing the open-ended search we looked at the weighted average of the number of bookmarked images over time and the total number of images of each category per iteration. The results are illustrated in Fig. 6. As we can see in Table 3, GP-SOM performs well in all categories but its performance is particularly high when it comes to finding *Excellent* images.

Table 3. The average number of bookmarked images per iteration for different algorithms in various types searches.

Algorithm	Target			Category			Open		
	q_E	q_G	q_S	q_E	q_G	q_S	q_E	q_G	q_S
Random	0.057	0.175	0.431	0.300	0.197	0.213	0.137	0.282	0.363
Exploitation	0.134	0.317	0.521	0.400	0.318	0.209	0.341	0.361	0.309
GP-SOM	0.151	0.346	0.493	0.680	0.297	0.212	0.484	0.380	0.364

4   Simulations

In the simulations we benchmark our system against two algorithms: LinRel [13], which forms an integral part of the *PinView* system [1], and the Gaussian Processes Bandit algorithm [16]. We assume that the choice of one of the presented images is a random process, where more relevant images are more likely to be chosen. In our simulation experiments, we applied the user model proposed in [13], which has been shown to be a close approximation of real user behavior. We assume a similarity measure $S(x_1, x_2)$ between images x_1, x_2, which also measures the relevance of an image x compared to an ideal target image t by $S(x, t)$. Let $0 \leq \lambda \leq 1$ be the uniform noise in the user's choice. The probability of choosing image $x_{i,j}$ is given by:

$$D\{x_i^t = x_{i,j} \mid x_{i,1}, \ldots, x_{i,k}; t\} = (1 - \lambda)\frac{S(x_{i,j}, t)}{\sum_{j=1}^{k} S(x_{i,j}, t)} + \frac{\lambda}{k}. \tag{2}$$

Assuming a distance function $d(\cdot, \cdot)$, a possible choice for the similarity measure is $S(x, t) = d(x, t)^{-a}$ with parameter $a > 0$. With the polynomial similarity measure, the user's response depends on the relative size of the image distances to the ideal target image. We use Euclidean norm as the distance measure between image x and the target image t. In all the experiments, the values of a and λ were kept constant at 4 and 0.1, respectively.

The reported results are averaged over 100 searches for randomly selected target images from the dataset. We simulated the retrieval process and we report the number of iterations it takes to find the hypothetical target. We also tested the influence of n, i.e. the number of images displayed at each iteration, on the performance of the algorithms. The results are summarized in Fig. 7. GP-SOM significantly outperforms LinRel and GP bandits for smaller values of n. For large values of n, there is no significant difference between LinRel and GP-SOM, however GP-SOM is more computationally efficient than LinRel. We believe this result is due to our enhanced multiple image selection policy with temporal pseudo feedback that allows us to select diverse images at every iteration and it has the most effect with small n.

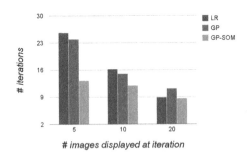

Fig. 7. Comparison of the performance of LinRel, GB-UCB and GB-SOM.

When designing the GP-SOM algorithm, we consider not only theoretical guarantees of the method but also its applicability to real-time systems. Below, we analyze the scalability issues of interactive Exploration/Exploitation algorithms. Let N denote the number of images in the dataset. At each iteration, we present n images where $n \ll N$. Let us consider the $i + 1$th iteration. The complexity of GP-SOM is $\mathcal{O}(\sqrt{N}i^2n^3)$ compared to $\mathcal{O}(N^3)$ of LinRel and $\mathcal{O}(Ni^2n^2)$ of GP (see [9] for a discussion on the computational efficiency of the initial version of algorithm). We can easily see the benefit of using Hierarchical GP bandits in Exploration/Exploitation trade-off. We decided not to compare GP-SOM to other bandit algorithms in real-life scenario due to this time-complexity limitations. Additionally, the SOM approach can be generalized into l level hierarchical bandits by introducing additional levels in the map and increasing the complexity only by a scalar factor. If we fix the number of arms in each run to be P and allow l level hierarchy, we can process P^l images with the complexity $l \cdot \mathcal{O}(Pi^2n^3)$. This means that introducing SOM in the preprocessing step brings a big advantage to the crucial for the user experience on-line step.

5   Conclusions and Discussion

The paper proposes a model of time-efficient relevance feedback mechanism with an Exploration/ Exploitation algorithm that balances between presenting images about which the system is uncertain and images about which the system has high confidence. We use clusters of images obtained through Self-Organizing Maps to serve as the first layer in GP bandits in order to overcome scalability issues in large unannotated databases.

The hierarchical technique used in GP-SOM facilitates running image searches much faster and can be easily applied to real-life systems. We created an image retrieval system *ImSe* incorporating the GP-SOM algorithm and tested its performance in different types of searches. We performed simulations and real-life experiments to test the performance of the proposed algorithm. Simulation results demonstrated that the proposed mechanism of selecting multiple images ensures their diversity and achieves better performance when a small number of images are presented at each iteration. The experimental results show that in all types of search, users of the GP-SOM algorithm were able to find more relevant images in a fewer number of iterations compared to other algorithms.

There are a number of interesting observations related to the experimental results. First, there we observed a clear difference between male and female behavior in image retrieval that we did not analyze yet in details: female participants tend to spend more time per iteration selecting images compared to male participants. Another interesting observation concerns performance of the GP-SOM algorithm in tasks with a varying level of "complexity". For instance, when the task is more difficult (few images in the database or the image contains very specific features, such as searching for a particular insect), GP-SOM gives us a larger improvement over a strategy based only on exploitation. We will conduct more tests in the future to fully confirm these observations.

An obvious direction for ongoing research is to learn individual feature kernel combination instead of relying only on one feature or uniform combination and incorporate this into a hierarchical scenario. Another possible direction is including feedback from eye-tracker to get another type of Relevance Feedback.

References

1. Auer, P., Hussain, Z., Kaski, S., Klami, A., Kujala, J., Laaksonen, J., Leung, A.P., Pasupa, K., Shawe-Taylor, J.: Pinview: implicit feedback in content-based image retrieval. JMLR **11**, 51–57 (2010)
2. Cox, I., Miller, M., Minka, T., Papathomas, T., Yianilos, P.: The Bayesian image retrieval system, pichunter: theory, implementation, and psychophysical experiments. Image Process. **9**(1), 20–37 (2000)
3. Datta, R., Li, J., Wang, J.: Content-based image retrieval: approaches and trends of the new age. In: Multimedia information retrieval, pp. 253–262. ACM (2005)
4. Hellinger, E.: Neue begründung der theorie quadratischer formen von unendlichvielen veränderlichen. Journal für die reine und angewandte Mathematik **136**, 210–271 (1909)
5. Huiskes, M., Lew, M.: The MIR flickr retrieval evaluation. In: MIR 2008 (2008)
6. Hussain, Z., Leung, A.P., Pasupa, K., Hardoon, D.R., Auer, P., Shawe-Taylor, J.: Exploration-exploitation of eye movement enriched multiple feature spaces for content-based image retrieval. In: Balcázar, J.L., Bonchi, F., Gionis, A., Sebag, M. (eds.) ECML PKDD 2010, Part I. LNCS, vol. 6321, pp. 554–569. Springer, Heidelberg (2010)
7. Kato, T., Kurita, T., Otsu, N., Hirata, K.: A sketch retrieval method for full color image database-query by visual example. In: Pattern Recognition. Computer Vision and Applications, IAPR, pp. 530–533. IEEE (1992)
8. Kohonen, T.: Self-organizing Maps, vol. 30. Springer Verlag, Heidelberg (2001)
9. Konyushkova, K., Glowacka, D.: Content-based image retrieval with hierarchical gaussian process bandits with self-organizing maps. In: ESANN (2013)
10. Kosch, H., Maier, P.: Content-based image retrieval systems-reviewing and benchmarking. JDIM **8**(1), 54–64 (2010)
11. Laaksonen, J., Koskela, M., Laakso, S., Oja, E.: Picsom-content-based image retrieval with self-organizing maps. Pattern Recognition Letters **21**(13), 1199–1207 (2000)
12. Manjunath, B., Ohm, J., Vasudevan, V., Yamada, A.: Color and texture descriptors. Circuits and Systems for Video Technology **11**(6), 703–715 (2001)
13. Hussain, Z., Auer, P., Leung, A., Shawe-Taylor, J.: Report on using side information for exploration-exploitation trade-offs, fp7-216529 pinview. Technical report, European Community's Seventh Framework Programme (2009)
14. Pandey, S., Agarwal, D., Chakrabarti, D., Josifovski, V.: Bandits for taxonomies: a model-based approach. In: SIAM International Conference on Data Mining (SDM) (2007)
15. Smeulders, A., Worring, M., Santini, S., Gupta, A., Jain, R.: Content-based image retrieval at the end of the early years. Pattern Analysis and Machine Intelligence **22**(12), 1349–1380 (2000)
16. Srinivas, N., Krause, A., Kakade, S., Seeger, M.: Gaussian process bandits without regret: An experimental design approach. In: CoRR (2009). arxiv.org/abs/0912.3995

17. Eickhoff, J.: Onboard Computers, Onboard Software and Satellite Operations. SAT, vol. 1. Springer, Heidelberg (2012)
18. Veltkamp, R.C., Tanase, M.: Content-Based Image Retrieval Systems: A Survey, pp. 1–62. Department of Computing Science, Utrecht University (2002). (preprint)
19. Zhou, X., Huang, T.: Relevance feedback in image retrieval: a comprehensive review. Multimedia Syst. **8**(6), 536–544 (2003)

Semantic Clustering of Russian Web Search Results: Possibilities and Problems

Andrey Kutuzov[✉]

Mail.ru Group, National Research University Higher School of Economics,
Moscow, Russia
akutuzov72@gmail.com

Abstract. The present paper deals with word sense induction from lexical co-occurrence graphs. We construct such graphs on large Russian corpora and then apply the data to cluster the results of Mail.ru search according to meanings in the query. We compare different methods of performing such clustering and different source corpora. Models of applying distributional semantics to big linguistic data are described.

1 Introduction

The presented paper deals with the problem of semantic clustering of search engine results page (SERP). The problem arises from the obvious fact that many user queries are ambiguous in some way. Thus, search engines strive to diversify their results and to present such results that are related to as many query interpretations as possible. For example, Google search for the Russian word '*максим*' returns:

1. five results related to a popular singer,
2. two results for a magazine,
3. one result for http://lib.ru, Maxim Moshkow's electronic library,
4. one result for a proper name.

However these results are not sorted by their meaning and are returned simply according to their relevance ranking, which for many of them seems to be almost equal. The obvious way to cluster the results is by the words their snippets share. Unfortunately, often snippets for results belonging to one query sense do not have a single content word in common (except for the query itself, which is useless). Cf. two snippets for the first query meaning from the example above:

1. '*МакSим начинает самостоятельно заниматься своей карьерой, пишет новые песни. В этот период певица выступает как малобюджетный проект, ...*'
2. '*МакSим презентовала видеоклип «Я буду жить», получивший широкую огласку еще до момента появления видео в сети.*'

© Springer International Publishing Switzerland 2015
P. Braslavski et al. (eds.): RuSSIR 2014, CCIS 505, pp. 320–331, 2015.
DOI: 10.1007/978-3-319-25485-2_12

They do not have a single common word, but still belong to one meaning (popular singer).

Moreover, snippets for different query senses can share some words. Cf. two snippets from the same search engine results page. They share the word *'автор'* ('author'), however the first snippet relates to the first meaning, while the second snippet shows the third one:

1. *'МакSим (Марина Абросимова) – одна из самых популярных и коммерчески успешных певиц в России, являющаяся **автором** и исполнителем...'*

2. *'Работает с 1994 года. Книги и тексты, разбитые по жанрам и **авторам**.'*

That means that there is a need for more sophisticated way to cluster search results. We should somehow learn which senses the query has and with which words these meanings are (probabilistically) associated. One of the possible ways to solve this problem is by extracting co-occurrence statistics from large corpora. The idea behind this is that word meaning is in fact the sum (or the average) of its uses. So, meaning is a function of distribution (cf. [1]). Thus, if we know with which words the query typically co-occurs and how these neighbors are related to each other, then we know the 'sense set' of the query. After that we can somehow measure semantic similarity of each search snippet on the SERP with each of the senses and map them to each other. This information can then be used to either rank the results, or mark them with appropriate labels.

The structure of the paper is as follows. In Sect. 2 we briefly overview work previously done on the subject. Section 3 describes the process of building co-occurrence graphs from large Russian corpora. In Sects. 4 and 5 we conduct an experiment on clustering SERPs with ambiguous queries from Mail.ru search engine with the help of the methods described before. The results are evaluated in Sect. 6. Section 7 draws conclusions concludes and provides suggestions for further research.

2   Related Work

As stated in the previous Section, we are inspired by a fundamental hypothesis than meaning depends on the distribution [1] and that frequency of linguistic phenomena (in our case, word co-occurrence) is important for determining these phenomena's place in the system of language [2]. Our work is also based on the idea that the senses of ambiguous lexical units should be induced from the data itself, not from a dictionary. No dictionary is perfect or comprehensive, because *'senses as identified in the dictionary identify points on a continuum of possibilities for how the word is used'* [3]. The only robust source of words' meanings in the text is the text itself. That's why we shift our focus away from selecting the most suitable senses from a pre-defined inventory towards discovering senses automatically from the raw data, which is natural text.

One of the first notes on practical application of this idea to word sense disambiguation and word sense induction is found in [4], where vector representations of word similarity derived from co-occurrence data are used. Broad review of contemporary (by 2012) state of the field is provided in [5].

The main source of methods for our present research is [6], which describes workflow for clustering web search results using graph analysis over co-occurrence networks. Specifically, we use the notion of query graph, consisting of query terms and words from search engine results page augmented with nearest neighbors and relations from a reference corpus. For partitioning query graph and clustering query senses we employed *Curvature* algorithm [6] and *Hyperlex* algorithm proposed in [7].

3   Building Co-occurrence Graph

The first thing we had to do was to select a text corpus to build the graph upon. It is well known that the larger the corpus is the more co-occurrence information it contains. However, increasing corpus size also leads to exponentially growing computation time. Thus, for the sake of time and because of the preliminary nature of our research, we restricted ourselves to three Russian corpora of smaller but still decent size:

1. Open Corpora[1] (1 million tokens), further *'OC'*;
2. Disambiguated fragment of Russian National Corpus[2] (1 million tokens), further *'RNC'*;
3. Corpus of random search queries from Mail.ru search engine[3] (2 million tokens), further *'QC'*.

The first two items are academic corpora of Russian texts, supposedly representing (written) language in general. They differ in that the first one consists of full texts published under various free and open licenses, while the second one is a random sample of sentences from the larger Russian National Corpus. Both of them come with morphological annotation.

The third corpus was taken for comparison. It is important in view of the aim of our research (to test semantic SERP clustering). Our intuition was that perhaps query corpus provides more 'real-life' sense inventory. It is two times as big as its counterparts, because 'connectivity' between its members is lower (see Table 2) and we had to compensate for this.

At the same time, it turned out that the first two corpora mixed into one give better results, thus below we will often refer to such 'meta-corpus' as *'Mix corpus'*.

---

[1] http://opencorpora.org

[2] http://ruscorpora.ru

[3] http://go.mail.ru

Table 1. Sizes of corpora participating in the experiment

Corpus	Size (tokens)
OC	490671
RNC	294849
Mix	785520
QC	1035483

Before constructing the graph itself, we preprocessed the corpora, namely:

1. Removed from QC all queries which did not contain Cyrillic characters (as apparently they are not Russian),
2. Processed QC with Freeling analyzer [8] to extract lemmas and morphological information for all tokens,
3. Removed stop words,
4. Removed all tokens except nouns, as we restrict ourselves to inducing only nominal senses (the same strategy was applied in [6]).

Sizes of preprocessed corpora are given in Table 1. Average query length in QC is 2.47 noun tokens per query.

After the corpus has been built, the process of constructing co-occurrence graph is rather straightforward: we create an empty graph and then populate it with vertexes denoting word types in the text (lemmas). After that for each lemma we find all its immediate neighbors in the corpus, that is, words to the left and to the right (sentence boundaries not crossed, queries considered to be 'sentences' as well). If two lemmas were neighbors at least one time, we draw an edge between corresponding neighbors.

Finally, we have an undirected graph in which noun lemmas are vertexes and co-occurrence relations are edges. For each edge we also calculate Dice coefficient [9]. It measures the 'strength' of the collocation, based on absolute frequency (c) of both words (w and w') and collocation (w, w'):

$$Dice(w, w') = \frac{2c(w, w')}{c(w) + c(w')} \tag{1}$$

One can also think about the graph as a matrix of Dice coefficient values for all possible pairs of lemmas in the corpus.

Table 2 gives an overview of the basic features of the graphs.

Table 2. Parameters of the graphs

Corpus	Vertexes	Edges	Average degree	Average path length	Clustering coefficient
OC	21881	257846	23.57	3.26	0.166
RNC	22467	163914	14.6	3.53	0.136
Mix	31984	395225	24.7	3.29	0.186
QC	85548	291033	6.8	4.07	0.16

One can see that the average degree of QC is lower in comparison with the other corpora (because queries are typically shorter that sentences in natural texts). That is one of the reasons for our decision to use a larger query corpus.

It should also be noted that all corpora comply to 'small world' definition [10], because their average path length is approximately the same as in a random graph with the same number of vertexes (N_V) and average degree (A_D), while clustering coefficient is significantly higher than it should be in the random graph. For example if *Mix* corpus were a random one, its average path length would be equal to 3.24 ($= \frac{log(N_V)}{log(A_D)}$), very close to the actual value. However, in this case, its clustering coefficient should be 0.0015 ($= \frac{2 \times A_D}{N_V}$), which is significantly lower than the actual value. The same is true for all other corpora.

'Small world' nature of our graphs means that vertexes in them tend to bundle into clusters, which is typical of many real-world networks. This finding supports the idea of extracting senses from such clusters. It also additionally proves the applicability of graph sense induction methods to our corpora, as English-language graphs in the related publications also showed such properties.

4   Building Query Graph

We experimented with clustering search engine results page on a set of sixty ambiguous one-word Russian queries, taken from *Analyzethis* homonymous queries analyzer[4]. *Analyzethis* is a search engines evaluation initiative, offering various search performance analyzers, including one for ambiguous or homonymous queries. We crawled Mail.ru search for these queries, getting titles and snippets (10 for each result).

The procedure of semantic clustering starts with building the so called query graph. Here we closely follow [6].

First, we lemmatize all snippets and titles and remove stop words and the query word itself. Then we construct a graph G_q with all nouns from snippets and titles as vertexes. Then we use one of the large corpora graphs (those that we built in Sect. 3) to find words strongly connected to the query word and add these words to the query graph. We consider a connection 'strong' if it falls under the following constraints:

$$\begin{cases} \dfrac{c(q,w)}{c(q)} \geq 0.01 \\ Dice(q,w) \geq 0.005 \end{cases} \qquad (2)$$

where c is absolute frequency in the corpus, q is the query and w is the word under analysis. Thresholds 0.01 and 0.005 were determined empirically while experimenting on the above mentioned ambiguous queries set. These thresholds produced most convincing sense clustering. However, the issue of choosing the thresholds is a subject for thorough evaluation in future.

---

[4] http://analyzethis.ru/?analyzer=homonymous

Thus, we now have G_q with no edges and vertex set consisting of words from the search result and strong neighbors of the query word. After that, for each pair of words (w,w') in G_q we check if they co-occur in the large corpus. If they do and $Dice(w,w') \geq 0.005$, we connect these words in G_q with an edge with weight = $Dice(w,w')$. Finally, we delete disconnected vertexes (those with the degree equal to 0).

5   Processing Query Senses and Results

With query graph at hand, we are ready to find which senses the query has. What we need is an optimal partition of the query graph, in which words related to different senses are in different parts of the graph. We apply two techniques for that, namely, *Curvature* from [6] and *Hyperlex* from [7].

5.1   Curvature

Curvature algorithm aims at finding vertexes from G_q with low local clustering coefficient. Our hypothesis is that these are words which serve as 'links' between different senses or 'uses' of the query. Then we remove vertexes with clustering coefficient below a certain threshold. It leads to the graph disjointing into several components related to different senses. Vertexes in these components represent lexical inventory of each sense. Disconnected vertexes are removed from the final graph.

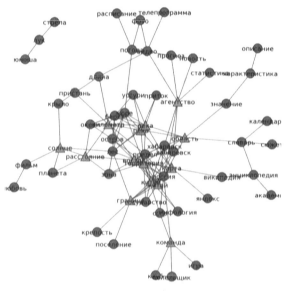

Fig. 1. Query graph for *'амур'* (*Curvature*)

Let us illustrate the process with the example of *'амур'* ('Amur') query. Figure 1 shows its query graph. It is already disconnected into two components and the meaning of love god (associated with words *'лук'*, *'стрела'* and *'юноша'*) is separated. However, other 'senses' of the query remain hidden in the giant component. Vertexes shown as triangles have low clustering coefficient and are thus marked for deletion.

So, we delete 'triangular' vertexes. Note that we chose threshold 0.3 — all vertexes with clustering coefficient below this are removed. It is also important that we do not delete vertexes with clustering coefficient = 0. This is because neighbors of such vertexes are not connected to anything except this vertex. If we remove it, a lot of disconnected vertexes will appear. Such clusters (consisting of only one word) do not make much sense. For example, the word '*лук*' on Fig. 1 is characterized by clustering coefficient = 0. If we remove it, then the whole component representing 'love god' meaning disappears.

Figure 2 shows the query graph after removing vertexes with low clustering coefficient. We now have 6 components (note that the labels for these clusters are introduced by us, not by the algorithm):

1. River (all vertexes except enumerated below)
2. Love god ('*юноша, лук, стрела*')
3. Hockey club ('*клуб, болельщик*')
4. Movie ('*любовь, фильм*')
5. Dictionary-1 ('*календарь, словарь, википедия, энциклопедия, академик, сюжет*')
6. Dictionary-2 ('*значение, описание, характеристика*')

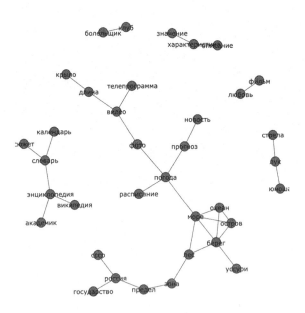

Fig. 2. Disjointed query graph (*Curvature*)

First 4 components clearly represent different meanings of the word '*амур*'. The last two are rather 'uses', typical contexts. However they can still be useful in clustering as they allow to keep encyclopedic results together.

5.2   Hyperlex

Hyperlex algorithm described in [7] introduces the notion of 'hubs' within the graph, meaning most inter-connected vertexes and employs the graph's maximum spanning tree. Just like the previous algorithm, it takes as an input the query graph G_q we prepared in Sect. 4 and the query itself.

First we create a list **L** with all vertexes from G_q sorted in decreasing order by their absolute frequency in the large corpus. Then for each item of this list we check if the corresponding vertex complies to the following constraints:

1. Vertex normalized degree is greater than or equal to 0.05,
2. Average Dice coefficient of vertex edges is greater than or equal to 0.007.

If the constraints are met, we add this word to the hub list, considering it to be a kind of a connector. Simultaneously, we remove this vertex and its neighbors from the list **L** and continue iterating. In case we meet a word which does not satisfy the requirements above, we check whether the list of hubs has at least two elements. If it does, we stop iterating, if not, we continue to the next item. Note that it differs from the original *Hyperlex* algorithm, where one should stop no matter how long the hub list is. In our Russian material it sometimes caused the hub list to remain empty or contain only one item, which is useless.

After we have the list of hubs, we augment G_q with query vertex and connect this vertex to all hubs putting infinite (or very high) Dice coefficient on the corresponding edges. Then, we produce a maximum spanning tree from this graph. Maximum spanning tree is an attempt to keep all the vertexes connected while eliminating cycles and using as few edges as possible with as high weights (in our case it is Dice coefficient) on them as possible. In the spanning tree, there is only one path between any two vertexes and this path lies through edges with maximum Dice. Because the query vertex and the hubs are connected by edges with infinite Dice, they are sure to be the center of the spanning tree and directly linked.

At last we remove the query vertex from the spanning tree, producing disjointed subtrees with hubs as roots. These subtrees represent query meanings. Note that we also delete all disconnected vertexes (those with degree = 0).

Let us present an example of Hyperlex at work with the same query '*амур*'. Our corpus is *Mix*. Initial state of the query graph G_q is the same as in Fig. 1.

We add the word '*амур*' to the graph G_q and connect it to vertexes selected as hubs:

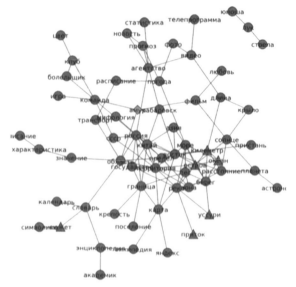

Fig. 3. Query graph for '*амур*' after augmenting it with the query vertex (*Hyperlex*)

'*область*, *фильм*, *команда*'. The result is presented on Fig. 3 with query vertex drawn as a diamond. For reference, vertexes which were introduced from the corpus and not from search results ('*сюжет*', '*океан*', etc.) are drawn as triangles.

Now we produce maximum spanning tree from G_q with Dice coefficient as weight measure. The tree is visualized on Fig. 4. Note that it has much fewer edges than the initial G_q.

Finally, we remove the query vertex and all vertexes that become disconnected after this removal. As a result, we have a disjointed graph shown in Fig. 5. The number of the components has grown from 2 to 4 (once again, labels are assigned by us):

1. Love god (*'юноша, лук, стрела'*)
2. Movie (*'любовь, фильм'*)
3. Hockey club (*'клуб, игра, болельщик, команда, цвет'*)
4. River (all the remaining vertexes)

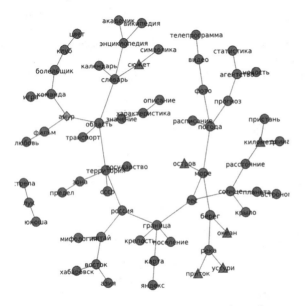

Fig. 4. Maximum spanning tree for *'амур'* query graph (*Hyperlex*)

One can see that *Hyperlex* successfully extracted the same four important meanings as the previous algorithm. At the same time, unlike *Curvature*, it managed to avoid two 'encyclopedic' clusters (obviously in common for too many queries) and leave their vertexes in the 'river' cluster. Also, *Hyperlex* is better because it describes 'hockey club' cluster in a richer way, using 5 relevant words instead of 2.

One can again note that in fact what we call 'senses' are not senses like meanings in the dictionaries. We agree with Jean Veronis who argues that co-occurrence networks reflect 'uses' rather than senses. So, what we have are typical environments where the word is used, and these environments are only loosely connected to what a lexicographer would call 'senses' or 'meanings'. However, we are fine with that, as we assume that clustering SERP according to 'typical uses' is at least equally important as clustering according to 'proper senses'. Perhaps, these senses are in fact less related to real-life, as even linguists sometimes have trouble matching the 'senses' found in a dictionary and the occurrences found in a corpus [7]. Additionally, as has already been stated, dictionary senses are always limited and by design cannot cover new semantic trends and subtle meanings quickly appearing and disappearing in the modern world. Thus, theoretically typical uses are more relevant for clustering than academic dictionary senses. To strictly prove it for the Russian material, one needs manually clustered data set (see Sect. 6), and we leave it for further research.

5.3   Mapping Results to Senses

Once we possess the sense inventory for the query, we can combine it with bags-of-words for each search result to finally perform SERP clustering. We do that in a rather straightforward way.

Given a set of senses represented by a lemma set each and a set of results (snippet and title) represented by lemma sets as well, for each pair of result (r) and sense (s) we calculate similarity measure sim. It is a simple number of lemmas in common for both sets divided by the number of lemmas in the result:

$$sim(r, s) = \frac{r \cap s}{length(r)} \qquad (3)$$

Then we choose the sense with maximum similarity and link this sense to the result. Thus, each result receives some sense, and is 'understood'.

In the future we plan to explore other means of calculating similarity measure as well, for example, counting tokens not types or considering weights on edges in the intersection.

6   Evaluation of SERP Clustering

Generally, evaluation of clustering is a rather harsh task. Perhaps, the best way to do this is to employ human assessment, but for the time being we limited ourselves to simple evaluation of the correctness of cluster number (that is, number of meanings).

Analyzethis service provides data about how many senses of an ambiguous query are there in the SERP. Thus we consider it to be an expert opinion and check how strong is our deviation from this 'gold standard'. For example, if *Analyzethis*

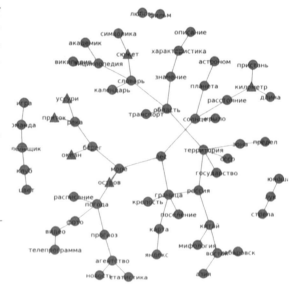

Fig. 5. Maximum spanning tree after removing query vertex (*Hyperlex*)

believes that there are three senses present on the SERP, and our clustering algorithm puts all the results into one cluster, this signals that the algorithm is not optimal. The same is true if the number of clusters is, for example, eight. The less our deviation from *Analyzethis* assessment is, the better. So, in fact we check that the employed algorithms do not produce senseless results (too many or too few meanings). We once again note that in order to evaluate the contents of the clusters themselves, one needs manually clustered SERPs for ambiguous queries. To our knowledge, there is no such a data set for Russian. We are working on creating it.

For the time being, we compared the number of clusters for each of ambiguous queries in four different settings (two corpora and two word sense induction methods). Then we calculated average deviation of our clustering number from that of *Analyzethis*. Table 3 provides the results of this comparison. Note that the average number of senses per query in *Analyzethis* data set was 2.65.

Table 3. Evaluation of SERP clustering (average deviation from *Analyzethis* assessment in number of senses and in percent from the average number of senses in the set)

Corpus	Curvature	Hyperlex
Mix	1.636 (61 %)	**1.288** (49 %)
Query	1.742 (66 %)	1.379 (52 %)

It is clear that *Hyperlex* consistently outperforms *Curvature*, and that *Mix* corpus does the same with the query corpus. *Hyperlex* victory comes as no surprise, as it uses maximum spanning tree notion, which seems to allow deeper grasping of graph structure. The victory of *Mix* corpus (which is smaller than the query corpus) is much less expected. We believe that there are two reasons for this:

1. As we have already mentioned, the query corpus is less 'dense' because of low length of queries. Thus, there are fewer edges and less data for algorithms.
2. Query corpus was lemmatized with Freeling while *Mix* corpus consists of manually annotated corpora. Glitches and outright errors of Freeling could impact graph quality. This can be fixed in the future either by improving Freeling or by using another lemmatizer.

Thus, at the moment, using *Mix* corpus and *Hyperlex* algorithm of word sense induction seems to be the best option. However, things surely can be different if we employ larger corpora (which we plan to do in the future).

7   Conclusion and Future Work

We showed that state-of-the-art methods of word sense induction and search results clustering based on semantic graphs do work for Russian data.

Application of such methods can lead to search engine results presentation getting closer to actual semantics of the results, not simply term frequency ranking. For a user, it would mean the possibility to immediately grasp which results in the SERP are actually related to the query sense, and which other senses exist. The power of this approach can be increased by wider employment of Semantic Web paradigm: semantically marked up web pages are represented by generally better and clearer snippets. Such snippets, in turn, should provide better data for graph-based word sense induction algorithms.

We plan to experiment on more types of query graph processing and launch a full-scale human evaluation of results. Also, it seems profitable to use not only separate words, but also compound phrases, as well as to construct graphs with not only immediate neighbors, but also with second-order co-occurrences (neighbors of neighbors). Additionally, experiments with larger query corpora may lead to new and inspiring insights in this field.

References

1. Harris, Z.S.: Distributional Structure. Springer, Heidelberg (1970)
2. Bybee, J.: Frequency of use and the Organization of Language. Oxford University Press, USA (2006)
3. Kilgarriff, A.: Dictionary word sense distinctions: An enquiry into their nature. Comput. Humanit. **26**(5–6), 365–387 (1992)
4. Schütze, H., Pedersen, J.O.: Information retrieval based on word senses. In: Proceedings 4th Annual Symposium on Document Analysis and Information Retrieval (SDAIR 1995), pp. 161–175 (1995)
5. Navigli, R.: A quick tour of word sense disambiguation, induction and related approaches. In: Bieliková, M., Friedrich, G., Gottlob, G., Katzenbeisser, S., Turán, G. (eds.) SOFSEM 2012. LNCS, vol. 7147, pp. 115–129. Springer, Heidelberg (2012)
6. Marco, A.D., Navigli, R.: Clustering and diversifying web search results with graph-based word sense induction. Comput. Linguist. **39**(3), 709–754 (2013)
7. Véronis, J.: Hyperlex: lexical cartography for information retrieval. Comput. Speech Lang. **18**(3), 223–252 (2004)
8. Padró, L., Stanilovsky, E.: Freeling 3.0: Towards wider multilinguality. In: Calzolari, N., Choukri, K., Declerck, T., Doğan, M.U., Maegaard, B., Mariani, J., Odijk, J., Piperidis, S. (eds.) Proceedings of the Eight International Conference on LanguageResources and Evaluation (LREC'12). Istanbul, Turkey, European LanguageResources Association (ELRA), May 2012
9. Smadja, F., McKeown, K.R., Hatzivassiloglou, V.: Translating collocations for bilingual lexicons: A statistical approach. Comput. Linguist. **22**(1), 1–38 (1996)
10. Watts, D.J., Strogatz, S.H.: Collective dynamics of 'small-world' networks. Nature **393**(6684), 440–442 (1998)

A Large-Scale Community Questions Classification Accounting for Category Similarity: An Exploratory Study

Galina Lezina[1]([⊠]) and Pavel Braslavski[2]

[1] Ural Federal University, Yekaterinburg, Russia
galina.lezina@gmail.com
[2] Ural Federal University/Kontur Labs, Yekaterinburg, Russia
pb@kontur.ru

Abstract. The paper reports on a large-scale topical categorization of questions from a Russian community question answering (CQA) service Otvety@Mail.Ru. We used a data set containing all the questions (more than 11 millions) asked by Otvety@Mail.Ru users in 2012. This is the first study on question categorization dealing with non-English data of this size. The study focuses on adjusting category structure in order to get more robust classification results. We investigate several approaches to measure similarity between categories: the share of identical questions, language models, and user activity. The results show that the proposed approach is promising.

Keywords: Question topic categorization · Community question answering · Question retrieval · Large-scale classification

1 Introduction

Community question answering (CQA) sites allow users to ask questions almost on every topic and get timely answers from other community members. Examples of general-purpose CQA platforms are Yahoo! Answers and its Russian counterpart Otvety@Mail.Ru (*otvety* means *answers* in Russian). Another popular CQA resource StackOverflow has a narrower scope – users ask there questions exclusively about software programming. Such services became a good complement of major web search engines such as Google and Bing. Users resort to their peers, when they have low search engine proficiency, encounter a complex search problem, or just want a more social search experience. CQA services have collected a vast amount of data and attract quite a big audience of users. Yahoo!

This work is partially supported by the Russian Foundation for Basic Research, project #14-07-00589 "Data Analysis and User Modelling in Narrow-Domain Social Media".

P. Braslavski et al. (eds.): RuSSIR 2014, CCIS 505, pp. 332–347, 2015.
DOI: 10.1007/978-3-319-25485-2_13

Answers claimed reaching one billion answers in May 2010[1]; Otvety@Mail.Ru has accumulated almost 80 million questions and more than 400 million answers by August 2012[2].

Topical classification of questions is an area of active research. Question classification can be helpful in several ways. First, category prompt for arriving questions makes question asking process easier for the user, maintains topical consistency within categories, and increases utility of categories for potential answerers (which again benefits questioners). Second, CQA archives contain a vast amount of topically labeled questions. Though partly noisy, these data can be still a valuable resource for question classification in external question answering tasks.

We describe an experiment on topical classification of a large data set of Russian questions originated from Otvety@Mail.Ru. The main purpose of this experiment is to learn to recommend appropriate category for the new arrived question. When posting a new question the user has to assign it to a category using drop-down lists; currently no hints are provided. By choosing the topically correct category for the posted question the user increases her chance of getting a good answer in the nearest future. In this paper we show that most users are not familiar with original category structure and rely on the experienced users is impractical.

In addition to inexperienced users problem we explore that Otvety@Mail.Ru categories structure has some drawbacks. This leads to a further category structure violations. Some categories are ambiguous to the user and overlaps with others. Again this leads question assignment to incorrect category.

The idea is to find similar categories and connect them together. These new categories can be accounted in question classification task. To do that we propose three different methods to calculate similarity of categories using the following features: sharing of identical questions, similarity distributions of words, and user activity.

Finally our contributions are threefold: (1) we describe a yearly non-English data set of questions that has not been previously used in research, (2) we perform a classification on a large data set that significantly exceeds in size data sets reported in the literature, (3) we investigate several approaches to category similarity, including users activity that can be helpful for category alignment in case of unbalanced and noisy label information.

The paper is organized as follows. The next section survey papers on question categorization within CQA context. Section 3 describes Otevty@Mail.Ru platform and the data set used in the study, including category structure, distribution of questions over categories, user activity throughout the year. The approaches to quantify closeness of categories are proposed in Sect. 4. Section 5 discuss classification methods and reports overall performance including our approach.

---

[1] http://yanswersblog.com/index.php/archives/2010/05/03/
1-billion-answers-served.
[2] http://otvet.mail.ru/news/#hbd2012 – accessed in July 2013.

2   Related Work

CQA data and tasks attract numerous researchers. Various methods for finding similar questions, search over large collections of questions and answers, experts search, etc. are proposed in the literature. Recent works made an attempt to organize (classify) CQA questions into an existing category hierarchy.

The task of determining CQA question topic has two goals. First is to facilitate browsing questions in CQA resources [1,5,11]. The category structure used in these papers resembles Yahoo! Answers in many ways, including user interface, rules, and incentives. In [1] authors proposed a kernalized framework to classify questions over hierarchical structure. Target category structure is a part of Yahoo! Answers structure: 6 top categories that includes the most popular and least popular categories. Totally they classified 11,354 questions from 127 leaf categories. In [5] authors randomly chosen 2057 Yahoo! Answers questions from 5 academic disciplines categories. Thus classified questions have less noise because they was asked in more formal categories. In [11] 3,900 questions from Yahoo! Webscope data set classified over 1,096 leaf categories. Authors compared different classification approaches using this data set. In [9] authors experimented with large-scale data set. They used more than 2 millons of questions for classification over Yahoo! Answers categories structure that includes 26 top-level and 1262 leaf level categories.

The second goal of CQA classification task is one of the question retrieval (QR) [3,7,8] problems.

The [3] proposes a category-based framework for search in CQA archives. Work conducts experiments with a data set that has 3,116,147 training and 127,202 test questions obtained from Yahoo! Answers. Authors build a classification model to classify a query question over structure that has 26 top-level and 1263 leaf level categories. In [7] authors determined question topic in question search task. Data set obtained from Yahoo! Answers includes 525,401 items from two categories which has 378 leaf categories. They used Yahoo! Answers taxonomy to get the specificity of topic terms. In [8] authors also exploited category information for improving performance of question retrieval. Experimental dataset includes 3,116,147 questions and 26 top-level and 1263 leaf level Yahoo! Answers categories.

The first problem solution would significantly improve user experience while the second makes possible to offer to the user similar questions from CQA archives and possibly avoid the user from posting the question. In our work we address to the first problem.

All those papers use Yahoo! Answers categories hierarchy as a target structure. Our data set differs in many ways from Yahoo! Answers. The most remarkable difference is the total number of categories. We describe in details our category structure in Sect. 3.1. Moreover some papers use not full Yahoo! Answers categories structure what probably should overestimate classification performance. Finally we have very large amount of source questions organized into much smaller number of categories.

All these papers deals only with data processing and classification method configuration and do not explore original category structure disadvantages. An adjacent to this is the problem named category hierarchy maintenance. Paper [12] propose a new approach to modify a given category hierarchy by placing documents into more topically suitable categories. Authors experiment with Yahoo! Answers, AnswerBag[3] and Open Directory Project[4] hierarchies. This work is built on the assumption that new topics arrive with a new documents and that semantics of the existing topics may change over time.

In our work we address to the problems that are not considered in previous works. First we address to the problem of target category structure disadvantages exploration. We explore its drawbacks through user experience. We do not try to discover new topics and to find their location in the category hierarchy but we try to find the most confusable to user categories and use this information about structure violations in the process of category prediction.

And second to our knowledge this is the first work that highlight the problem of classification of large-scale Russian-language questions by CQA categories hierarchy.

3   Otvety@Mail.Ru Data Collection

In this section we overview Otvety@Mail.Ru service structure and present the data collection that we use in our experiments.

3.1   Categories

In Otvety@Mail.Ru all questions are organized in categories hierarchy that has 28 top-level nodes and 186 leaf nodes.[5] Figure 1 shows part of the Otvety@Mail.Ru categories hierarchy.

Some categories are fine grained in the subcategory level: the largest categories are "Food, cooking" and "Legal advice" since they have 14 subcategories which encompass a wide range of sub-topics. The smallest are "Science, Technology, Languages" and "Style & Fashion" as they have 4 subcategories which are quite coarse. Also some top-level categories such as "Humor", "Adult" and "Other" do not branch.

Generally topics of the categories represent the interests of the community. Common quite understandable "seasonal fluctuations" on some topics could be traced over time. Figure 2 shows an examples of user activity in four subcategories. In the "Education – Homework" ("edu_homework") category the number of question decreases in summer starting from June till September. Percentage of questions about homework per month range from 0.26 % to 3.8 % throughout the year. The maximum of asked questions in the "Travel, Tourism – Holidays Abroad" ("travel_abroad") category asked in the July - usually the holiday

---

[3] http://www.answerbag.com.

[4] http://www.dmoz.org.

[5] See http://otvet.mail.ru/categories for a full list of categories.

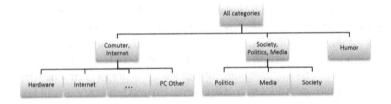

Fig. 1. Part of Otvety@Mail.Ru categories hierarchy.

season - and the minimum is in the December. Percentage of questions varies from 0.18 % to 0.45 %. Questions about holidays are asked 2.5 times more often in July than in December.

Subcategories "Food, Cooking – Other Cooking" ("food_other") and "pc_other" have no such fluctuations. "food_other" and "pc_other" subcategories has small changes throughout the year - from 0.225 % to 0.258 % and from 2.0 % to 2.4 % respectively.

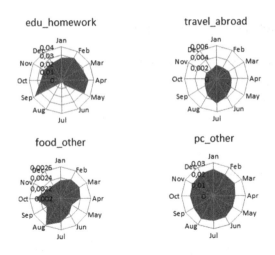

Fig. 2. Distribution questioners activity in categories "edu_homework", "travel_abroad", "food_other" and "pc_other" in 2012.

Almost every category has "other" subcategory (like "PC other" in the "IT" category) which itself are noisy because they contain all questions that possibly have no suitable subcategory or could be assigned to more than one subcategory. This drawback heavily violates categories structure, makes them coarse and indistinguishable between each other.

Another major problem is that people often ask at Otvety@Mail.Ru exactly the same and very similar questions in different subcategories, so categories and subcategories overlaps. All this make categories structure hard to use for both questioners and answerers.

On the one hand user may be confused at the level of subcategories. In the Example 1 user asks question about graphics card in "Computer, Internet – Other Computer" ("pc_other") subcategory while the similar question is asked in "Computer, Internet – Hardware" ("hardware") category (Example 2).

Example 1. "What graphics card is better? GTX 560 or GT 630" [6]

Example 2. "What graphic card is better?" [7]

On the other hand user may confuse top-level categories. For example the question from Example 3 is asked in the "Animals and Plants – Wildlife" [8], "Animals and Plants – Houseplants" [9], "Family, Home, Kids – Housekeeping" [10] and "Animals and Plants – Gardening" [11] categories. This question is related to different top-level categories "Animals and Plants" and "Family, Home, Kids".

Example 3. "what is the name of the flower on the picture?"

For some sort of questions user assumes some categories to be synonymous. In the current Otvety@Mail.Ru categories structure some questions could be assigned to more than one category.

This violates categories structure and makes user experience with the CQA service much worse. The classifier trained on this data set will probably confuse the categories that confuses the user. Our goal is to find similar subcategories to modify original structure. The approaches of categories structure modifications are described in details in Sect. 4.

3.2   Experimental Data Collection

To modify categories structure and train classifier we use all questions asked in 2012. The data set was obtained through Otvety@Mail.Ru API [12]. This data set contains 11,170,398 questions from different categories and subcategories some of which are not used in the service anymore. Examples of such useless categories are "Beauty and Health – Doctor" and "Newcomers". So we do not use this categories in our predictions.

Category named "Golden" is useless because it is not topical. According to formal definition the "golden" category is a special one and it includes selected questions about some facts which may be of interest to a wide range of users. We also do not use this category in our experiments.

We removed all questions asked in these three categories and finally we get 10,739,727 questions asked in 186 categories for experiments.

---

[6] http://otvet.mail.ru/question/167517346.
[7] http://otvet.mail.ru/question/83696264.
[8] http://otvet.mail.ru/question/69108691.
[9] http://otvet.mail.ru/question/69166385.
[10] http://otvet.mail.ru/question/69656908.
[11] http://otvet.mail.ru/question/69709407.
[12] http://otvet.mail.ru/api/v2/question?qid=24141950.

Most of the removed questions was asked in the "Newcomers" category. Figure 3 shows 10 most popular categories in 2012. Almost 10 % of the total number of questions in the data set was asked in the "Humor" category. These top 10 subcategories comprise 40 % of all questions and the other 60 % are asked in the rest 176 (!) subcategories.

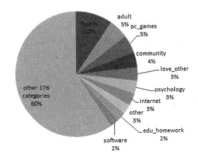

Fig. 3. The most popular categories in 2012.

This percentages changes slightly from month to month, but top categories remain the same. We used questions asked in 11 month to find similar categories and to train classifier. Questions asked in the December of 2012 are used to evaluate classification results. Originally December data set had 989,521 questions but after removing redundant categories we get 939,472 questions.

We did lexical pre-processing of questions before experiments. We perform data pre-processing in three steps:

1. Remove punctuation and lowercase questions.
2. Lemmatize words using AOT[13]. AOT is a software for automatic text processing and is intended mainly for the analysis of the Russian language.
3. Remove stopwords.

3.3 Users

In 2012 at Otvety@Mail.Ru 2,287,417 unique users asked at least one question. More than half (1,406,132) of all active users asked question in the service only once. Figure 4(a) shows dependence of number of questions on the number of users who asked this number of questions.

More frequently users ask questions in one or two categories. The Fig. 4(b) shows that 236,670 users ask more than two questions in one category (but possibly different subcategories). In this figure we do not take into account the users who ask only one question. Most frequently users ask questions in two different categories and only one user asked questions in each of 28 categories[14].

---

[13] http://www.aot.ru/.
[14] http://otvet.mail.ru/profile/id9112629.

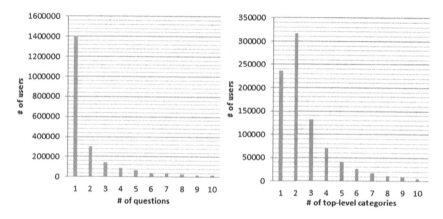

Fig. 4. Distribution of the number of questions depending on the number of users who ask this number of questions (a). Number of categories depending on the number of users who ask questions only in this number of categories (b)

The same situation is typical for subcategories where user ask questions in a limited number of subcategories - mostly in two subcategories. This limited set of categories of the user possibly is an area of her interests but there is another way to explain this behaviour of the user.

For users who ask questions only in two subcategories we assume that they might not be sure what subcategory best suits the question. Some users might post one question in two different but topically similar subcategories. We check this assumption in the Sect. 4.3.

4   Finding Similar Categories

In this paper we learning to predict the most probable category for the question. Section 5.1 describes baseline classification category prediction approach. For clarity, presenting classification results we also provide examples of categories which confuses the baseline classifier.

Our goal is to improve classification performance by finding categories that often confuses users. Regardless of the baseline classification results we try to find these confusable categories. Further in the paper we show that our approach allows to connect categories that often confuses baseline classifier. We assume that categories that confuses users probably will confuse the classifier too so we modify categories structure to make it more unambiguous and less confusable to the classifier and as a consequence to the user.

We find semantically similar subcategories and connect them so these connections form the new categories from the old one so ambiguous questions could be assigned to them. We use three similarity measures to find similar subcategories.

4.1 Connecting Subcategories Using Similar Questions

At Otvety@Mail.Ru some similar questions belong to different categories because sometimes it is hard for the user to determine which category is more topically appropriate to the question. We assume that subcategories are similar if they share many similar questions. We denote this method of finding similar question by $QSim$.

Question similarity $QSim$ is calculated as follows:

$$QSim(S_i, S_j) = \sum_{q \in Q_{ij}} \frac{min(S_i(q), S_j(q))}{S_i(q) + S_j(q)}, \tag{1}$$

here Q_{ij} is the set of questions that is assigned to the S_i and S_j; $S_i(q)$ and $S_j(q)$ are numbers of times question q was assigned to the S_i and S_j respectively.

To find similar and duplicate questions we use simhash [2] algorithm. Simhash is based on comparison of bags of words and gives the same hash values for the same and similar questions. In our application questions are similar if they have the same vocabulary but may have different set of particles and stopwords. Questions from Examples 4 and 5 in Russian language have the same meaning but they differs lexically. Simhash can handle this case because we remove stopwords and particles before calculating hash values.

Example 4. *"Who has any plans for today?"*[15]

Example 5. *"What are your plans for today?"*[16]

As an example this measure gives a strong connection between "pc_other", "Computers, Internet – Software" ("software"), "Computers, Internet – Internet" ("internet") and "hadrware" subcategories. These 4 subcategories share common top-level "Computer, Internet" category in original Otvety@Mail.Ru hierarchy. $QSim$ also connects "pc_other" subcategory with the subcategories "Science, Technology, Languages – Technology" ("technics") and "Goods and Services – Mobile devices" ("mobiles") from different top-level categories. Indeed in "technics" subcategory users ask many questions about computers and hardware like in the Example 6.

Example 6. *"Hp laptop speakers are hissing, what I should I do?"*[17]

According to $QSim$ the subcategories "technics" and "mobiles" has weak connection but they are connected too. Example 7 shows the question that is more suitable to the "mobiles" category but was asked in the "technics" subcategory.

Example 7. *"What is better to buy HTC One Mini Silver or Iphone 4s 8 GB"*[18]

---

[15] http://otvet.mail.ru/question/76074787.
[16] http://otvet.mail.ru/question/75570807.
[17] http://otvet.mail.ru/question/167836262.
[18] http://otvet.mail.ru/question/167848364.

4.2   Connecting Subcategories Using Vocabulary

Another approach is to find similar subcategories using vocabularies. We assume that similar subcategories have similar set of words because users ask similar questions. The Kullback-Leibler Divergence (KL-divergence) is a good measure to find subcategories that are lexically similar.

KL-divergence can be calculated as follows:

$$D_{kl} = \sum_{w \in W_{ij}} \log \left(\frac{P_{S_i}(w)}{P_{S_j}(w)} \right) P_{S_i}(w), \tag{2}$$

here $P_{S_i}(w)$ is the probability that word w occurs in the S_i subcategory; W_{ij} is the set of words that occur both in S_i and S_j subcategories.

KL-divergence is an asymmetric measure: $D_{kl}(S_i||S_j) \neq D_{kl}(S_j||S_i)$ so to calculate distance between two subcategories we use the sum of these measures. We denote this measure by $KLSim$ and calculate it as follows:

$$KLSim(S_i, S_j) = D_{kl}(S_i||S_j) + D_{kl}(S_j||S_i) \tag{3}$$

$$KLSim(S_i, S_j) = \sum_{w \in W_{ij}} (P_{S_i}(w) - P_{S_j}(w)) \log \left(\frac{P_{S_i}(w)}{P_{S_j}(w)} \right) \tag{4}$$

According to Eq. 2 the KL-divergence operates with an intersection of vocabularies W_{ij} of two subcategories S_i and S_j whence the KL-divergence cannot be computed if this intersection is small or empty. To overcome this drawback we use smoothing that was proposed in [4]. Instead of $P_{S_i}(w)$ probability we use smoothed $D_{S_i}(w)$:

$$D_{S_i}(w) = \begin{cases} \gamma P_{S_i}(w) & \text{if } w \in W_i \\ \beta & \text{otherwise} \end{cases}, \tag{5}$$

here W_i is the set of words occurring in S_i subcategory; the parameter β is a positive number smaller than the minimum word probability occurring in either S_i or S_j subcategories and γ is a normalization coefficient and it is based on the requirement:

$$\sum_{w \in W_i} \gamma P_{S_i}(w) + \sum_{w \in W_i, w \notin W_j} \beta = 1 \tag{6}$$

The parameter γ is calculated as follows:

$$\gamma = 1 - \sum_{w \in W_i, w \notin W_j} \beta \tag{7}$$

The parameters γ and β are calculated for each pair of subcategories independently.

As a result the set of connected using $KLSim$ categories pairs is very similar to the set of pairs obtained using $QSim$ described in the previous section. We give a short comparison of these measures in the Sect. 4.4.

4.3   Connecting Subcategories Using User Activity

Recall that users who ask more than one question in Otvety@Mail.Ru more often assign them only two different subcategories. We motivated by the assumption that users who are confused between two semantically similar categories ask question in two similar categories. We use this assumption to compute categories similarity. We call this measure User similarity and denote it by $USim$. User similarity is calculated as follows:

$$USim(S_i, S_j) = \frac{U_{ij}}{U_i + U_j},   \tag{8}$$

here U_{ij} is the number of users who asks questions both in S_i and S_j; U_i and U_j are the total number of users who ask questions in the S_i and S_j subcategories respectively.

As the result $USim$ measure connects subcategories from one common category of the original Otvety@Mail.Ru categories hierarchy. It gives only two pairs of connected subcategories which subcategories is assigned to different categories in the original structure. These pairs of connected subcategories are "music"/"drama" and "drama"/"internet".

4.4   Similarity Thresholds

Original 186 Otvety@Mail.Ru subcategories produce 17,205 pairs and it makes no sense to connect all subcategories so we have to choose the most similar subcategories. We select thresholds for all three measures independently and if pair similarity value does not pass the threshold's value we connect them in new one. In the Section. 5.3 Fig. 6 shows the performance of classifier depending on the selected threshold of similarities for all measures. Empirically selected threshold values corresponds to the moment where classifier performance begins to sharply increase.

Finally we take 106 pairs of similar subcategories for $QSim$ similarity result, 78 pairs for $KLSim$, and 40 pairs for $USim$.

Table 1 lists the number of connected subcategories pairs for each measure.

Table 1. Connected with $QSim$, $KLSim$ and $USim$ pairs.

Similarity measure	# of connected pairs	# of connected subcategories from different categories	Total # of categories in the new structure
$QSim$	107	62	218
$KLSim$	78	41	217
$USim$	40	2	202

Modified structures built using $QSim$ and $KLSim$ measures are very similar to each other because they connect 56 same subcategories. $USim$ and $KLSim$ share only 22 pairs, while $USim$ and $QSim$ have 31 pairs in common.

Generally *USim* connects subcategories from one common category. This means that users who ask only two questions is interested in one common topic and address one question within one top-level category. *QSim* and *KLSim* connect subcategories both from same and different top-level categories.

5   Classification

In this section we describe classification approach and evaluation methods. Table 2 presents evaluation results for different classification tasks and in Sect. 5.2 we describe all methods with its notations.

5.1   Baseline Approach

Our baseline is a standard approach for text classification tasks - support vector machine with bag of words features vector. It classifies questions by original Otvety@Mail.Ru hierarchy.

Figure 5 shows the Hinton diagrams of baseline classifier's confusion matrices for flat top-level and lower-level classification. Figure 5(b) shows the part of confusion matrix obtained for flat top-level categories classification. It is interesting that generally humor is the most confusable category and it is less often confused with technical categories than with non-tech (more frequently it is confused with "society", "philosophy", "love" and "adult" categories).

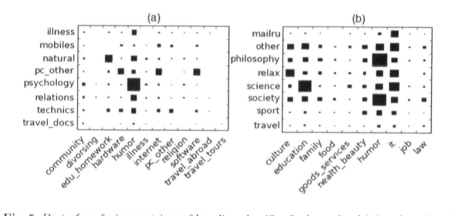

Fig. 5. Part of confusion matrices of baseline classifier for lower-level (a) and top-level (b) categories

Figure 5(a) show the part of confusion matrix of original lower-level categories classification result. In this figure we can see confusions between subcategories at leaf level. Here we can see that classifier frequently confuses subcategories from one common category like "hardware" and "pc_other", "religion" and "pshychology", etc.

5.2   Methods

We evaluate question classification performance over original Otvety@Mail.Ru categories hierarchy and three modified categories structures. Evaluation methods of classification over modified categories structures in Table 2 is denoted similarly with its measures: $QSim$, $KLSim$ and $USim$ respectively. We independently classify questions over top-level (TLC) categories and over lower-level (LLC) subcategories of original Otvety@Mail.Ru hierarchy.

Otvety@Mail.Ru categories hierarchy is useful resource not only for internal question category recommendation task. It also can be used to determine topic of the question from external resource - a search engine query subject for example. Recall that query topic identification is actively used in the question retrieval task. Some categories from original structure is not useful for topic prediction task. These categories are "humor", "other" and "about Mail.Ru project". "humor" and "other" is not objective while "about Mail.Ru project" is meaningful only for Otvety@Mail.Ru users. The category "other" has questions on many topics as well as "humor". We exclude these three categories from original Otvety@Mail.Ru categories structure and evaluate classification performance over top-level (TLC*) categories and lower-level (LLC*) subcategories.

Hierarchical classification is an effective approach in hierarchical classification task. We denote it by TLC/LLC. In hierarchical classification approach we build one classifier to predict top-level category and classifiers for every top-level category to predict subcategory.

Recall that we have 10,739,727 questions asked in 2012 and test data set includes 939,472 questions. After removing questions from useless categories we have 8,456,252 questions for training and 815,170 for testing.

All methods use the same baseline classifier. They differs only in evaluation approach.

5.3   Classification Performance

Table 2 presents evaluation results in terms of accuracy.

Accuracy is calculated as follows:

$$Accuracy = \frac{TP + TN}{TP + FP + FN + TN},\tag{9}$$

here T means True, F is False, P is Positive and N is Negative.

In the case of top 3 evaluation we take three most probable categories predicted by the classifier and see whether correct category are in the predicted. So we give the user an opportunity to choose between recommended categories.

Classifier needs relatively small amount of data for training. Recall that we have about 10 million of questions for training but the accuracy stops growing after 500 thousands of training samples. Figure 7 shows accuracy values depending on the size of training data set for different classification tasks.

Table 2. Classification results

Evaluation method		# of classes	Accuracy	
			Top 1	Top 3
Baseline	TLC	28	0.56	0.79
	LLC	186	0.40	0.63
	TLC*	25	0.61	0.83
	LLC*	171	0.42	0.65
	TLC/LLC	183	0.66	0.91
	QSim	218	0.57	0.80
	KLSim	217	0.52	0.76
	USim	202	0.49	0.70

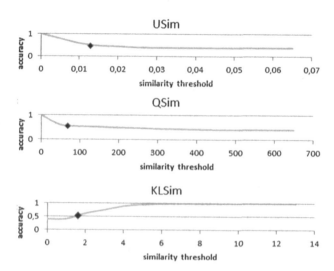

Fig. 6. Classifier accuracy depending on the selected measure threshold

We evaluating classification by modified structures for $QSim$, $KLSim$ and $USim$ measure. Evaluation on structure built with $USim$ measure gives us the lowest accuracy. Generally $USim$ connects subcategories belonging to the one common category in the original Otvety@Mail.Ru hierarchy while $QSim$ and $KLSim$ connect subcategories from different categories that users often confuse. $USim$ possibly reflect an areas of user's interest and not subcategories that users often confuse within one common category because they do not know which subcategory is more appropriate for a given quesion.

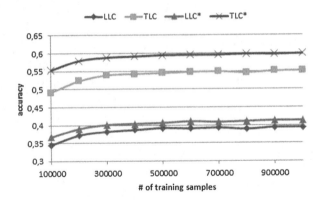

Fig. 7. Accuracy over the number of training samples.

6   Conclusion

Top-level classification by the structure without general categories like "humor" and "other" and specific "about Mail.Ru Project" category exceeds the classification by original categories structure results by 5 %. "humor" and "other" categories itself are noisy because they have questions from all possible categories. In the "humor" category users can post jokes of any topic. Determining jokes is another scientific problem and it is not addressed in this paper. Performance of classification by subcategories (LLC*) without categories "humor" and "other" does not differ from classification by subcategories (LLC) of original structure. Classifier is often confused between subcategories of different top-level categories. The same is relevant to the users.

Hierarchical classification is an effective approach in such problems. In 91 % cases a correct category is in top 3 predicted categories. In this case we even do not take into account the similar categories.

Another CQA question classification problem is that questions itself are short sparse texts while sparse text classification is another well known problem. For example this problem is described in [6]. We do not handle question text sparseness in our paper.

An open-ended question is how to choose similarity measures' thresholds. Recall that we selected it empirically and we do not provide clear guidelines how to choose it. Anyway the more similar pairs coincided with the most confusable by baseline classifier categories. But it could just be the feature of our data set.

References

1. Chan, W., Yang, W., Tang, J., Du, J., Zhou, X., Wang, W.: Community question topic categorization via hierarchical kernelized classification. In: Proceedings of the 22nd ACM International Conference on Conference on Information and Knowledge Management, pp. 959–968 (2013)

2. Charikar, M.S.: Similarity estimation techniques from rounding algorithms. In: Proceedings of the Thiry-Fourth Annual ACM Symposium on Theory of Computing, pp. 380–388 (2002)
3. Cao, X., Cong, G., Cui, B., Jensen, C.S., Zhang, C.: The use of categorization information in language models for question retrieval. In: Proceedings of the 18th ACM Conference on Information and Knowledge Management, pp. 265–274 (2009)
4. Bigi, B.: Using Kullback-Leibler distance for text categorization. In: Sebastiani, F. (ed.) ECIR 2003. LNCS, vol. 2633, pp. 305–319. Springer, Heidelberg (2003)
5. Blooma, M.J., Coh, D.H.-L., Chua, A.Y.: Question classification in social media. Int. J. Inf. Stud. **1**(2), 101–109 (2009)
6. Li, B., King, I., Lyu, M.R.: Question routing in community question answering: putting category in its place. In: Proceedings of the 20th ACM International Conference on Information and Knowledge Management, pp. 2041–2044 (2011)
7. Duan, H., Cao, Y., Lin, C.Y., Yu, Y.: Searching Questions by Identifying Question Topic and Question Focus. In: ACL, pp. 156–164 (2008)
8. Cao, X., Cong, G., Cui, B., Jensen, C.S.: A generalized framework of exploring category information for question retrieval in community question answer archives. In: Proceedings of the 19th International Conference on World Wide Web, pp. 201–210 (2010)
9. Cai, L., Zhou, G., Liu, K., Zhao, J.: Large-scale question classification in CQA by leveraging Wikipedia semantic knowledge. In: Proceedings of the 20th ACM International Conference on Information and Knowledge Management, pp. 1321–1330 (2011)
10. Broder, A.Z., Fontoura, M., Gabrilovich, E., Joshi, A., Josifovski, V., Zhang, T.: Robust classification of rare queries using web knowledge. In: Proceedings of the 30th Annual International ACM SIGIR Conference on Research and Development in Information Retrieval, pp. 231–238 (2007)
11. Qu, B., Cong, G., Li, C., Sun, A., Chen, H.: An evaluation of classification models for question topic categorization. J. Am. Soc. Inf. Sci. Technol. **63**(5), 889–903 (2012)
12. Yuan, Q., Cong, G., Sun, A., Lin, C.Y., Thalmann, N.M.: Category hierarchy maintenance: a data-driven approach. In: Proceedings of the 35th International ACM SIGIR Conference on Research and Development in Information Retrieval, pp. 791–800 (2012)

Towards Crowdsourcing and Cooperation
in Linguistic Resources

Dmitry Ustalov[1,2,3](✉)

[1] Krasovsky Institute of Mathematics and Mechanics, Ekaterinburg, Russia
dau@imm.uran.ru
[2] Ural Federal University, Ekaterinburg, Russia
[3] NLPub, Ekaterinburg, Russia

Abstract. Linguistic resources can be populated with data through the use of such approaches as crowdsourcing and gamification when motivated people are involved. However, current crowdsourcing genre taxonomies lack the concept of cooperation, which is the principal element of modern video games and may potentially drive the annotators' interest. This survey on crowdsourcing taxonomies and cooperation in linguistic resources provides recommendations on using cooperation in existent genres of crowdsourcing and an evidence of the efficiency of cooperation using a popular Russian linguistic resource created through crowdsourcing as an example.

Keywords: Games with a purpose · Mechanized labor · Wisdom of the crowd · Gamification · Crowdsourcing · Cooperation · Linguistic resources

1 Introduction

Crowdsourcing has become a mainstream and well-suited approach for solving many linguistic data gathering problems such as sense inventory creation [1], corpus annotation [2], information extraction [3], etc. However, its most effective use still remains a problem because human annotators' motivation and availability are tantalizingly constrained and it is crucial to get the most of performance from the effort interested people can make.

Another extremely popular term nowadays is *gamification*. The origin of the gamification concept is, of course, video game industry. The idea of gamification is in embedding interactive and game-based techniques into application to strengthen user engagement and increase the time spent annotating. Due to the insufficiency of exploration, gamification is more rarely used in academia when compared to the industry.

Cooperation is a major, if not principal, element of today's video games, which is confirmed by the presentations made in recent years at E3 — the largest video game exposition and event. Initially, multiplayer mode in video games was focused on *player versus player* competitions, but a few years ago the focus has changed to *cooperated human players versus AI* and *guild versus guild* games.

© Springer International Publishing Switzerland 2015
P. Braslavski et al. (eds.): RuSSIR 2014, CCIS 505, pp. 348–358, 2015.
DOI: 10.1007/978-3-319-25485-2_14

The work, as described in this paper, makes the following contributions: (1) it presents a survey on crowdsourcing taxonomies and cooperation in linguistic resources, (2) makes recommendations on using cooperation in existent genres of crowdsourcing, and (3) provides an evidence of the efficiency of cooperation represented by a popular Russian linguistic resource created through crowdsourcing.

The rest of this paper is organized as follows. Section 2 focuses on related work towards crowdsourcing genres and cooperation in linguistic resources. Section 3 is devoted to the cooperative aspect of crowdsourcing. Section 4 discusses cooperation using OpenCorpora as the example, which is a sufficiently popular Russian linguistic resource created through crowdsourcing. Section 5 interprets and explains the obtained results. Section 6 concludes with final remarks and directions for the future work.

2   Crowdsourcing Genres & Activities

Early studies on crowdsourcing genres in their wide definition were conducted in 2009. Quinn and Bederson in their technical report [4] proposed the term of *distributed human computation* along with the taxonomy of seven different genres of these computations such as games with a purpose, mechanized labor, wisdom of crowds, crowdsourcing, dual-purpose work, grand search, human-based genetic algorithms, and knowledge collection from volunteer contributions.

In the same year Yuen et al. also presented [5] another taxonomy of five crowdsourcing genres: initiatory human computation, distributed human computation, social game-based human computation with volunteers, paid engineers and online players, which is similar to the previously mentioned.

Many studies following the early ones are focused on classification of whether a crowdsourced project belongs to a specific class of the given taxonomy. For instance, Sabou et al. study of correlation between crowdsourcing genres [6], quality assessment [7], and guidelines on corpus annotation through crowdsourcing [2] align various best practices among the established genres.

There are other attempts to create a taxonomy of crowdsourcing genres. Zwass investigated the phenomena of *co-creation* [8] and proposed a taxonomy of user-created digital content which includes the following: knowledge compendia, consumer reviews, multimedia content, blogs, mashups, virtual worlds. The resulted taxonomy appears to be too general and, since it was not intended, does not fit the natural language processing field perfectly.

Erickson presented *four quadrant model* [9] composed of two orthogonal dichotomies to classify crowdsourcing projects: "same place–different places" and "same time–different times". The resulted taxonomy tends to assign all the mentioned above crowdsourced projects to the "different places–different times" quadrant also called *Global Crowdsourcing*.

Some studies propose much narrower dichotomies. This is the case of the research conducted by Suendermann & Pieraccini [10], which introduces a concept of *private crowd* being a trade-off between two extremes: an inexpensive, highly available yet uncontrolled *public crowd* such as the Amazon's one, and the expensive to hire, high-quality and professional expert annotators. The *private crowd* term can be referred to as *controlled crowd*.

2.1 Three Genres of Crowdsourcing

In 2013, Wang et al. aggregated most of the previous studies in their very well-done survey. The mentioned work emphasizes three intuitive and well-separated genres of crowdsourcing [11]:

Games with a purpose (GWAPs), when a player without any special knowledge is put into a gaming environment and have to make right decisions to win the game under the pressure of time or any game mechanics' constraints. Phrase Detectives[1] and JeuxDeMots[2] can be considered as good examples of such games.

Mechanized labor (MLab), when an annotator who meet the preliminary requirements is asked to answer a questionnaire on a centralized platform and is rewarded for their work by micropayments. The most well-recognized examples of MLab are Amazon Mechanical Turk[3] and CrowdFlower[4].

Wisdom of the crowd (WoTC), when motivated volunteers share their knowledge on the given topic in the free form in order to answer some question, to explain something to other people, and so on. The obvious examples of WoTC are Wikipedia[5] and Yahoo! Answers[6].

Observations reveal that research papers often do not specify the exact crowdsourcing genre and treat the crowdsourcing term as a synonym to MLab due to extreme popularity of the Amazon's product.

2.2 Cooperation in Linguistic Resources

Cooperation, derived from *to cooperate*, is to work actively with rather than against others [12, p.435]. Unfortunately, cooperative crowdsourcing in linguistic resources is less explored in the literature, but present studies show that considering the concept of cooperation in crowdsourcing is a trending topic deserving attention.

An early study of Wikipedia and its quality by Wilkinson and Huberman [13] found a statistically significant correlation between page edits, talkpage conversations and the quality of these pages. The study revealed the fact that pages with more intense discussion activity often have better quality than less discussed ones.

---

[1] https://anawiki.essex.ac.uk/phrasedetectives/.
[2] http://www.jeuxdemots.org/.
[3] http://mturk.com/.
[4] http://www.crowdflower.com/.
[5] http://wikipedia.org/.
[6] https://answers.yahoo.com/.

A study by Arazy and Nov [14] pays a special attention to *local inequality* — inequality of editors' contribution in a particular article, and *global inequality* — inequality in overall Wikipedia activity for the same set of editors. As a result, they found that global inequality has an impact on local inequality, which influences editors' coordination in a positive way, which in its turn contributes to quality.

Budzise-Weaver et al. [15] consider several cases of multilingual digital libraries and their collaboration both with state institutions and crowdsourced projects in order to provide multilingual information access for users. The paper does not describe how exactly crowdsourcing can help digital libraries in doing their job, but does demonstrate significant interest to crowdsourcing from an interdisciplinary point of view.

Ranj Bar and Maheswaran [16] in their case study on Wikipedia concluded that new mechanisms are needed to coordinate the activities in crowdsourcing due to the fact that high quality articles are controlled by small groups of permanent editors, and supporting these articles is a huge burden for the editors.

3 Crowdsourcing Genres and Cooperation

Each of the three crowdsourcing genres has its own identities; and the principle of paritipants' cooperation changes with each particular crowdsourcing instance. However, it seems possible to denote three common points:

- **attractiveness**, the degree of how a participant can find a crowdsourcing process attractive,
- **usefulness**, the degree of how a participant can find his activity results useful to their own purposes,
- **difficulty**, the degree of how it is difficult to embed cooperative elements into a process.

When specific case studies are available, the correspondent details are provided.

3.1 Games with a Purpose

The main advantage of GWAPs is their *high attractiveness*, because people love video games and it is easier to get new participants than in other genres of crowdsourcing. One may find *low usefulness* in these games, but the more attractive the game is, the less other factors are becoming important.

It is necessary to mention that video games are a very costly kind of software and producing GWAPs requires not only creating a game, but also designing innovative game mechanics allowing a player to both enjoy the game and to implicitly produce valuable data. Thus, games with a purpose have *high difficulty* to be realized.

Authors of Phrase Detectives say that the cost of data gathering using their means is lower than using other approaches [17], but they did not consider the total cost of the game design and development. Elements of real-time players' cooperation may enhance GWAPs attractiveness even more. The evidence of this is the fact that modern cooperative multiplayer video games like Dota 2 or Destiny have substituted traditional *free for all* (deathmatch) multiplayer games.

3.2   Mechanized Labor

Since MLAB projects are often deployed on specialized platforms available on the World Wide Web, the main advantage of MLAB is its *low difficulty*: cooperative elements may be embedded supplementarily to the annotation process through allowing annotators to join teams and making them participate in the team-based activity.

In order to cover as much domains as possible, platforms' owners provide only very utilitarian and generic interfaces allowing one to answer a questionnaire without exposing them to any domain-specific features.

Since MLAB participants are often rewarded for their work that may be or may not be interesting for them, the mechanized labor projects have *medium usefulness* and usually *low attractiveness*.

3.3   Wisdom of the Crowd

The strong side of WOTC projects is, indeed, *high usefulness* due to the fundamental principle of such a genre, when volunteers make efforts to make their resource better for everyone. WOTC have *low attractiveness* for the same reasons, however it depends on every particular instance.

The above mentioned study by Arazy and Nov also touches upon a typical regulation problem called "edit warring" in Wikipedia [14], when "editors who disagree about the content of a page repeatedly override each other's contributions, rather than trying to resolve the disagreement through discussion".

The phenomena of "edit warring" was later studied by Yasseri et. al [18]. Such a problem may be partially resolved by using the controlled crowd instead of the public one when volunteers have a mentor and responsibility for their actions [19]. Therefore, such projects have *medium difficulty*.

4   Evidence

An evidence that cooperation does work and really stimulates participants to do more assignments is the case of OpenCorpora, which is a project focused on creation of a large annotated Russian corpus through crowdsourcing [20].

Currently, OpenCorpora participants have to annotate morphologically ambiguous examples in the MLAB manner. One can annotate examples individually, but has an opportunity to join teams and annotate examples in cooperation

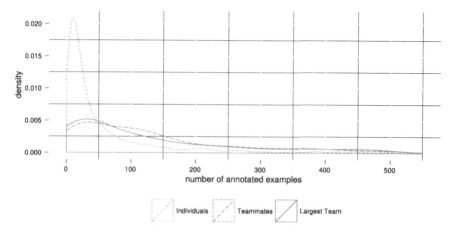

Fig. 1. Annotated examples' densities by three user groups: the individuals, the team-mates, and the largest team; the highlighted region corresponds to the interval between 50 and 350 examples.

with their teammates. A team can be created and joined by everyone, and teams challenge each other by means of active collaborators, annotated examples, and error rates.

As according to the full-scale pilot study conducted on one of the largest Russian information technologies' websites[7], volunteers were very positive about their participation in the cooperative annotation. The study was followed by the creation of the largest team uniting 170 participants. The team got the 2nd place in the total rank[8] based on the number of the annotated examples.

The possible explanation of such a result would be found in what have driven the participants' motivation. It was not only their altruism and readiness to help, but the possibility for their team to get the leading places in the total rank, as well as their personal participation being one of the keys to the team's possible success.

4.1   "Is There a Relationship?"

To make it possible to study the present result more thoroughly, the Open-Corpora team has kindly provided us with the dataset consisted of user ID, the group's name, and various activity information including total number of the annotated examples per user. Hereafter participants who joined a team are referred to as *teammates*, and those who did not join a team are referred to as *individuals*.

The initial dataset contains information on 2642 users: 2219 of them are individuals and 423 are teammates. The distributions' densities are depicted at

---

[7] http://habrahabr.ru/post/152799/#comment_5315923.
[8] http://opencorpora.org/?page=stats.

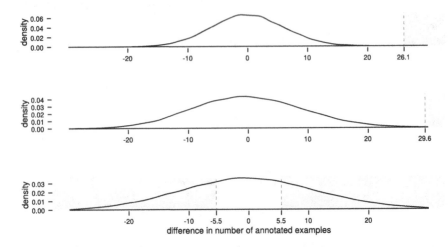

Fig. 2. Simulated differences in number of annotated examples, the vertical dashed lines represent observed differences: the upper plot (a) corresponds to H, the middle plot (b) corresponds to H', and the lower (c) corresponds to H''.

Fig. 1 and seem to be right-skewed. In order to remove outliers from the dataset, users who annotated less than 50 examples or more than 350 examples have been excluded. As a result, the dataset has been reduced to 579 individuals and 195 teammates, 71 of those are the members of the largest team.

In general, the individuals annotated 801 531 examples and the teammates annotated 970 650, while in the dataset the individuals annotated 71 150 examples and the teammates annotated only 29 049 examples.

Hence, the research question is *"Is there a relationship between being a team member and the number of annotated examples for a regular OpenCorpora user?"*

4.2   Inference

Since the dataset is right-skewed and such hypothesis tests as *t-test* may be unreliable, a randomization test was implemented in the R programming language and executed for 25 000 times under the significance level of $\alpha = .05$ in order to estimate the unbiased *p-value*.

The true difference in means of the numbers of annotated examples among the teammates (μ_T) and the individuals (μ_I) has been examined. The following hypothesis H was evaluated in order to find a relationship between being a team member and the number of annotated examples:

$H_0 : \mu_T - \mu_I = 0$, the teammates and the individuals on average have no difference in their annotation activity,

$H_A : \mu_T - \mu_I > 0$, the teammates tend to annotate more examples on average than the individuals.

The density of differences in the number of annotated examples is demonstrated at Fig. 2(a): the observed difference in means for this one-tailed test is $\bar{x}_T - \bar{x}_I = 26.085$, and the p-value is $p = 0$. Thus, $p < \alpha$ and the null hypothesis H_0 has been rejected, suggesting that $\mu_T > \mu_I$: the teammates tend to annotate more examples than the individuals.

5   Discussion

The obtained result can also be explained by a teammate being more loyal and attached to the resource than an individual. Therefore, it is reasonable to study the performance of a particular team.

5.1   The Largest Team Vs. the Individuals

In order to compare the behavior of the individuals and the largest team members instead of all the teammates, the true difference in means of the numbers of annotated examples among the teammates of the largest team (μ_H) and the individuals (μ_I) was examined. The following hypothesis H' was evaluated in the similar way as the previous one:

$H'_0 : \mu_H - \mu_I = 0$, the teammates of the largest team and the individuals on average have no difference in their annotation activity,

$H'_A : \mu_H - \mu_I > 0$, the teammates of the largest team annotate more examples on average than the individuals.

The simulation results for this one-tailed test are presented at Fig. 2(b): the observed difference in means is $\bar{x}_H - \bar{x}_I = 29.552$ and the p-value is $p = .001$. Since $p < \alpha$, the null hypothesis H'_0 has been rejected, suggesting that $\mu_H > \mu_I$: the teammates of the largest team annotate more examples than the individuals.

This result agrees well with the H_0 hypothesis and can be explained by the fact that the largest team is still relatively small and consists of only 170 teammates who were highly motivated for a short time due to news rotation on the website where they came from. Their activity decreased significantly when the announcement of the OpenCorpora disappeared from the news headline. Their team took the 2nd place on the leaderboard; they lost to the the leading team as the latter had annotated approximately seven times more examples (501 963 versus 76 559).

5.2   The Largest Team Vs. Other Teams

Statistical testing of teams' performance based on comparison of their impact is complicated due to lack of participants in other teams. For instance, the 2nd largest team is comprised of 36 users only, the 3rd largest — 24, the 4th — 13, which is insufficient for any meaningful test. However, it is indeed possible to compare the performance of the largest team with the performance of other teams considered together.

The true difference in means of the numbers of annotated examples among the teammates of the largest team (μ_H) and other teams (μ_R) was examined, and the following hypothesis H'' has been evaluated:

$H_0'' : \mu_H - \mu_R = 0$, the teammates of the largest team and other teammates on average have no difference in their annotation activity,

$H_A'' : \mu_H - \mu_R \neq 0$, the teammates of the largest team and other teammates on average have the difference in their annotation activity.

The simulation results for this two-tailed test are presented at Fig. 2(c): the observed difference in means is $\bar{x}_H - \bar{x}_R = 5.453$ and the *p-value* is $p = .629$. Since $p > \alpha$, the null hypothesis H_0' has not been rejected, suggesting that $\mu_H = \mu_R$: the teammates of the largest team annotate the same number of examples as other teammates do.

6   Conclusion

According to the obtained results, there *is* a correlation between being a team member and the number of annotated examples for a regular OpenCorpora user. The use of team-based cooperation can improve the user activity on crowdsourced linguistic resources. However, since the study is observational, it was impossible to establish causal relationships between the variables.

When organized in teams, users do provide more annotations comparing with those who are not organized in teams. Thus, it is highly recommended for a crowdsourced resource to provide users with the opportunity to join teams and annotate examples in cooperation with their teammates.

The statistical hypotheses have been evaluated with use of the randomization test with the significance level of .05. The present dataset is available[9] in an depersonalized form under the Creative Commons Attribution-ShareAlike 3.0 license. The source code of the above mentioned simulation program is included under the MIT License.

Further work may be focused on assessing the quality of team-based cooperation results and on studying the patterns of cooperation and the efficiency of their use in other popular linguistic resources created through crowdsourcing.

Acknowledgments. This work is supported by the Russian Foundation for the Humanities, project no. 13-04-12020 "New Open Electronic Thesaurus for Russian", and by the Program of Government of the Russian Federation 02.A03.21.0006 on 27.08.2013.

The author would like to thank Dmitry Granovsky for the extended statistical information collected from http://opencorpora.org/. The author is also grateful to the anonymous referees who offered very useful comments on the present paper.

[9] http://ustalov.imm.uran.ru/pub/opencorpora-cooperation.tar.gz.

References

1. Biemann, C.: Creating a system for lexical substitutions from scratch using crowd-sourcing. Lang. Resour. Eval. **47**(1), 97–122 (2013)
2. Sabou, M., Bontcheva, K., Derczynski, L., Scharl, A.: Corpus annotation through crowdsourcing: towards best practice guidelines. In: Proceedings of the Ninth International Conference on Language Resources and Evaluation (LREC 2014), European Language Resources Association (ELRA), pp. 859–866 (2014)
3. Lofi, C., Selke, J., Balke, W.T.: Information extraction meets crowdsourcing: a promising couple. Datenbank-Spektrum **12**(2), 109–120 (2012)
4. Quinn, A.J., Bederson, B.B.: A Taxonomy of Distributed Human Computation. Human-Computer Interaction Lab Tech Report, University of Maryland (2009)
5. Yuen, M.C., Chen, L.J., King, I.: A survey of human computation systems. In: International Conference on Computational Science and Engineering, 2009. CSE 2009. vol. 4, pp. 723–728. IEEE (2009)
6. Sabou, M., Bontcheva, K., Scharl, A.: Crowdsourcing Research Opportunities: Lessons from Natural Language Processing. In: Proceedings of the 12th International Conference on Knowledge Management and Knowledge Technologies, pp. 17:1–17:8. ACM (2012)
7. Sabou, M., Scharl, A., Michael, F.: Crowdsourced knowledge acquisition: towards hybrid-genre workflows. Int. J. Semant. Web Inf. Syst. **9**(3), 14–41 (2013)
8. Zwass, V.: Co-Creation: toward a taxonomy and an integrated research perspective. Int. J. Electron. Commer. **15**(1), 11–48 (2010)
9. Erickson, T.: Some thoughts on a framework for crowdsourcing. In: CHI 2011 Workshop on Crowdsourcing and Human Computation (2011)
10. Suendermann, D., Pieraccini, R.: Crowdsourcing for Industrial Spoken Dialog Systems. In: Eskénazi, M., Levow, G.A., Meng, H., Parent, G., Suendermann, D. (eds.) Crowdsourcing for Speech Processing: Applications to Data Collection, pp. 280–302. Transcription and Assessment. John Wiley & Sons, Ltd (2013)
11. Wang, A., Hoang, C.D.V., Kan, M.Y.: Perspectives on crowdsourcing annotations for natural language processing. Lang. Resour. Eval. **47**(1), 9–31 (2013)
12. Kohn, A.: No Contest: A Case Against Competition. Houghton Mifflin Harcourt, New York (1992)
13. Wilkinson, D.M., Huberman, B.A.: Cooperation and quality in wikipedia. In: Proceedings of the 2007 International Symposium on Wikis, pp. 157–164. ACM (2007)
14. Arazy, O., Nov, O.: Determinants of wikipedia quality: the roles of global and local contribution inequality. In: Proceedings of the 2010 ACM Conference on Computer Supported Cooperative Work, pp. 233–236. ACM (2010)
15. Budzise-Weaver, T., Chen, J., Mitchell, M.: Collaboration and crowdsourcing: the cases of multilingual digital libraries. Electron. Libr. **30**(2), 220–232 (2012)
16. Ranj Bar, A., Maheswaran, M.: Case study: integrity of wikipedia articles. In: Confidentiality and Integrity in Crowdsourcing Systems. SpringerBriefs in Applied Sciences and Technology. Springer International Publishing, pp. 59–66 (2014)
17. Poesio, M., Chamberlain, J., Kruschwitz, U., Robaldo, L., Ducceschi, L.: Phrase Detectives: Utilizing Collective Intelligence for Internet-scale Language Resource Creation. ACM Trans. Interact. Intell. Syst. **3**(1), 3:1–3:44 (2013)
18. Yasseri, T., Sumi, R., Rung, A., Kornai, A., Kertész, J.: Dynamics of conflicts in wikipedia. PLOS ONE **7**(6), e38869 (2012)

19. Braslavski, P., Ustalov, D., Mukhin, M.: A spinning wheel for YARN: user interface for a crowdsourced thesaurus. In: Proceedings of the Demonstrations at the 14th Conference of the European Chapter of the Association for Computational Linguistics, Association for Computational Linguistics, pp. 101–104 (2014)

20. Bocharov, V., Alexeeva, S., Granovsky, D., Protopopova, E., Stepanova, M., Surikov, A.: Crowdsourcing morphological annotation. In: Computational Linguistics and Intellectual Technologies: papers from the Annual conference "Dialogue", RGGU, pp. 109–124 (2013)

Author Index

Braslavski, Pavel   332

Caragea, Cornelia   3

Fonarev, Alexander   253

Gareev, Rinat   263
Giles, C. Lee   3
Głowacka, Dorota   307
Gollapalli, Sujatha Das   3

Hofmann, Katja   21

Ignatov, Dmitry I.   42, 276
Inselberg, Alfred   142
Ivanov, Vladimir   263

Karpov, Nikolay   297
Konyushkova, Ksenia   307
Kutuzov, Andrey   320

Lai, Pei Ling   142
Lezina, Galina   332
Li, Xiaoli   3

Malioukov, Alexander   276
Mikhailova, Maria   276

Nakov, Preslav   185

Porshnev, Alexander   297

Redkin, Ilya   297
Rosso, Paolo   229

Ustalov, Dmitry   348

Zakirova, Alexandra Yu.   276

Printed in the United States
by Baker & Taylor Publisher Services